Integrated Functional Sanitation Value Chain
The Role of the Sanitation Economy

Integrated Functional Sanitation Value Chain

The Role of the Sanitation Economy

Peter Emmanuel Cookey
Rivers State College of Health Science and Management Technology, Port Harcourt, Nigeria

Thammarat Koottatep
Asian Institute of Techology, Bangkok, Thailand

Walter Thomas Gibson and
Bear Valley Ventures, Cheshire, United Kingdom

Chongrak Polprasert
Thammasat University, Bangkok, Thailand

Published by   **IWA Publishing**
                                Unit 104–105, Export Building
                                1 Clove Crescent
                                London E14 2BA, UK
                                Telephone: +44 (0)20 7654 5500
                                Fax: +44 (0)20 7654 5555
                                Email: publications@iwap.co.uk
                                Web: www.iwapublishing.com

First published 2022
© 2022 IWA Publishing

Disclaimer
The information provided and the opinions given in this publication are not necessarily those of IWA and should not be acted upon without independent consideration and professional advice. IWA and the Editors and Authors will not accept responsibility for any loss or damage suffered by any person acting or refraining from acting upon any material contained in this publication.

British Library Cataloguing in Publication Data
A CIP catalogue record for this book is available from the British Library

ISBN: 9781789061833 (Paperback)
ISBN: 9781789061840 (eBook)
ISBN: 9781789061857 (ePub)

Doi: 10.2166/9781789061840

This eBook was made Open Access in July 2022.

Contents

Chapter 1
Re-conceptualizing the sanitation value chain. *1*
Peter Emmanuel Cookey, Thammarat Koottatep and
Chongrak Polprasert

Chapter 2
Product design and development. *19*
Walter Gibson

Chapter 3
Product/equipment manufacturing **37**
Peter Emmanuel Cookey, Thammarat Koottatep and
Chongrak Polprasert

Chapter 4
Facility integration, installation and construction **75**
Harinaivo A. Andrianisa, Asengo Gerardin Mabia and
Peter Emmanuel Cookey

Chapter 10
Governance and enabling systems............................ *265*
Peter Emmanuel Cookey, Mayowa Abiodun Peter-Cookey,
Thammarat Koottatep and Chongrak Polprasert

Meet the editors

Dr. Peter Emmanuel Cookey is a Chief Lecturer with the Rivers State College of Health Science and Management Technology, Port Harcourt, Nigeria. He received his PhD in Environmental Management from the Prince of Songkla University, Hat Yai, Thailand. For the past 28 years, Peter has served as a higher education lecturer, trainer, speaker, magazine publisher, writer and sanitation professional, and co-authored the IWA Publications book *Regenerative Sanitation: A New Paradigm for Sanitation 4.0* and also contributed a chapter (Safe-Sanitation: Adaptive-Integrated Management Systems (SAIMS): A Conceptual Process Tool for Incorporating Resilience) in the book *Water, Climate and Sustainability* published by Wiley. He teaches postgraduate classes and mentors research as Adjunct Lecturer with the Institute of Geoscience and Environmental Management, (IGEM), Rivers State University, Nigeria; worked as a Senior Lecturer/Researcher on non-sewered sanitation with the IHE Institute of Water, Delft, Netherlands and a Senior Research Specialist with the Asian Institute of Technology (AIT), Bangkok, Thailand. Peter also trains future sanitation professionals at bachelor level on water, sanitation and hygiene courses to prepare them for work, entrepreneurships and research in the sector. His interdisciplinary academic background, covering health, environment, science, technology and governance, supports a versatile work experience in the private and public sectors covering governance, regulations, impact assessments, monitoring and evaluation, research, academic work, consulting/training and on global water and sanitation entities such as the Water Supply and Sanitation Collaborative Council (now the Sanitation and Hygiene Fund) in Geneva as well as being a pioneer member of the International Standard Development Committee of the Alliance for Water Stewardship team that developed and delivered the first global water stewardship standards. He also initiated and facilitated the Rivers State Water and Sanitation Sector reforms and co-chaired the drafting of legal and governance instruments for water and sanitation in the State. All of these experiences, along with being the co-Founder of the EarthWatch Research Institute (a non-profit) of Nigeria, co-Publisher of the *EarthWatch Magazine* and co-Convener of the EarthWatch Conference on Water and Sanitation, give Peter an in-depth and comprehensive perspective on the core issues of sanitation with personal and rich insights. You can find Peter's many academic articles such journals as *Humanities and Social Science Communications* (Nature); *Journal of Water Sanitation and Hygiene for Development* (IWAP); *Journal of Ocean and Coastal Management* (Elsevier) and *Ecological Indicators* (Elsevier); *Lakes & Reservoirs: Research and Management* (Wiley); *Hydrology* (MDPI); and *Journal of Water Resource and Hydraulic Engineering* (World Academic Publishing).

Dr. Thammarat Koottatep is Professor of the Environmental Engineering Management in the Asian Institute of Technology (AIT), Thailand. He is an internationally recognized professional on faecal sludge management, sustainable sanitation systems, nature-based technology for waste and wastewater treatment, and water reuse technology. His major scholarly contributions include publications of more than 60 refereed international journal papers, 3 books, and 9 book chapters. He has invented several toilet and sanitation technologies, one of which is patented to his credit, and several are filing. He has jointly developed professional master's degree programs in regenerative sanitation and marine plastics abatement. He has secured significant funded projects including, research and training grants, and most notably, the Bill & Melinda Gates Foundation grants on "Decentralized Wastewater Management in Developing Countries: Design, Operation and Monitoring" and "Development of Fecal Sludge Management Toolbox". He has contributed significantly to capacity building in faecal sludge management and decentralized wastewater treatment systems in Thailand and abroad, including capacity strengthening for experts, planners, and policy makers.

Dr Walter Thomas Gibson studied Natural Sciences at Corpus Christi College, Cambridge and went on to take a PhD in Medical Biochemistry at the University of Manchester. He subsequently had a long career in industrial R&D with Unilever, mostly at the interface between research, product development and marketing, as a scientist, R&D leader and resource manager. He worked at the Colworth and Port Sunlight laboratories in the UK, as well as the Hair Innovation Centre in Compiegne, France, and Unilever Technology Ventures in San Francisco. His major achievements were building a hair growth research programme, helping to launch the organics brand, reshape Unilever's corporate research strategy and establish an effective research capability in China. In these roles he worked closely with Unilever brand and technical leaders across the world and was also closely involved in major external partnerships with leading universities. His final role in Unilever was as Head of Biosciences for Home and Personal Care, during which time he interacted closely with the hygiene group at the London School of Hygiene and Tropical Medicine (LSHTM) and became closely involved in Unilever's external partnership with UNICEF. These experiences generated a strong desire to use what he had learnt in consumer-led innovation to help address the lack of new products in hygiene and sanitation, and in 2007 he set up Bear Valley Ventures to help catalyse innovations in these areas. He became a visitor at LSHTM under the guidance of the late Professor Val Curtis and this led to him directing a major research and innovation project in on-site sanitation (funded by a grant from the Bill & Melinda Gates Foundation to LSHTM) in 2009–2012. This resulted in two innovations to improve sanitation in low-income settings: the Tiger Toilet and the use of black solider fly (BSF) larvae to treat human faecal waste. He has since worked closely with a range of organisations in several countries to help develop and take these innovations to market.

Dr. Chongrak Polprasert is currently Professor of Civil and Environmental Engineering at the Thammasat School of Engineering, Thammasat University, Thailand. He received his Ph.D. in civil and environmental engineering under a Fulbright scholarship from the University of Washington, Seattle, USA, and began his career as Research Assistant at the International Development Research Centre in Ottawa, Canada, working on a low-cost sanitation and resource recovery project with the World Bank. Professor Polprasert was a faculty member at the Asian Institute of Technology (AIT) during 1997–2009, held the AEON Group Chair of Environmental Engineering during 1991–1995 and was Dean of the School of Environment, Resources and Development from 1996 to 2005. He was appointed as Director and Professor at the Sirindhorn International Institute of Technology (SIIT) of Thammasat University, Thailand, during 2009–2012. His research during the past 40 years has been in the areas of sanitation, waste reuse and recycling, and hazardous waste engineering and management. He has supervised over 120 master's theses and 23 doctoral dissertations conducted by international students. Professor Polprasert has published more than 120 papers in international refereed journals, in addition to conference and workshop proceedings. He is the author of the IWA Publishing book *Organic Waste Recycling – Technology and Management*, now in its third edition. Professor Polprasert has served in the US National Academy of Sciences panel on productive utilization of wastes in developing countries and the Water Environment Federation task force on natural systems for wastewater treatment. Professor Polprasert was the recipient of the Biwako Prize for Ecology (Japan), Outstanding scientist award (Thailand), Thailand Research Fund publication award, Outstanding researcher award in engineering and industry (Thailand), elected fellow of the Graduate School of Engineering, the University of Tokyo, Japan, fellow of the Royal Institution, Thailand, and listed in World's Top 2% Scientists (Career category). He has served as visiting professor/scholar at the United Nations University; the UNESCO-IHE Institute for Water Education, Delft, the Netherlands; Lulea University, Sweden; Tohoku University and Kyoto University, Japan. The above works have led Professor Polprasert to a number of appointments to committees and advisory bodies in both Thailand and abroad.

Meet the contributors

Dr. Olufunke Cofie is the West Africa Regional Representative for the International Water Management Institute (IWMI), responsible for leading IWMI's research in the sub-region. She is also the Flagship Lead for Water and Land Solutions, CGIAR Research Program on Water, Land, and Ecosystems. Her research is on water and sanitation linkages to agriculture and smallholder agricultural water management, counting over 20 years of field experience in international development research. She co-developed a waste-based soil ameliorant marketed in Ghana. In addition to her scientific research capacity, she has held research leadership positions for about 15 years. She has also consulted for international organizations, including the World Bank, UNEP, IFAD, and International NGOs, and serves on several Scientific Committees and Management Boards. A former university don, she holds a PhD in Soil Science and a Master's degree in Business Administration. She has published over 100 scientific articles through diverse outlets.

Jack Sim, a.k.a. "Mr Toilet", founded the World Toilet Organization in 2001. In 2013, 19 November was designated as UN World Toilet Day – the day is now celebrated globally each year to raises awareness of the need for action to end the sanitation crisis. Jack Sim became a Schwab Fellow of the World Economic Forum in 2005, Ashoka Global Fellow in 2006, and named a Heroes of the Environment for 2008 by Time Magazine. Jack broke the global taboo around toilets and sanitation by bringing it to centre stage with his unique mix of humor and serious facts.

Mayowa Abiodun Peter-Cookey With a childhood poring over encyclopaedias and ancient history books as her leisure activity, Mayowa Abiodun Peter-Cookey has been on a quest for knowledge and what it is capable of for a long time. Years later, a Bachelor in Philosophy degree gave her the foundation she needed for this quest for knowledge and learning; pairing this with her passion for writing and knowledge transfer, she launched the EarthWatch Research Institute, Nigeria with her husband and others, where she served as Administrator of the EarthWatch Conference on Water and Sanitation and co-Publisher/Chief Editor of the *EarthWatch Magazine* and co-Convener of the EarthWatch Conference on Water and Sanitation. The focus of most of her work has been water and sanitation and after 20 years of consulting and training in the sector and a Master's of Science in Human and Social Development that explored the concept of skill upgrading for informal workers, she has turned her attention to knowledge and skill capacity upgrades to improve sanitation conditions as her doctorate pursuit. Mayowa is a part-time lecturer with the School of Foundational Studies, Rivers State College of Health Science and Technology, Nigeria where she taught students of the School of Medical Sciences introductory courses in Philosophy and Logic. She has published articles on skill upgrading and co-authored book chapters on resilient sanitation and lake basin management and conference paper on sanitation education; and is a silent partner in many other writing projects upon which she hones her skills. She is the patient and content wife of Peter Cookey and proud mother of Hadassah, their precocious two-year-old.

Dr Harinaivo Anderson Andrianisa is Associate Professor of Water and Environmental Engineering with 22 years of experience in project management, R&D, consultancy, education and capacity building related to water, sanitation and environmental issues in sub-Saharan French speaking countries, principally West Africa and Madagascar. He joined the International Institute of Water and Environmental Engineering (2iE) in Burkina Faso in 2012 as Senior Lecturer of Urban Water and Sanitation. Now Head of the Water, Sanitation and Hydro-Agricultural Development Engineering Department, his interests are focused on pro-poor water/wastewater technologies and sustainable services, water and soil pollution in watershed affected by artisanal mining, circular economy and urban mining. Prior to joining 2iE, he was the Managing Director of a consulting company in Madagascar. Dr Andrianisa was an ERASMUS+ Fellow in 2019, a Japan Government Mobukagakusho Fellow from 2002 to 2008, and a Japan Society of Civil Engineers Research Prize winner in 2006. He is actively involved in various professional network including the Global Sanitation Graduate School initiative. He is married with three children.

Asengo Gerardin MABIA is both a Water and Sanitation and Agronomist Engineer. Junior Water and Sanitation Consultant at the International Institute for Water and Environmental Engineering (2iE) since 2021, He obtained his Master's degree in Water, Sanitation and Hydro- Agricultural Development Engineering at 2iE in 2021. He is currently working in technical support and organizational capacity building of small and medium enterprises in the WASH sector in Burkina Faso, at 2iE Incubator. He has been involved in many studies on faecal sludge management in Burkina Faso using a value chain approach and inclusive urban sanitation such as "Scoping study and rapid needs assessment for the establishment of a Capacity Hub network for inclusive urban sanitation in four (4) West and Central African Francophone countries: case of Burkina Faso" and "Reconceptualization and rethinking of the Facility Integration, Installation and Construction stage of the sanitation value chain (SVC): a case study in Burkina Faso". One of his main drivers is to advance access to sustainable sanitation in developing countries.

Mahugnon Samuel AHOSSOUHE is a Water and Sanitation Engineer. He obtained his Master's degree in Water, Sanitation and Hydro-agricultural Development Engineering from the International Institute of Water and Environmental Engineering (2iE) of Burkina Faso in 2021. Mr. Ahossouhe worked as a Technical Assistant Trainee at the Drinking Water Treatment Department of the National Office of Water and Sanitation (ONEA) in Burkina Faso. He did his end of training internship at the Department of Water, Sanitation and Hydro-Agricultural Development Engineering at 2iE and wrote his master's thesis on the 'Reconceptualization and Rethinking of the "Sanitation Services" stage of the "Sanitation Value Chain (SVC)": a case study in Ouagadougou, Burkina Faso'.

Foreword

2021 was the 10-year anniversary of the International Faecal Sludge Management (FSM) conference. What started off as a small conference in Durban, South Africa has become a growing field focused on non-sewered sanitation. Seeing the progress that has been made over the past 10 years is something to be celebrated – the advances in research, tools, frameworks, and knowledge exchange have shown us the need for expanding the types of sanitation technologies, but also that in 2022 two-thirds of the world's population are still without access to safely managed sanitation.

The initial deadline for the UN's Sustainable Development Goals was set for 2030, which is now just 8 years away. As the sanitation field has grown and adapted, so must our reference point and research. This book helps to bridge the complex nature of the sanitation value chain to bring a holistic and integrated view of the multiple and different actors involved in creating safely managed non-sewered sanitation. We have so often approached development issues as singular issues, creating silos.

The recent global COVID-19 pandemic caused a lot of the world to quickly shift their thinking, and what we at the FSM Alliance observed was how many of our established ways of thinking and our approaches showed the limitations of the silo approach. While Deepak Chopra has been viewed as controversial at times in the medical community, I found this quote on how we move forward very relevant to how we have viewed sanitation in the past:

> . . . *We use reductionist mental models that break up complexity into small pieces to examine the components of things at ever finer levels of granular detail – hoping we can put them back together coherently. But escalating crises prove we have exhausted the usefulness of this paradigm. Almost every major challenge humanity is facing, from cancer and climate change to food and consciousness, needs complex systems thinking to solve.* Chopra, D. (4 May, 2021) To survive our technological transformation, civilization needs a cognitive revolution. Yahoo Life. https://www.yahoo.com/lifestyle/deepak-chopra-survive-technological-transformation-civilization-needs-cognitive-revolution-154120760.html?guccounter=1

The current sanitation problem has largely been approached as a technical problem and it has been largely engineers approaching the issue from that perspective. Our failure to provide improved access to sanitation has been mostly viewed as not having the right technical solutions.

Scientific American recently published an article written by an engineering student, Grace Wickerson, on the limitations of looking at the world's problems from only an engineering perspective or viewed from the framework of

"*technical-social dualism, the idea that the technical and social dimensions of engineering problems are readily separable and remain distinct through-out the problem-definition and solution process.*" Wickerson, G. (24 February, 2022) The Culture of Engineering Overlooks the People It's Supposed to Serve. Scientific American. https://www.scientificamerican.com/article/the-culture-of-engineering-overlooks-the-people-its-supposed-to-serve/

There are numerous nontechnical parameters that must be considered when thinking about sanitation planning.

The City-Wide Inclusive Sanitation (CWIS) framework has also pushed our sector to expand out from purely an engineering point of view.

"*With its focus on equity, a CWIS approach challenges investment and service delivery norms that have excluded many communities and marginalized groups from safe sanitation facilities and services. A CWIS approach includes their interests and voices as core objectives of and resource for planning, design, and implementation of services.*" The Bill & Melinda Gates Foundation. (Undated). https://cwiscities.com/#cwismle https://s3.amazonaws.com/resources.cwis.com/learning/88/BMGFCWISFactsheet.pdf

All three of these examples represent vastly different industries, but the core theme is the need to expand our approach and use a more complex systems thinking approach. This type of integrated systems thinking will be necessary as the threats to our modern world continue to ignore the neatly drawn paradigms we've built in the past to categorize and respond to societal challenges.

This new publication and the new Integrated Functional Sanitation Value Chain (IFSVC) helps to address this issue by expanding the paradigm that we have used to approach sanitation solutions, with a focus on the role that both public and private entities will need to play. This book expands and operationalizes the CWIS framework by helping to translate into a new approach that looks at stakeholders that have not been included in the past. Chapter Two also begins with an emphasis on starting with consumer insights and how this has been overlooked in the past, particularly for low–middle income countries, and that not following human-centred design principles can have serious consequences, such as "*[w]ithout following a process like this there is a real risk that solutions will fail to have the impact they are intended to deliver, either through lack of demand or lack of use or both, even if they function well technically*". It also identifies that this is a non-linear approach, just like much in our human daily lives. The new IFSVC also includes chapters on sanitation advocacy, management systems, incorporates the role of the private sector, and the circular sanitation economy. Expanding to include these topics helps everyone involved in sanitation to shift their thinking and therefore designing solutions for an ecosystem and not just a singular technical issue.

The FSM Alliance is optimistic about the changes in our thinking that this book advocates for and is committed to helping advance this new expanded paradigm as we draw closer to 2030. We would also like to congratulate and thank the lead authors and all the contributing authors involved in the creation of this new approach to the sanitation value chain.

Jennifer Williams
CEO, FSMA

Foreword

When Peter Cookey and the author team reached out to the Toilet Board Coalition with the concept of this book in 2020, we were delighted to support. We can't help but think back to the humble launch of the sanitation economy framing in 2017 at SIWI's World Water Week in Stockholm. Our members, including leading businesses, investors and development organizations in the sector had pioneered this thinking in the founding of the Coalition in 2015 and to see it represented at World Water Week felt catalytic. Since that day, we have been fortunate to welcome thousands of innovative thinkers across our platforms to learn about the untapped value and opportunities of the sanitation economy.

In 2020, at the onset of the global COVID19 pandemic we sat (virtually) with Peter to walk through the vision of this book, the voices and breakthroughs that could feature in it and the impact we could jointly envision this piece to realise. To see and read the full piece in its fruition is an immense credit to the remarkable efforts of Peter Emmanuel Cookey, Thammarat Koottatep, Walter Thomas Gibson, Chongrak Polprasert and their contributing authors. Here, we again feel an electric buzz, this time of a tipping point being reached in our sector.

We are thrilled at the global uptake of the sanitation economy framing over the last five years. This book, illustrating the stakeholders, roles and processes of the sanitation economy adds rich detail for local and national advocates interested to grow thriving sanitation economies in their communities. The authors demonstrate wonderful breadth of understanding and expertise on the sanitation economy and the power of an integrated, functional approach to the value in sanitation systems. We anticipate this book to rapidly be looked to as foundational literature for those pursuing careers in the sanitation landscape.

What we see when we look at sanitation, much like the authors, is a landscape of value – anchored in consumer relationships, renewable resources, impactful data and information about public health. We see an industry centred on human interaction, protecting and nurturing our planet. This is the philosophy behind why we were founded, to uncover and showcase the value in sanitation services and product provision. The framing of business opportunities and a global marketplace of sanitation-related services and products – the sanitation economy - was an organic next step.

The book's objective, to direct attention towards building an expansive Integrated Functional Sanitation Value Chain, giving birth to sanitation economies around the world, is superbly aligned with our values. Since the introduction of the sanitation economy framing, we have seen a dynamic shift in the understanding and value of market-based approaches to sanitation. Individuals in the important roles of advocacy, knowledge management and the enabling environment around these markets have increasingly understood the importance of this approach. Consequently, support from leaders and the donor and investment sector is now coming into alignment. The 2021 WHO/Unicef JMP report outlined that a quadrupling of progress is needed to achieve SDG6 on time – acceleration likely only to be achieved through strategic catalytic investment that moves beyond traditional funding models and brings innovative financing and private capital into the sanitation economy. As we globally learn how to work better with private sector stakeholders of all sizes in this sector, we will come to embrace not only their financial support but also their insights, wisdom and very importantly, their skills. The outlining in this book of roles and potential influences scoped for each stakeholder group of the Integrated Functional Sanitation Value Chain is an important step towards recognising and harnessing the unique value and impact that each individual in the IFSVC can bring.

We know Peter Emmanuel Cookey, Thammarat Koottatep, Walter Thomas Gibson, and Chongrak Polprasert's work here will further advance understanding and alignment across the sector. The concept of IFSVC enables academics and practitioners of a breadth of professions and studies to grasp the vast opportunities of the sanitation economy and an Integrated Functional Sanitation Value Chain. The inclusive and integrated approach presented in the following pages sets the stage for improved operations and facilitated relationships across and throughout the value chain, thus inviting greater engagement and accelerated progress.

We are routinely encouraged and emboldened by the success we see emerging around the world; the businesses at the heart of the sanitation economy that are growing two, three, four times faster than they were five years ago; the investment we see flowing into the sanitation economy in a more catalytic, informed and sustainable way; the re-prioritisation among leading donors to focus on market-based approaches and private sector engagement. We are learning and we are accelerating.

Alexandra Knezovich
Managing Director
Toilet Board Coalition

doi: 10.2166/9781789061840_xxiii

Preface

The concept of the integrated functional sanitation value chain (IFSVC) was proposed from our previous book on *Regenerative Sanitation: A New Paradigm for Sanitation 4.0* (section 6.3), based on the need to track the sanitation value chain (SVC) from stages of product design and development, manufacturing/production, facility integration, installation and construction, sanitation services, sanitation biomass recovery and conversion, marketplace and sales, sanitation advocacy, sanitation management knowledge, and governance as well as a system of enabling environment. The need to rethink the SVC has become necessary in order to provide better information and comprehension of the firms, businesses, enterprises and organizations that operate within the sanitation industry from input suppliers to the end market buyers. Through the development of efficient and effective IFSVC, the sanitation sector could enhance the achievement of the SDG 6 and foster the generation of additional finance, income, and employment for different groups who engage along the value chain, and in so doing contribute to other SDGs. IFSVC development building on entrepreneurial dynamics could improve safely managed sanitation services, competitiveness and value addition.

While integrated and functional safely managed sanitation services and poverty reduction constitute the goals of the IFSVC, clearly achieving this will require improvements in the operation and interactions of sanitation businesses, firms and enterprises. Production and service delivery is the philosophy that guide the objectives of this book. Thus, the IFSVC enables businesses and enterprises to evaluate their processes so that they can provide the greatest opportunities to reduce operational costs, optimise efforts, eliminate waste, and improve health and safety, as well as increase profitability. The concept of IFSVC enables the academicians, professionals, practitioners, businesses and entrepreneurs to see mixed economic, environmental and social gains not realized by the traditional sanitation value chain, thus bringing a much wider range of companies and other stakeholders into active engagement with sanitation systems and services. With this approach sanitation can also provide some solutions for water security, energy security, food security and health.

This book explores concepts, frameworks, principles and practical case studies that support and represent an IFSVC for sewered and non-sewered sanitation systems. The authors and contributors identified and examined practical and operative linkages within a systemic loop that capture various functions at different stages of activities within different enterprises and production/service processes related to sanitation management and economy from design and production to final market and user. This involves enterprises and ventures within each stage of the IFSVC as well as those businesses directly and/or indirectly involved with providing safely managed sanitation services/products across the local, national, regional and global supply chain and in particular communities where they operate. The focused intention is to direct thinking towards building an expansive SVC that supports the growth of the sanitation economy. Also, this book showcases up-to-date research findings to support the concepts, frameworks, and principles presented therein, and also applicable cases that highlight leading sanitation and related businesses, education and research organizations as well as global supply chain ventures involved in the provision of safely managed sanitation products, services and facilities; and do so from the IFSVC perspective.

The structure of the book is inter- and trans-disciplinary and is made up of ten chapters aiming to come in handy as a tool that provides guidance on defining elements necessary for the development and upgrading of SVC to IFSVC. This book provides effective and efficient learning material and the book is active in its presentations in that it helps the user/reader to practice what was learnt. The **Chapter Objective** introduces the user/reader to the focus of the chapter. The **Take Action** section challenges the user/reader to do something with the knowledge gained. The section on **Journal Entry** helps the user/reader to review issues in an objective manner and to make notes that can be referred to at a later date. The **Reflection** section focuses on the ability of the user/reader to analyze and interpret the issues raised based on facts and the real situation on the ground. The **Guiding Questions** section provides learning exercises for the user/reader to further improve their knowledge, skills and competency. It is hoped that all these sections will challenge the minds of users/readers of this book enough to find answers and solutions in areas where they never existed before.

Chapter 1 introduces the readers to the concept of the integrated functional sanitation value chain (IFSVC) and its relationship to sanitation economy. Chapter 2 highlights the need to identify innovations to generate higher value end-products that would transform the economics of the final stage of the chain and drive closer integration. Chapter 3 raises the case for manufacturers in developing and developed countries to become more integrated into the sanitation manufacturing value chain in order to drive demand and affordability. Chapter 4 provides a detailed insight into the complexity of integration, installation and construction services. Chapter 5 emphasizes the importance of extending public–private partnerships, and improved coordination of all the actors in service delivery. Chapter 6 focuses on sanitation biomass recovery and transformation by enhancing resource recovery and reuse in a circular sanitation bioeconomy. Chapter 7 gives an indication of the size of the sanitation market and requirement for growing the sanitation economy. Chapter 8 focuses on sanitation advocacy business models that support a common cause by creating high social impacts with a sizeable market for sanitation products and services. Chapter 9 reveals how building a strong foundation in sanitation knowledge management could cope with the complex dynamics inherent in the sector, as this is vital to unlocking value across the IFSVC. Finally, Chapter 10, stresses the importance of strengthening linkages between all actors and stakeholders, along with the role that governance can play. Attempts to address these challenges will require the transformation of the sector to a more economically viable state, especially for the non-sewered parts of the sanitation spectrum. In conclusion, companies, entrepreneurs, enterprises and all stakeholders need to map and understand their sanitation business value chain to be able to make the best of the sanitation economy.

The principal audiences for this book include undergraduate and graduate students of sewered and non-sewered sanitation, environmental and biological sciences, environmental and public health, community health, engineering, and emerging sanitation management as well as researchers, professionals, practitioners, advocacy-agents, regulators, knowledge providers, policy makers and solution providers in related fields in both developed and developing countries. This book can be used as a textbook to teach; a resource for research; a reference for professionals and practitioners; and for planning and implementation of sanitation solutions; advocacy and intervention; developing, sharing and managing sanitation knowledge, as well as developing and implementing institutional and regulatory frameworks.

This work was written within a year and half and relied on the direct and indirect support of many people. We are grateful to Jennifer Williams of the Faecal Sludge Management Alliance (FSMA) for their supporting this book for online open access and agreeing to do write a Foreword for this book. We are also grateful to Daigo Ishiyama of LIXIL, Babitha George of Quicksand, Marc Lewis of the BioCycle, David Auerbach and Sheila Kibuthu of Sanergy, Eduardo Perez of Global Communities, Alison Parker of Cranfield University, Sejal Tembwalkar of 3S India, Claire Furlong of IHE-Delft, Geoff Revell of WaterSHED, Atitaya Panuvatvanich of Asian Institute of Technology (AIT), John Sauer of PSI, Jim Lane of Sanivation, and Ling Tao of US National Renewable Energy Laboratory (NREL) for their input, comments and image provision. Significant contributions by Mayowa Abiodun Peter-Cookey, EarthWatch Research Institute, Port Harcourt, Nigeria are also appreciated.

We would like to express our sincere appreciation to Mark Hammond of IWA Publishing, London, and the entire IWA staff for their professional support in the publication of this first edition.

Peter Emmanuel Cookey
Thammarat Koottatep
Walter Thomas Gibson
Chongrak Polprasert
Port Harcourt, Nigeria
February 2022

Abbreviations and symbols

Abbreviations	Meanings
ABHS	Alcohol-based hand sanitizer
ASPs	Activated Sludge Plants
AD	Anaerobic digestion
AW	Agricultural waste
ASSC	Augmented sanitation service chain
BoP	Base-of-Pyramid
BMGF	Bill & Melinda Gates Foundation
BOL	Beginning of life
CAGR	Compound annual growth rate
CEPTs	Combined Effluent Treatment Plants
CBS	Container-based sanitation
CoPs	Communities of practice
CSR	Corporate social responsibility
CSE	Circular sanitation economy
CE	Circular economy
CBE	Circular bio-economy
Covid-19	Coronaviruses
DBO	Design-build-operate
EWP	End Water Poverty
ETPs	Effluent Treatment Plants
EOL	End-of-life
EU	European Union
FPW	Fermentation processing waste
Forestry residue	Forestry residue
FW	Food waste
FS	Faecal sludge

FSTPs	Faecal sludge treatment plants
FSM	Faecal sludge management
GHGs	Greenhouse gases
GoI	Government of India
GVCs	Global value chains
HTL	Hydrothermal liquefaction
HTC	Hydrothermal carbonization
IFSVC	Integrated functional sanitation value chain
ICT	information communication and technology
IC	Intellectual capital
IYS	International Year of Sanitation
KA	Knowledge Application
KAD	Knowledge Acquisition/Capturing
KE	Knowledge Evaluation
KSR	Knowledge Storage and Retrieval
KCD	Knowledge Creation and Development
KD	Knowledge Dissemination
KP	Knowledge Protection
KM	knowledge management
KMS	Knowledge Management System
KR	Knowledge resources
KI	knowledge innovations
KE	Knowledge Economy
KIFs	Knowledge intensive firms
KIOs	Knowledge intensive organisations
KIBSFs	knowledge intensive business services firms
KISA	Knowledge Intensive Services Activities
KIS	Knowledge Intensive Services
KIAs	Knowledge Intensive Activities
KVC	Knowledge value chain
KExps	Knowledge experts
KEnts	Knowledge entrepreneurs
KWers	Knowledge workers
LB	Lignocellulosic biomass
LM	Livestock manure
MTA World	Mondragon Team Academy
MPW	Marine processing waste
MDGs	Millennium Development Goals
MSEaP	Manufactured sanitation equipment and allied products
MOL	Middle-of-life
MNCs	Multinational corporations
NGOs	Non-governmental organizations

NLB	Non-lignocellulosic biomass
OHS	Occupational health and safety
O/M	Operations and maintenance
PPP	Public private partnership
ReGenSan	Regenerative Sanitation
ROI	Return-on-investments
SA	Sanitation advocacy
SAC	Sanitation advocacy campaign
Sani-K	Sanitation knowledge
Sani-KMart	sanitation knowledge market
SWA	Sanitation and Water for All
SDC	Sanitation advocacy campaign
SDOs	sanitation advocacy organizations
SS	Sewage sludge
SIVC	Sewer Infrastructure Value Chain
STPs	Sewage Treatment Plants
SDGs	Sustainable Development Goals
SAMaT	Sanitation advocacy management tool
SACV	Sanitation advocacy value chain
SDPs	Sanitation-derived-products
SGVC	Sanitation global value chain
SFIIC	Sanitation facility integration, installation and construction
SFIICVC	Sanitation facility integration, installation and construction value chain
SMPE	Sanitation manufactured products and equipment
SMPVC	Sanitation manufactured products value chain
SMVC	Sanitation manufacturing value chain
SBRC	sanitation biomass recovery and conversion
SBRCVC	sanitation biomass recovery and conversion value chain
SSVC	Sanitation services value chains
SaniM-KVC	Sanitation Management Knowledge Value Chain
SaniM-K	Sanitation management knowledge
Sani-KM	Sanitation knowledge management
Sani-Kmart	Sanitation knowledge market place
Sani-KRs	Sanitation knowledge resources
Sani-KWers	Sanitation Knowledge Workers
Sani-KExps	Sanitation Knowledge Experts
Sani-ERT	Sanitation Education, Research and Training
SSC	Sanitation service chain
S-SP	Small-scale producers
SPVC	Sanitation production value chain
SHPE	Sanitation/hygiene products enterprises

SSE	Smart sanitation economy
SSC	Sanitation service chain
SVC	Sanitation value chain
SUSANA	Sustainable Sanitation Alliance
TBC	Toilet Board Coalition
UN	United Nations
UNIDO	United Nations Industrial Development Organization
USAID	United State Aids for International Development
UCD	User- Centred Design
UDDT	Urine diverting dehydrating toilet
VC	Value chain
VIP	Ventilated improved pit latrine
WSSCC	Water and Sanitation Collaborative Council (now Sanitation and Hygiene Fund)
WC	Water closet
WTE	Waste-to-energy
WSP	Water and Sanitation Program
WHO	World Health Organization
WTO	World Toilet Organization

doi: 10.2166/9781789061840_0001

Chapter 1

Re-conceptualizing the sanitation value chain

Peter Emmanuel Cookey, Thammarat Koottatep and Chongrak Polprasert

Chapter objectives

The objective of this first chapter is to present the integrated functional sanitation value chain (IFSVC) map to get an overview of the stages of the IFSVC, the actors and their functions in the value chain, and also the flow of products and services through the chain. The general IFSVC map also provides information on governance and enabling systems.

1.1 INTRODUCTION

The value chain (VC) system is a key way to address important sanitation technological and institutional gaps in production and service delivery (Drost *et al.*, 2012) and could constitute a natural platform for development actions and also serve as a market systems approach to improve access to safely-managed sanitation (Springer-Heinze, 2018a). The value chain concept is used to gain a better understanding of how and where enterprises and institutions are positioned within a chain and identify opportunities and potential leverage points for improvement (Rawlins *et al.*, 2018). Sanitation value chain (SVC) actors and/or enterprises have several interests in common and all depend on the same end-markets to be successful whereby it is necessary for them to interact with each other and the same enablers and supporters to reach the market. The SVC provides the sustainable market that enables more customers and entrepreneurs to exchange products and services, thereby increasing market depth and reducing the burden on public finance. VC also optimizes the finance, products and information flow that enterprises can identify and exploit for new opportunities and to reduce external threats (Springer-Heinze, 2018a; USAID, 2018).

This book considers the sanitation value chain (SVC) to be the full range of activities that are required to bring a product and/or service from conception through the different phases of production to delivery to final consumers and disposal after use (Kaplinsky, 2000, 2002; M4P, 2008). In a narrow sense, this includes the range of activities performed within a firm to produce a certain output (Porter, 1985), while in the broad approach, it is a complex range of activities implemented by various actors (primary and secondary producers, processors, traders, service providers) to bring products and/or services from conceptualization through chains to the sale of product and/or provision of services (M4P, 2008). This approach does not only look at the activities implemented by a single enterprise, but also includes all its backward and forward linkages, until the products and services are linked to the end-users in an 'integrated-functional' system (Kaplinsky, 2000, 2002; M4P, 2008). Thus, this concept encompasses the issues of organization and coordination, and the strategies and the power relationship of the different actors in the chain (M4P, 2008). This, however, is different from the usual perspective of the value chain in the sanitation sector whereby the SVC is depicted as actors and businesses within the sanitation service chain (SSC) and other enterprises involved in faecal sludge and sewage management (FSM) which are aspects of the sanitation value chain (SVC) (Strande *et al.*, 2014; WWAP, 2017). The SSC and FSM are descriptive frameworks with distinct technological steps showing the flow of sanitation service provision to end-users and do not really represent the whole picture of the sanitation value chain (SVC) (Hyun *et al.*, 2019; Osann & Wirth, 2019). This is a restrictive use of the concept of value chain because it is focused within the SSC actors and local businesses involved in FSM activities alone, but these are just stages/levels in the sanitation value chain (see Figure 1.1).

Using the SSC as the SVC may only capture the value-added activities at the sanitation services stages of the integrated functional sanitation value chain (IFSVC), which has probably led to misconceptions by some practitioners that the value that could generate additional financial flow in the sanitation system is at the back end of the SSC (Murray & Ray, 2010). This misleading perspective limits the efficiency and practicality of the complex sanitation value chain and could interfere with the circular bioeconomy potential of the sanitation economy (Akinsete *et al.*, 2019; Koottatep *et al.*, 2019). The reconceptualization of the SVC is supported by Hyun *et al.* (2019) when they provided an augmented sanitation service chain (ASSC) and called for the redesigning of sanitation systems that could contribute to better health and cleanliness, climate change adaptations, support for sustainable food systems, and human rights for the poorest communities. The ASSC concept expands upon the traditional SSC materials and social functions, such as the decision makers, key financial actors and how they affect other entities within the sanitation system. However, various materials flows, social functions,

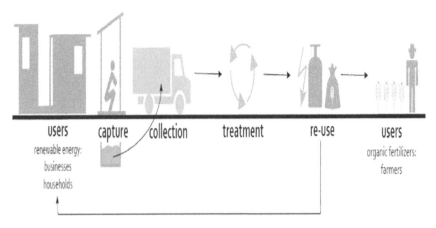

Figure 1.1 Sanitation service chain (SSC) used to illustrate sanitation value chain (SVC) (from van Welie and Romijn, 2018).

and the roles of actors in the chain that determine the goals were not originally part of the sanitation service and/or value chain (Hyun *et al.*, 2019). On the other hand, the ASSC also obscures other players in the SVC that do play active roles in the sanitation economy, and this actually provides more evidence for new perspectives of the SVC. The WWAP (2017) also aligned a financial flow model of faecal sludge management with the SSC to indicate where a utility achieves full cost recovery through discharge fees and revenues from selling treated faecal sludge (Strande *et al.*, 2014), but this did not properly identify the specific value-added activities that could drive the business model responsible for the financial flows and account for failure in most practical applications. It is evidence that there are missing links in the SVC as it is perceived currently, and that the complete linkages and interconnection within the sanitation markets could be explored and mapped to really understand, appreciate and properly activate the sanitation economy, particularly in developing countries.

Therefore, re-thinking the sanitation value chain (SVC) becomes necessary in order to provide better information and understanding of the firms that operate within the sanitation industry, from input suppliers to end market buyers, and the support markets that provide technical, business and financial services (Market Links, 2021). Such a broad scope for analyzing the sanitation industry is needed because the principal constraints to competitiveness may lie within any part of its stages and/or levels as well as the market system or the environment in which it operates (Market Links, 2021). Thus, the SVC should capture all the value-added activities and enterprises within the entire sanitation economy (Koottatep *et al.*, 2019), and should not be restricted to the SSC, but directed towards building an expansive SVC that could transform the sanitation economy. This perspective is critical for the innovation and great opportunities to drive design, production and services that not only match individual, societal and cultural expectations, but also activate cultural and behavioural changes that support socioeconomic well-being, provide employment, business and investment opportunities, as well as reducing human exposure to sanitation matter. One major obstacle in modernizing sanitation practices and integrating resource recovery is the need to ensure human health and safety throughout the sanitation value chain (Bischel *et al.*, 2019), and all of these will in turn boost the possibilities of achieving the SDG 6 sanitation targets and other related SDGs.

The SVC could enable businesses to evaluate their processes so that they could reduce operational costs, optimise efforts, eliminate waste, and improve health and safety, as well as increase profitability (Reese *et al.*, 2016). It also conceptualizes activities needed to provide products or services to customers and depicts the way a product gains value (and costs) as it moves along the path of design, production, marketing, delivery and service to customers (Springer-Heinze, 2018a, 2018b). Furthermore, it includes all producing and marketing enterprises operating in the entire sanitation sector that create and deliver services in the sanitation market such as product and service design and development (Chapter 2); manufacturing of infrastructure (fixtures and fittings) as well as consumables (like hand sanitizer, toilet paper, disinfectant) (Chapter 3); facility, integration, installation and construction services (Chapter 4); sanitation services within the sanitation service chain (SSC) (Chapter 5); and sanitation-biomass transformation enterprises (Chapter 6) that use the biomass generated from the sanitation service chain (SSC) for safe recovery of water, nutrients, organic matter and energy which are finally delivered to customers (Koottatep *et al.*, 2019); SVC as a market-based approach where value is added at different stages of the sanitation sector including marketing (Chapter 7); advocacy for policy and behaviour change (Chapter 8); and sanitation management knowledge (Chapter 9) to deliver innovative and fit-for-purpose products and services. Thus, the sanitation value chain depicts how customers' value can accumulate along the chain of activities that lead to an end-product or service (Koottatep *et al.*, 2019).

The expansive transformation of the SVC has the capacity to address the current global sanitation backlog, estimated at 2.3 billion people without access to any sort of improved sanitation facility and 4.5 billion – more than half the world's population – that still lack access to safely managed sanitation (TBC, 2019a). These concerns and the need to shift the SVC from an initial focus on product/service design/development and follow Regenerative Sanitation (ReGenSan) principles are the reasons why an integrated-functional version of the SVC was proposed by Koottatep *et al.* (2019) and is expanded in this publication: an integrated functional sanitation value chain (IFSVC).

1.2 PORTER'S VALUE CHAIN MODEL

This concept was introduced by Michael Porter of Harvard Business School and has gained wide popularity for understanding the fundamentals of any development business (McLeod, 2012). Porter (1985) viewed the organization as a chain or series of processes that each add value to the products and/or services for the customer. Thus, the framework promotes firm competitiveness by directing attention to the entire system of activities involved in producing and consuming a product (Inomata, 2017). Porter (1985) considering competitive advantage has used the value chain framework to assess how a firm should position itself in the market and in the relationship with suppliers, buyers and competitors. The idea of the competitiveness of an enterprise can be summarized as follows: how a firm provides customers with a product or service of equivalent value compared with competitors, but at lower cost; or alternatively, how the enterprise can produce a product or service that customers are willing to pay a higher price for (M4P, 2008). This provides a tool that firms can use to determine their source of competitive advantage, as Porter argued that the source of competitive advantage cannot be detected merely by looking at the firm as a whole; rather the firm should be broken down into series of activities where competitive advantages could be found in one or more of the activities (M4P, 2008; Porter, 1985, 2008). Porter's Value Chain Model defines the value chain into two distinct types of activities (see Figure 1.2):

(1)   Primary activities which include inbound logistics (getting materials in for adding value by processing it); operations (which are all the processes within the manufacturing); outbound logistics (distribution to the points of sale);

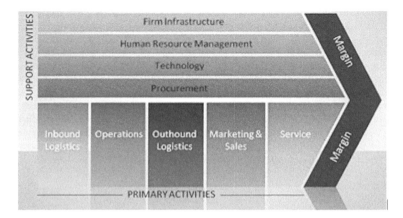

Figure 1.2 Porter's value chain competitive advantage (Porter, 1985, 2008). (image credit: Michael porter)

marketing and sales (selling, branding and promoting); and service (maintaining the functionality of the product, post sale) (Porter, 1985, 2008).

(2) Support activities that feed into all the primary activities, such as the firm infrastructure, human resources, procurement to buy or source goods at the right price and technology (Inomata, 2017; Porter, 1985, 2008).

Porter (1985) argues that a company's business is best described as value change. Managers can win the competition by targeting elements of the value chain through their company for specific purposes (Sutarmin, 2016).

A company's value chain activities make up part of a larger stream of activities in an industrial sector, which Porter referred to as a value system; either upstream (i.e. suppliers) and/or downstream (i.e. distribution system). This includes the suppliers that feed the companies with the input materials used to create and produce/provide products and services and the network of organizations that interact to meet the demand of the market and customer (Porter, 1985, 1990, 2008).

1.3 GLOBAL VALUE CHAIN AND/OR NETWORK APPROACH

More recently the concept of value chains has been applied to the analysis of globalization (Kaplinsky, 2000, 2002; Kaplinsky & Morris, 2001; Springer-Heinze, 2018a, 2018b), using the framework of the value chain to examine the ways in which firms and countries are globally integrated to assess the determinants of global income distribution (M4P, 2008). By mapping the range of activities along the chain, a value chain analysis breaks down the total value chain earnings into the rewards that are achieved by different parties in the chain. This approach can also show how firms, regions and countries are linked to the global economy, especially in the case of sanitation, where the critical products and services are not produced in most developing countries where SDG 6 targets are not being met. SVC may be national, regional, or global, depending on the goods and services in question, as activities may take place in several parts of a country, regions, different countries and on different continents, although some cases may be more limited, involving only a few locations within country across the globe (Springer-Heinze, 2018a, 2018b). For instance, world ceramic sanitaryware production sources raw materials from several countries while the factories or industries are located in different countries and the products are distributed and marketed all over the world (see chapter 3). The 'value network' and/or 'integrated functional' approach extends the idea of the value chain to social networks and views enterprises as being embedded in a complex of horizontal and vertical situations involving multiple players and where processes are not actually linear (McLeod, 2012; Trienekens, 2011).

The SVC incorporates business models for toilet provision, products and services, re-usable water and nutrients, data and information and is designed to provide new benefits across the economy and society (Akinsete *et al.*, 2019). Thus, the SVC can be defined as a socioeconomic system that includes all enterprises cooperating to serve the sanitation market. The enterprises forming the SVC interact constantly – buying and selling products and services, exchanging information, supporting each other, and cooperating to pursue shared interests (Springer-Heinze, 2018a; Trienekens, 2011). These enterprises are the core of a wider value chain community that consists of private associations, specialized service providers and industry-specific public organizations that provide support. In essence, improving the value chain builds on collaborations between partners in the sanitation industry at large (Springer-Heinze, 2018a). The network of value chains includes market outlets from local, regional, national and international situations that focus on vertical and horizontal relationships (formalizing business linkages through written agreement and contracts) between actors in the chain (Trienekens, 2011).

1.4 RELATIONSHIP BETWEEN THE SANITATION ECONOMY AND VALUE CHAINS

Integrating the sanitation economy with value chain systems presents vast potential for global economic growth and has the ability to transform businesses. The integrated functional sanitation value chain (IFSVC) system brings together all the designers, producers, processors, buyers/users and sellers in integrated-systemic functional networks and/or chains that add value to goods and services as they pass from actors along the spectrum of conception to the final consumers which include knowledge management and advocacy actors in the sanitation economy. This is excellently captured by the concept of the sanitation economy as postulated by the Toilet Board Coalition (TBC) as a 'robust marketplace of products and services, renewable resource flows, data and information that could transform cities, communities, and businesses. It creates a new trajectory for sanitation management by addressing it as a 'solution provider for sectors and governments facing constraints on essential resources such as water, nutrients, energy and proteins' that could also provide insightful and innovative data for public health and consumer behaviour and the invention of integrated and novel solutions and technologies for all-round sanitation systems (see Figure 1.3). In addition, it provides another option for tackling sanitation challenges and its impact on other SDGs and socioecological systems by translating global sanitation needs into sustainable business solutions with outstanding value of multi-billion dollars a year in the marketplace (TBC, 2019a). The sanitation market, where transactions and trade of sanitation products and services take place, consists of three interconnecting sections as described below.

- **The Toilet Economy:** this encompasses both the products and services that provide safe toilet access and maintenance (as well as related products/services) for all at both private and public levels across centralized and decentralized and/or

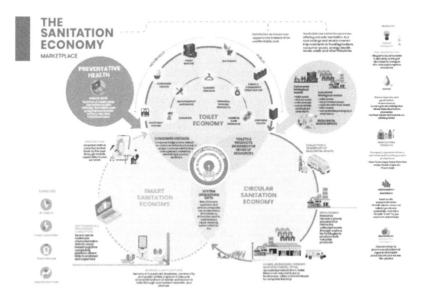

Figure 1.3 Illustration of sanitation economy (Source: Toilet Board Coalition, 2019a, 2019b).

sewered and non-sewered systems within contexts of high and low water tables, high and low income, as well as rural, urban and peri-urban areas (e.g. household toilets, community toilets, public toilets, auxiliary products). Also, there are a variety of sanitation solutions, both dry and wet, such as full waterborne systems, septic tanks, biodigesters, container-based systems, pit and improved ventilated pit latrines, and composting toilets. These systems operate differently according to resources required, energy used, and most importantly how the waste produced is processed (Frost & Sullivan, 2021). Resources are collected from these systems and are used to add value in the Circular Sanitation Economy, while data about consumer usage are captured by the Smart Sanitation Economy to provide knowledge for users, operators, and other businesses, which could be used to improve user experience, operating efficiency, and range of potential products and services (Frost & Sullivan, 2021). Also, institutions across the globe have 'reinvented' the toilet and produced a variety of alternatives to the conventional flush toilet that many of use today. New innovative systems have been developed to use less or zero water, to be more energy efficient, and if waste is produced, to be treated and reused or transformed into products such as fertilizer. These solutions can provide full on-site sanitation solutions that include generation and containment, as well as treatment of waste in areas without access to sewer systems and water supplies; and they can bring sanitation to the world's poorest communities and contribute to the sustainable development goals (Frost & Sullivan, 2021).

• **The Circular Sanitation Economy:** this alternative to a linear economy is one in which resources are reused as much as possible to minimize the generation of waste. The aim is to maximise the value that can be extracted from them during their lifetime and the products or materials are recovered or regenerated at the end of their lifecycle (Frost & Sullivan, 2021). The circular sanitation bioeconomy moves away from the traditional waste management view of human waste as having no value; that is toilet resources (i.e. human waste and perhaps farm animals' too) are recovered, recycled and reused to produce other organic products, often in combination with solid waste within and around a circular bioeconomy system. Nutrients, water and other matter are recovered and treated as sanitation resources to create value-adding products (e.g. biogas, electricity, biochar, organic fertilisers, proteins, animal feed, etc).

• **The Smart Sanitation Economy:** this will be present in smart cities where the sanitation system is digitized. In this economy, data of consumers and service providers are collected to monitor, improve and maintain systems within the sanitation economy. This includes collecting data on consumer health, the usage of public toilets, and sewage treatment facilities (Frost & Sullivan, 2021). For example, sanitation intelligence is provided by gathering data and information through technologies that capture usage, sewage treatment, health indicators, maintenance and repair systems to produce knowledge for informed decisions by governments, businesses, citizens and other stakeholders (e.g. health data, smart technologies, smart logistics, data analytics and applications) (Akinsete *et al.*, 2019; TBC, 2019b). A sanitation economy with a re-designed SVC presents vast potential for global economy growth while addressing one of the most urgent major challenges of our time, achieving universal access to improved safely managed sanitation (SDG6). It monetizes toilet provision, products and services, biological resources, data and information to provide benefits across the economy and society (Akinsete *et al.*, 2019).

1.5 INTEGRATED FUNCTIONAL SANITATION VALUE CHAIN (IFSVC)

Value chain integration is a process by which multiple enterprises within a shared market segment of the sanitation sector collaboratively loosely to tightly plan, implement and manage the flow of goods, services and information along the value system in a way that increases customer value and optimizes the efficiency of the chain (Dobbs, 1998; Papazoglou *et al.*, 2002). The integration shows the 'extended enterprises' that are creating and enhancing customers-perceived value by means of cross-enterprise collaboration (Papazoglou *et al.*, 2002). On the other hand, 'functional integration' aims at more efficient linkages of elements in the sanitation supply chain, namely to ensure that suppliers closely meet the requirements of customers in terms of costs, availability, and time (Rodrigue, 2006). Thus, 'integrated-functional' complementarity is established through a set of supply/demand relationships involving physical and information flows among the value chain actors and stakeholders, thereby ensuring that efficiencies and economies are achieved through the principles of flow (Rodrigue, 2006). Since sanitation activities are not necessarily locked inside a single organization (or even a business unit within an organization), but are more of a large-scale set of interactions between players in multiple industries, companies and/or countries, the sequence of value-adding activities is not particularly linear or constrained within a given sequence.

The value chain 'network' and/or 'integrated functional' system differs from the Porter value-chain concept by shifting the focus from firms to the configuration of business activities (Inomata, 2017). Thus, a corporate entity is first decomposed into a set of businesses with individual functions that constitute analytical units (business activities) – such as product design, materials procurement, material marketing and distribution – that tend to be defined in a way to pursue the individual objective of that particular unit, which may or may not conflict with the objective of other units. This implies that all activities in the value chain are collectively organized to ensure optimization of the functioning of the corporate entity as a whole (Inomata, 2017; Porter, 1985). The IFSVC is then conceptualized as production and service networks of horizontally and vertically related enterprises and/or businesses loosely or/and tightly joined together locally, regionally, nationally and internationally with the aim of working towards providing products or services to the sanitation market (Trienekens, 2011).

This IFSVC concept focuses on the operational functions within the sanitation sector in combination with sanitation enterprise operators, external actors such as supporters of the sector (professional associations, communities of practice, etc.) that provide support to the growth of the SVC, and enablers that govern and regulate (government) SVC activities (Koottatep *et al.*, 2019; Springer-Heinze, 2018a, 2018b). It expands the sanitation value chain activities beyond the SSC and ensures comprehensive and integrated-functional solutions by taking the total system approach of the sanitation economy, which fundamentally realigns flows of products, services and information as well as recovery of nutrients, water, energy, data, and finance within the economy; acting as a root cause solution for a number of areas beyond sanitation itself (Akinsete *et al.*, 2019). The IFSVC pre-supposes that the SVC (and sanitation economy) cannot be effective unless all stages are working connectively in an integrated synergistic systemic manner (Koottatep *et al.*, 2019) across local, regional and international arenas focusing on vertical and horizontal relationships between actors (Trienekens, 2011). Traditional value chains are linear processes, where upstream suppliers provide material for products or services, and downstream units provide distribution and point of sale (Chofreh *et al.*, 2019; Reese *et al.*, 2016). However, the IFSVC is a non-linear system that links all enterprises, businesses and actors across the sanitation spectrum from local, regional, national and international systems involved in adding value to each stage of the sanitation value chain until the products and/or services reach the hands of the end-users.

The value chain is made up of eight main value-chain stages (product design and development; manufacturing services; facility integration; installation and construction; sanitation services; sanitation biomass recovery; marketplace and sales; sanitation advocacy and sanitation management knowledge) as well as governance and business enabling systems that link in a closed loop and represent the 'enterprisation' of the entire sanitation sector, including the SSC. This implies that the IFSVC explores all enterprises, ventures and activities within the whole system by identifying the functional linkages within a systemic loop that captures the various operations and different stages related to sanitation management from conceptualization to the final market, as well as the enterprises and ventures within each stage. In short, value-added activities within each stage and the way the stages link to each other through enterprise interactions depict the integrated functional value chain system. Also, added value is created at different stages and by different actors throughout the IFSVC. The value added may be related to quality, cost, delivery times, delivery flexibility, innovation, and so on. (Trienekens, 2011), while the size of the value added is decided by the end-customer's willingness to pay. Opportunity for an enterprise to add value depends on a number of factors, such as market characteristics (size and diversity of market) and technological capabilities of the actors (Kaplinsky, 2000).

All the stages are anchored (throughout the chain) by the supporting and enabling structures and mechanisms (Koottatep *et al.*, 2019), see Figure 1.4, and must be closely interconnected.

Details of the stages are provided below:

(I) *Product design and development* – the initial stage of planning and conceptualization of sanitation products and services

(II) *Manufacturing* – the stage of primary and secondary production of user products, supply and distribution to the market and service delivery such as toilet combo, faucets, pipes, and so on.

(III) *Facility integration, installation and construction* – the stage of installation, construction, connection, operation and fabrication of sanitation facilities at point of use (such as septic tanks, public toilets, treatment facilities, pipe-laying, etc).

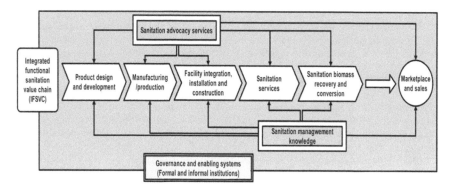

Figure 1.4 Nine stages of the integrated functional sanitation value chain (Source: authors).

(IV)   *Sanitation services* – the stage of activities after the user interface from collection to disposal and/or treatment (such as emptying, transportation, disposal, etc.)

(V)   *Sanitation biomass recovery* – the recovery of organic nutrients, biogas, manure by various processes and their transformation for broader applications through the utilization of changed products for other purposes such as agriculture, aquaculture, horticulture, and so on.

(VI)   *Marketplace and sales* – the final market for all products and services as well as the sanitation materials obtained from the recovery and transformation processes. The major markets for sanitation products and services include (but are not limited to): the healthcare industry, transportation industry (air, land, water), rural and urban households, hospitality industry, agro- and allied industries, educational industry, businesses providing access to safely managed sanitation for their employees, across supply chains and in communities where they operate, and so on. The market provides feedback to the designers and developers at the conceptualization stage for appropriate, acceptable and marketable products/services and innovations

(VII)   *Sanitation advocacy* – building a critical mass of people to support a common cause by creating high social impact with a sizable market for sanitation products and services

(VIII)   *Sanitation management knowledge* – this involves managing knowledge sourcing, acquisition, creation, transformation, dissemination and usage as a key to developing innovations and competitive advantage as value is added to or created by available knowledge capital/assets to produce improved performance, capabilities and competences in individuals, organisations and industrial sectors

(IX)   *Governance and enabling systems* – enhancing and regulating services to all the operators in the value chain. Typical support services include setting of professional standards; provision of information; trade fairs and export marketing; research on generally applicable technical solutions; professional, vocational training or political advocacy. These services are often provided by business associations, chambers of commerce or by specialized public institutions. Typical enabling services are provided by relevant government institutions, major providers of public utilities, educational and research institutions, civil societies and intervention organizations, governance and regulatory services of the value chain.

The sanitation economy is dependent on the value-added activities across all stages of the IFSVC. To this effect, the IFSVC identifies the functional linkages that capture all the business activities at different stages in the sanitation economy (Koottatep *et al.*, 2019), as shown in Table 1.1, and also describes the socio-economic reality of the sanitation sector.

All chain actors, private enterprises in particular, need to understand the value chain they are a part of, its functioning and failure, and their own position in it. The results are used to prepare decisions on objectives and strategies (Springer-Heinze, 2018a). Sanitation enterprises can develop a vision of change and determine collaborative strategies based on a shared view on the state of the value change (Springer-Heinze, 2018a). Government and public actors could use analysis of the IFSVC to identify and plan supportive actions and monitor impacts of their policies on the value-chain players.

Table 1.1 Integrated functional sanitation value chain (IFSVC) activities and processes.

Stages	Enterprises/businesses
Product design and development services	Research and development, education and training institutes, wastewater treatment plant designers, home, commercial and toilets/accessories designers, architectural and interior designers, sanitary engineering consultants, sanitary wares and accessories designers, recovery, recycling and reuse system designers, town and urban planners, public and environmental health consultants, and so on.
Manufacturing services	Sanitary wares and accessories, treatment plant/accessories, toilets/accessories, recovery, recycling and reuse accessories, cement manufacturers, importers, suppliers, retailers, distributors, plumbing materials/ accessories, cleaning and hygiene products, metal, cements and wood work, suppliers/distributors, and so on.
Facility integration, installation and construction services	Fabrication and installation, logistics and transportation, architecture and real estate, sanitary engineering, town and urban planners, wastewater treatment plant installation, faecal and sewage sludge treatment plant installation, recovery, recycling and reuse, home and commercial building toilet installation, plumbing and cement works, public and mobile toilet installation, microenterprise services, sales, installation and construction, local mason, metal, woodwork and concrete building contractors, testing and quality control laboratories, and so on.
Sanitation service	Logistics and transportation, wastewater treatment plant operations and maintenance, faecal and sewage sludge collection, emptying and treatment, public and private utilities, environment and public health consultants, operations and maintenance, sanitary engineering, town and urban planners, cleaning and hygiene, testing and quality control laboratories, public, home, commercial building, mobile, toilet maintenance, plumbers, local artisan and masons, education, training and capacity development, and so on.
Sanitation biomass recovery services	Resource recycling and recovery, composting and organic matter recovery, wastewater resource recovery, faecal and sewage sludge treatment recovery, public and private utilities, certification and verification, health, safety, quality control and assurance, training/capacity building and so on., Also, organic fertilizers and manure, recycled water suppliers, biogas and energy producers and suppliers, aquaculture, horticulture, animal feeds producers, parks and gardens management, farmers' cooperative organizations, aquaculture cooperative organizations, certification and verification, health, safety, quality control and assurance, and so on.
Marketplace and sales	This is where demand meets supply, where buyers or customers meet suppliers and a transaction related to sanitation provision takes place. This is conducted by the sanitation enterprise (or in some cases a public utility) which arranges for promotion, production, distribution, sale, and delivery of the goods or services through its operations.
Sanitation advocacy services	These are enterprises involves in public education and influencing public opinion; research for interpreting problems and suggesting preferred solutions; constituents' actions and public mobilization; agenda setting and policy design; lobbying; policy implementation, monitoring, and feedback, and so on.
Sanitation management knowledge services	This is concerned with the generation, capture, storage and sharing of knowledge with an intent to take timely actions for increasing competitive advantage
Governance and enabling systems	Policy, legislators, regulators, guidelines and standards developers, land-use planning, sustainable financing, investors and banks, policymakers, and so on. Also, includes: ministries/departments/agencies responsible for environment, health, water resources, economic planning and cooperatives, natural resources, agriculture, fisheries and aquaculture, trade, commerce, industry, gender and development, education, information and communication, financial and insurance institutions, public and private investors, marketing and advertising, multilateral organizations, research community, international, regional, national and local non-governmental organizations, community-based organizations, and so on.

1.6 VALUE CHAIN ANALYSIS

Mapping the IFSVC will stand as the initial process in analysing the value chain so as to identify the main supportive business activities, and all related components and the relationships between them (Chofreh *et al.*, 2019). This will not only provide an overview of the system by identifying the position of value-chain actors but could also help to visualize key topics for the value-chain analysis and then structure the information according to the functions and stages of the chain. A mapping like this will reduce the complexity of economic reality into a comprehensive visual model (Springer-Heinze, 2018a, 2018b), see Figure 1.5.

The value-chain map can be characterized by eight generic elements of which five constitute the basic value chain maps: (i) marketed products/services or group of products/services that define the value chain, (ii) an end market in which the products/services are sold to customers, (iii) a series of value-chain stages through which the products/services reach the end market, (iv) enterprises or chain operators conducting the business operations, (v) business linkages between these operators, (vi) selected business linkages with sub-contractors and operational service providers, (vii) support service providers, and (viii) public agencies performing regulatory functions (Springer-Heinze, 2018a).

The basic value-chain map shows the micro level of the value chain, that is the value chain stages, the different types of operators and their relation to the end market. The value-chain operators are the owners of the merchandise along the chain. They buy the main raw materials, perform the productive/creative process and pass on the semi-finished and final commodities to the consumers in the end market. It is important to note that business linkages in the value-chain map always refer to the interactions between the value-chain operators (Springer-Heinze, 2018a).

Also, the concept of business models occupies a key role in the value chain analysis. To define a specific type of operator in the value chain map, analysts look for its business

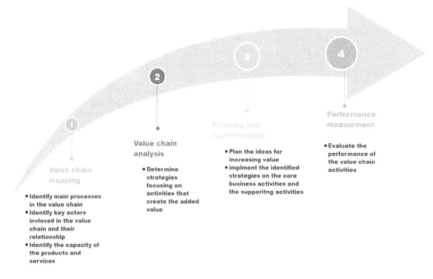

Figure 1.5 Steps in a value chain analysis (Source: Authors).

model. For instance, enterprises of similar sizes and with similar business models are grouped together, while enterprises that have a business model in common are classified as a particular category of operator. The business models of the operators constitute the backbone of the value-chain map (Springer-Heinze, 2018b). The methodology combines analytical and procedural aspects and could be structural, economic, environmental, social and poverty analyses:

(I)   The structural analysis involves mapping, which is a visual representation of the value chain system. Value chain maps identify products and end markets, business operations (functions), chain operators, and their linkages, as well as the chain supporters. The basic value chain map is a descriptive conceptual model. The value chain mapping is most essential to the chain and is the core of any value chain analysis (Springer-Heinze, 2018a, 2018b).

(II)   Economic analysis indicates market prices, quantifying the volume of product and the market shares of particular segments in the chain. The data are used to determine the value added along the stages and chain competitiveness which include cost of production and marketing (Springer-Heinze, 2018a, 2018b).

(III)   The environmental analysis determines the impact of sanitation systems on the ecosystems, natural resource and climate change as well as the contribution of IFSVC's of sanitation biomass recovery and sanitation services to the circular bioeconomy. The main tool is the construction of an impact matrix that allows identifying impacts of the chain on the environment (Springer-Heinze, 2018a, 2018b).

(IV)   The social and poverty analysis focus on the vulnerable groups that do not have access to safely managed sanitation services and how the chain can use to provide these essential services to the poor of the society (Springer-Heinze, 2018a, 2018b; Sutarmin, 2016).

In conclusion, this book explores concepts, frameworks, principles and practical case studies that support and represent an *integrated functional sanitation value chain (IFSVC)*. Authors and contributors identified and examined practical and operative linkages within a systemic loop that capture various functions at different stages of activities within different enterprises, and production/service processes related to sanitation management and economy from design and production to final market and user. We also explored enterprises and ventures within each stage as well as those businesses directly and indirectly involved with providing safely managed sanitation services and ancillary businesses across the global supply chain (in the particular communities where they operate).

The intention is to direct thinking towards building an expansive *IFSVC* that could give birth to a sanitation economy. The contributors to this book connected the missing links between the *IFSVC* and the sanitation economy, thus providing a new lens with which to translate global sanitation needs into sustainable business solutions. The book showcases up-to-date research findings to support the concepts, frameworks, and principles presented therein, and also relevant cases that highlight leading sanitation and related business, education and research organizations, as well as global supply chain ventures involved in the provision of safely managed sanitation products, services and facilities, and do so from the IFSVC perspective. Consequently, the concept of the IFSVC should enable academicians, professionals, practitioners, businesses and entrepreneurs to see mixed economic, environmental and social gains not realized by traditional sanitation value chain considerations, thus bringing a much wider range of companies and other stakeholders into active engagement with sanitation systems and services. With this approach sanitation can also provide a solution for water security, energy security, food security and healthcare. The book also intends to create academic interest in exploring and describing the IFSVC as well as research studies into the IFSVC that could be operating within local, regional, national and global sanitation management and economic settings.

1.7 Take Action

(I) Take an inventory of sanitation and related enterprises in your city and show how they interact with others, if they do.

(II) Develop promotional materials for IFSVC in your city, town and municipality to inform others about the need to develop and/or strengthen IFSVC; and send out to sanitation-related professional bodies and chambers of commerce.

(III) Organize an event to bring together sanitation businesses/enterprises and other stakeholders to discuss the existing sanitation value chain (SVC) and how to upgrade and develop it to the IFSVC.

1.8 Journal Entry

(I) Make a table indicating the different activities of the IFSVC and related players in your own area.

(II) What is the sanitation value chain and how is it relevant to SDGs 1, 6, 9, 11, 12, 2, 3, and 7?

1.9 Reflection

(I) Write a short essay on understanding the agricultural value chain, how it operates and how this knowledge could be used to create and/or upgrade existing SVCs.

1.10 Guiding Questions

(I) What does it mean to reconceptualize the sanitation value chain and why is it necessary?

(II) What was the gap in previous perspectives of using the sanitation service chain (SSC) alone to represent the sanitation value chain?

(III) What is the relationship between the sanitation economy and the integrated functional sanitation value chain (IFSVC)?

(IV) How does Porter's value chain differ from the integrated functional sanitation value chain (IFSVC)?

(V) How can the integrated functional sanitation value chain (IFSVC) be described?

(VI) Who are the actors that participate in the IFSVC? What functions do they perform? How many exist within the chain?

(VII) What are the factors that determine the capacity of an enterprise to add value in the IFSVC?

(VIII) With the aid of a diagram, give a detailed description of the value chain stages of the IFSVC.

(IX) What is IFSVC mapping and analysis? What information does it provide and what is the usefulness of such data?

REFERENCES

Akinsete A., Bhagwan J., Hicks C., Knezovich A., Naidoo D., Naidoo V., Zvimba J. and Pillay S. (2019). The Sanitation Economy Opportunity for South Africa: Sustainable Solutions for Water Security & Sanitation – A Business Perspective. Water Research Commission & Toilet Board Coalition, Pretoria, South Africa. https://www.wrc.org.za/mdocs-posts/the-sanitation-economy-opportunity-in-south-africa/ (accessed 4 March 2020)

Bischel H. N., Caduff L., Schindelholz S., Kohn T. and Julian T. R. (2019). Health risks for sanitation service workers along a container-based urine collection system and resource recovery value chain. *Environmental Science & Technology*, **53**, 7055–7067. https://doi.org/10.1021/acs.est.9b01092

Chofreh A. G., Goni F. A., Zeinalnezhad M., Navidar S., Shayestehzadeh H. and Klemeš J. J. (2019). Value chain mapping of the water and sewage treatment to contribute to sustainability. *Journal of Environmental Management*, **239**, 38–47, https://doi.org/10.1016/j.jenvman.2019.03.023

Dobbs J. H. (1998). Competition's New Battleground: The Integrated Value Chain. Cambridge Technology Partners, Cambridge, MA, USA. Available on line from https://www.scirp.org/(S(351jmbntvnsjt1aadkposzje))/reference/ReferencesPapers.aspx?ReferenceID=856972

Drost S., van Wijk J. and Mandefro F. (2012). Key Conditions for Successful Value Chain Partnerships: A Multiple Case Study in Ethiopia. Working Paper Series 033. The Partnerships Resource Centre, Rotterdam, The Netherlands,. ISSN 2211-7318. https://repub.eur.nl/pub/77626/ (accessed 27 December 2021)

Frost & Sullivan (2021) Sanitation economy value chain opportunity assessment. *Gates Open Research*, **5**, 68. Available from: https://gatesopenresearch.org/documents/5-68 (accessed 27 December 2021)

Hyun C., Burt Z., Crider Y., Nelson K. L., Prasad C. S. S., Rayasam S. D. G., Tarpeh W. and Ray I. (2019). Sanitation for low-income regions: a cross-disciplinary review. *Annual Review of Environment and Resources*, **44**(1), 287–318, https://doi.org/10.1146/annurev-environ-101718-033327

Inomata S. (2017). Analytical frameworks for global value chains: an overview. In: Measuring and Analyzing the Impact of GVCs on Economic Development. Global Value Chain Development Report 2017, IBRD and OECD (eds.), International Bank for Reconstruction and Development/The World Bank, Washington DC, pp. 15–35. https://digitallibrary.un.org/record/3829265?ln=en (accessed 22 December 2021)

Kaplinsky R. (2000). Globalisation and unequalisation: what can be learned from value chain analysis? *The Journal of Development Studies*, **37**(2), 117–146. https://doi.org/10.1080/713600071 (accessed 22 December 2021)

Kaplinsky R. (2002). Gaining from global value chains: the search for the nth rent. In: Who Gets Ahead in the Global Economy? Industrial Upgrading, Theory and Practice, G. Gereffi (ed.), Johns Hopkins Press, New York, pp. 240–254.

Kaplinsky R. and Morris M. (2001). A Manual for Value Chain Research, http://asiandrivers.open.ac.uk/documents/Value_chain_Handbook_RKMM_Nov_2001.pdf (accessed 22 December 2021)

Koottatep T., Cookey P. E. and Polprasert C. (eds.) (2019). Resource system. In: Regenerative Sanitation: A New Paradigm for Sanitation 4.0, IWA Publishing, London, UK, pp. 209–282. https://doi.org/10.2166/9781780409689_0209

Market Links (2021). Value Chain Development. https://www.marketlinks.org/good-practice-center/value-chain-wiki/overview-value-chain-approach (accessed 28 December 2021)

McLeod G. (2012). Value Chain/Network Improvement. Inspired 3Q2012. https://static1.squarespace.com/static/55df0347e4b064f219e05fb4/t/5792056ccd0f686a939c5b99/1469187496779/Value-Chain-Network-Improvement.pdf (accessed 06 March 2020)

Murray A. and Ray I. (2010). Commentary: back-end users: the unrecognized stakeholders in demand-driven sanitation. *Journal of Planning Education and Research*, **30**(1), 94–102. https://doi.org/10.1177/0739456X10369800

M4P (2008). Making value chain work better for the poor: a toolbook for practitioners of value chain analysis, version 3. Making Market Work Better for the Poor Project UK. https://www.rflilc.org/wp-content/uploads/2020/08/making_value_chains_work_better_for_the_poor_a_to_14413.pdf (accessed 27 December 2021).

Osann E. R. and Wirth J. (2019). Non-Sewered Sanitation Devices: A New ISO Standard for the Reinvented Toilet. Presentation at the National Environmental Health Association (NEHA) Annual Education Conference Nashville, TN, 10 July 2019. https://www.nrdc.org/sites/default/files/non-sewered-sanitation-devices.pdf (accessed 06 March 2020)

Porter M. E. (1985). Competitive Strategy. Macmillan Publishing Co., Inc., USA.

Porter M. (1990). The Competitive Advantage of Nations. MacWilliam Press, London. http://www.economie.ens.fr/IMG/pdf/porter_1990_-_the_competitive_advantage_of_nations.pdf (accessed 12 January 2022)

Porter M. E. (2008). Competitive Advantage: Creating and Sustaining Performance. Amazon Digital Services, Incorporated, New York.

Papazoglou M. P., Yang J. and Tsalgatidou A. (2002). The role of eServices and transactions for integrated value chains. In: Business to Business Electronic Commerce, M. Warkentin (ed.), Mississippi State University, USA, Idea Group Publishing, pp. 141–170. https://doi.org/10.4018/978-1-930708-09-9.ch011 (accessed 22 December 2021)

Rawlins J. M., De Lange W. J. and Fraser G. C. G. (2018). An ecosystem service value chain analysis framework: a conceptual paper. *Ecological Economics*, **147**, 84–95. https://doi.org/10.1016/j.ecolecon.2017.12.023 (accessed 22 December 2021)

Reese J., Gerwin K., Waage M. and Koch S. (2016). Value Chain Analysis: Conceptual Framework and Simulation Experiments. Nomos Verlagsgesellschaft, Baden-Baden, Germany.

Rodrigue J.-P. (2006). Transportation and the geographical and functional integration of global production networks. *Growth and Change*, **37**(4), 510–525. https://doi.org/10.1111/j.1468-2257.2006.00338.x. https://www.researchgate.net/publication/4990757_Transportation_and_the_Geographical_and_Functional_Integration_of_Global_Production_Networks (accessed 22 December 2021)

Springer-Heinze A. (2018a). ValueLinks 2.0: Manual on Sustainable Value Chain Development. Vol. 1: Value Chain Analysis, Strategy and Implementation. Eschborn, Germany. https://beamexchange.org/uploads/filer_public/f3/31/f331d6ec-74da-4857-bea1-ca1e4e5a43e5/valuelinks-manual-20-vol-1-january-2018_compressed.pdf (accessed 26 April 2021)

Springer-Heinze A. (2018b). ValueLinks 2.0: ValueLinks 2.0: Manual on Sustainable Value Chain Development, vol. **2**. Value Chain Solutions, Eschborn, Germany. https://beamexchange.org/uploads/filer_public/d3/a4/d3a4882e-eb14-4c30-8f7e-6ba4f51f6ec9/valuelinks-manual-20-vol-2-january-2018_compressed.pdf (accessed 26 April 2021)

Strande L., Ronteltap M. and Brdjanovic D. (eds.) (2014). Faecal Sludge Management: Systems Approach for Implementation and Operation. IWA Publishing, London.

Sutarmin J. D. P. (2016). Value chain analysis to improve corporate performance: a case study of essential oil export company in Indonesia. *Investment Management and Financial Innovations*, **13**(3-1), 183–190. https://doi.org/10.21511/imfi.13(3-1).2016.04

TBC (2019a). The Sanitation Economy at Sector Scale: Transformative Solutions for New Business Value. https://www.toiletboard.org/media/53-The_Sanitation_Economy_at_Sector_Scale_FINAL.pdf (accessed 04 March 2020)

TBC (2019b). Scaling up the Sanitation Economy 2020–2025. https://www.toiletboard.org/media/52-Scaling_the_Sanitation_Economy.pdf (accessed 05 March 2020)

Trienekens J. H. (2011). Agricultural value chains in developing countries a framework for analysis. *International Food and Agribusiness Management Review*, **14**(2), 51–82. https://edepot.wur.nl/189057 (accessed 29 December 2021)

USAID (2018). Scaling Market Based Sanitation: Desk Review on Marketbased Rural Sanitation Development Programs. USAID Water, Sanitation, and Hygiene Partnerships and Learning for Sustainability (WASHPaLS) Project, Washington, DC. https://pdf.usaid.gov/pdf_docs/PA00T52M.pdf (accessed 27 December 2021)

van Welie M. J. and Romijn H. A. (2018). NGOs Fostering transitions towards sustainable urban sanitation in low-income countries: insights from Transition Management and Development Studies. *Environmental Science & Policy*, **84**, 250–260. http://doi.org/10.1016/j.envsci.2017.08.011 (accessed 04 March 2020)

WWAP (United Nations World Water Assessment Programme) (2017). The United Nations World Water Development Report 2017. Wastewater: The Untapped Resource. UNESCO, Paris.

doi: 10.2166/9781789061840_0019

Chapter 2

Product design and development

Walter Gibson

Chapter objectives

The aim of this chapter is to consider how value is added through product design and development, give examples at different points in the IFSVC, and examine the constraints and challenges to be overcome if the full potential of the IFSVC is to be realized.

2.1 INTRODUCTION

A key premise of this book is that realizing the sanitation economy will demand new enterprises creating new products and services which offer greater value for their customers than those currently available. Although the value chain involves many different company activities, technology development and product design are vital elements (Porter, 2001). In this chapter we will consider how value is added through product design and development, give examples at different points in the IFSVC, and examine the constraints and challenges to be overcome if the full potential of the IFSVC is to be realized. As we will see, progress is being made but there is scope to go further and there are exciting prospects on the horizon. Throughout the chapter 'product' should be taken to mean both products and services: they are often interlinked and interdependent in a new customer offering and underpinned by similar processes. Product design and development involves a number of different processes, each of which offers an opportunity to create value for the customer and the business. These will be considered in turn, highlighting ways in which value can be created and giving examples of how this is being applied to sanitation.

2.1.1 Starting from customer insights

Successful businesses in other sectors invest heavily in understanding what kind of products their customers want, as opposed to need, and are willing to pay for. They seek insights into habits, attitudes, and behaviours that will drive better product appeal, competitive advantage and ultimately value. Technological advances are important, but they are not sufficient on their own to drive customer adoption: the new product must satisfy some desire for improvement, which could be functional and/or emotional, ideally both. Because sanitation has historically been viewed, particularly in developing countries, as a public good, the focus of new product design and development has often been on recipients as end-users or beneficiaries, rather than as customers (Mulumba *et al.*, 2014). With the growing role of the private sector in delivering improved sanitation, this is starting to change. Indeed, we argue that value creation for the customer must be central to product design and development in sanitation if the sanitation economy is to flourish. And the converse is true also: product design and development are central to value creation.

While this may already be the case for some parts of the sanitation market, for example sanitary ware, historically it has not always been the case for basic sanitation provision in low-to-middle income countries where the starting point for product development has tended to be functionality, for example containment of faeces, and public health, for example prevention of contamination. With this approach, usage of toilets has been lower than desired if the full public health benefits are to be obtained (Chambers & Myers, 2016). Furthermore, without toilet usage for waste collection the IFSVC cannot function fully. To drive change, the critical distinction that must be made increasingly in future is between what end-users *need* in terms of better sanitation products and services and what customers *want* and are willing to buy – and use. And it is important to widen horizons in terms of customers: in the IFSVC these go beyond toilet users and include for example those who buy high value products at the end of the value chain, such as farmers buying fertilizer.

By focusing on what the customer wants, as opposed to what they need, value creation through product design begins by understanding at a deep level the desires and aspirations of the customer, their existing habits and attitudes, and their likes and dislikes of current products. It uses a variety of techniques such as ethnographic research to gain insights into the kind of product that customers would value and willingly purchase. It also tries to understand the context in which the product will be used and its constraints. Context can include cultural factors, the environment, social and economic conditions. Without following a process like this there is a real risk that solutions will fail to have the impact they are intended to deliver, either through lack of demand or lack of use or both, even if they function well technically. For example, a study of the barriers to sustainability and scale of household water treatment practices concluded that user preferences need to be addressed in order to achieve sustained demand (Ojomo *et al.*, 2015). While efficacy is important this may not be the basis on which a product is chosen; again, in the case of water treatment, convenience and product aesthetics play a major part (Ojomo *et al.*, 2015). Albert *et al.* (2010) noted that for household water treatment and storage 'product dissemination at scale to the poor will not occur until we better understand the preferences, choices, and aspirations of the at-risk populations.' This is undoubtedly also true for sanitation and all development sectors.

A key sign that customer preferences have not been sufficiently well understood during the development of a product is that it fails to be used or is used differently to the intended purpose. In sanitation, a prime example of this is the pit latrine, the most basic form of storing human excreta. Although these can be beneficial and be the basis of safely managed sanitation if emptying services are available, the latter is not always the case and then the user has the anxiety of what to do when the latrine is full. A study in India concluded that there is a strong preference for a latrine to be (impractically) much larger than the typical government-recommended size because of concerns about how it will be emptied (Coffey *et al.*, 2017). Many different factors are at play here including beliefs and values related to purity and pollution, as well as the caste system. Open defecation in parts of rural India remains an issue despite increased ownership of latrines (Gupta *et al.*, 2020).

Time spent gaining customer insights is thus vital: insights are the foundation for subsequent steps in the innovation and product development process including concept generation, prototyping and willingness-to-pay studies with constant evaluation and feedback until it is clear that there is a product which customers want to buy, which works as intended and which is ready to launch.

Typically this process is called user-centred design or human-centred design (Text Box 2.1) and there are a number of well established methods which can be followed (see note at end of this chapter e.g. of tools available).

In 2014 Mulumba *et al.* observed that there is a lack of sanitation products poor people wish to buy and that this is a barrier to scaling up private sector provision of sanitation. The application of UCD is beginning to change that: over the past 5–10 years UCD principles have been used to develop innovative sanitation products and services along the IFSVC. Examples are shown in Table 2.1 below, with some of the products being illustrated in Figure 2.2.

Text Box 2.1: What does a User-Centred Design (UCD) process involve?

Although there are different toolkits available, they all have many features in common. They all begin with an attempt to discover what the customer really wants, through deep immersion in their lives, the products they use, how they behave, what problems and challenges they face. This is a search for inspiration, for insights that can form the basis of new ideas and new solutions that customers will want to use and buy. It is followed by an idea generation phase, which will typically involve reframing the problem based on new insights, getting further external stimuli, and co-creation with users. Some early prototyping may be involved. Lastly there is an implementation phase when solutions are made real, tested and refined until they are ready to be launched in the market.

It is important to stress that this is not a linear process. It is likely to be highly iterative. For example, putting a prototype into the hands of a consumer can reveal new insights and start the process all over again. There is always room for improvement and learning.

Quicksand, a design agency based in India, applied these principles to the issue of urban sanitation in India. Over a period of 10 months they conducted in-depth research into the sanitation experience of low-income families in five different cities, using observation, interviews, participatory research and rich media documentation. The insights this generated provided direction for innovations which could improve urban sanitation for the poor (Quicksand, 2011a; Quicksand Sammaan Brief, n.d.). These were later brought to life and implemented in new community toilet designs through Project Sammaan in partnership with the Bhubaneswar and Cuttack municipalities. The scope of innovations included not just the design and construction of the physical facilities, but also operations and maintenance, the business model and branding and communications (Surfaces Reporter, 2020)

Experiencing what your target customer experiences every day can be powerful and influential in terms of the final outcome. Daigo Ishiyama, Leader of Innovation at SATO and a member of the team that developed the SATO pan, had such an experience in Bangladesh during a 'deep dive' to generate disruptive innovations in collaboration with iDE (MacArthur *et al.*, 2015). He recalled seeing a full latrine through the open hole in the slab and thinking 'I've got to close that hole' (Ishiyama, 2017), see Figure 2.1. This was a key insight and imperative for the later design steps.

Figure 2.1 User-centred research in Bangladesh (image credit, LIXIL).

These innovations reveal the scope of possibilities to bring improved benefits to a range of different customers along the length of the value chain. In this way they enhance the value proposition at each stage and make the whole chain more robust. The extent to which they are willing and able to pay for those benefits will determine the success of a given venture and that element of the chain, which may in turn influence and enable other stages. This emphasizes the importance of conducting willingness-to-pay studies prior to a new product launch with the desired target market, but the ultimate test is indeed to present it to the consumer and see if they will pay for it.

While the benefits of sanitation can be viewed in terms of cost-effectiveness, where health, social and economic indicators can be used to judge its merits as a public health intervention (Hutton *et al.*, 2007), the customer may make a different set of judgements based on the information presented: they may want to know how easy it is to build and where they can purchase it from; they may want to know how long it will last and what maintenance it will need; they may want to know if it will smell like some of the toilets they have used in the past. The advantage of UCD is that these concerns and wants should have been factored into the design and thus increase the chances of a positive customer response.

Many of the innovations mentioned in Table 2.1 have passed that test: as examples, SATO has shipped over 5.1 million units across 41 countries, improving sanitation for approximately 25 million people (LIXIL, 2021), iDE have impacted more than 1 m households (iDE, 2021), more than 30 000 Digni-Loos have been sold (E. Perez, 2021, personal communication), and Sanergy is serving more than 120 000 residents every day in Nairobi (Sanergy, 2021). While the number and success of innovations in the market is encouraging, there is still scope to deepen our understanding of all the different possible customers along the value chain. In particular, a more comprehensive understanding of customers for current and future biomass/biotransformation products is vital to drive the whole chain to a new level. Further, it should not be considered that because the products in Table 2.1 have been developed, there is no further need for

Table 2.1 Customer insights behind recent sanitation innovation.

Stage	Enterprise	Customer	Example Insights	Product
Design and Product Development Services	Quicksand	Communal toilet users, India	Facility design establishes rules of use more clearly (Quicksand, 2011b)	Gender-inclusive toilet
	Clean Team (design involved Ideo/WSUP/ Unilever)	Off-grid urban poor, Ghana	No one wants to 'slosh a bucket of waste through their home' (Ideo, n.d.)	Branded Toilet and complete sanitation service
	Cranfield University Centre for Competitive Creative Design	Off-grid, water stressed communities	Users liked the mechanical flush and preferred a deeper bowl (Hennigs *et al.* 2019)	Nano Membrane Toilet
Equipment Manufacturing Services	LIXIL (user research with iDE)	Rural low-income families in Bangladesh	Existing toilets unhygienic, suboptimal in function (MacArthur *et al.* 2015)	SATO Pan
	Duraplast (facilitated by Global Communities)	Rural poor households Ghana	Key design factors included smell, flies, ease of cleaning, colour (Borkowski and Perez, 2019)	Digni-Loo
	Sanergy (Auerbach, 2015)	Urban slums, Kenya	Ease of cleaning, small footprint enables installation near home	Fresh Life Toilet
Facility Integration, Installation and Construction Services	iDE (Pedi *et al.* 2012), WaterSHED (Pedi *et al.* 2011)	Rural Cambodian families	Both studies revealed a strong preference for price-prohibitive 'permanent' pour-flush solutions	Easy Latrine, Unbranded toilet package
Sanitation Services	Saraplast, 3S	Urban women outside the home, India	Lack of clean, safe public toilets for women (Economic Times, 2020)	Ti bus
	WSUP/BoP Innovation/ UX (Kisker and Drabble 2017)	Emptying business owners	Very little active customer acquisition	Pula App
Sanitation Biomass Recovery and Conversion Services	Sanergy (Auerbach, 2015)	Farmers, Kenya	Dissatisfied with current animal feeds, difficult to source organic fertilizer domestically	KuzaPro Evergrow
	Sanivation	Large companies using solid fuel boiler, Kenya	Businesses looking for more sustainable fuel supply (J. Lane, 2020, personal communication)	Superlogs

Figure 2.2 Some products developed with UCD principles. Clockwise from top left: Sammaan Community Toilet (*image credit, Quicksand*); Digni-Loo (*image credit, Alberto Wilde, Global Communities*); Ti Bus (*image credit, 3S*); SuperLogs (*image credit, Sanivation*); KuzaPro (*image credit, Sanergy*); Fresh Life toilets (*image credit, Sanergy*).

customer research: there is always scope to gain new insights and improve the product and enhance its value. In a competitive environment this is vital.

2.1.2 The importance of a brief and a plan

Customer insights and product ideas are necessary but not sufficient: turning these into a fully fledged product ready to put into the hands of the customer is equally important. This part of the process should be guided by a brief setting out all the elements that should be considered in developing the final product or service. Who will use it? How will it deliver what they want? What functionality is needed and is it technically feasible? What material and component options are available? How will it be made and reach the market? How much will it cost? These and other questions must be addressed if the potential of the new product is to be realized. The added value in this stage is derived from considering and solving all the detailed practical questions that make the final product possible and deliver its promised benefits at an affordable price.

A good design brief goes beyond a description of the concept or physical design and will define the objectives and key criteria for all the different dimensions required to create a successful new product. Preparing a brief ensures that all these different factors are considered from the outset and that there are no surprises later which could mean delays or even failure. For example, considering the manufacturing process will have a bearing on the types of materials used and their functionality, as well as the compatibility of different potential components. Social and environmental impact factors should be considered. Figure 2.3 summarises some of the key elements to be considered in preparing a design brief.

Planning determines how the brief is to be implemented and the new solution delivered. It defines key steps, activities, their duration and sequence, and team roles and resources. The value of planning goes beyond saying how long it will take and how much it will cost, important though those are. Planning helps to anticipate issues. It brings the team together to ensure no misunderstanding about who does what and when. Time spent at the outset of a product development project in planning with all the members of the team present can save time and resources later and increase the chances of a successful outcome. It generates milestones, which if used correctly are a valuable aid in tracking progress and deciding on when it is appropriate to move from one phase to the next.

A good plan may not need too much adjustment as work progresses but if the team is trying to do something very novel there may well be uncertainties in terms of timings and resource requirements which require continuous monitoring and adjustments to be made. Further, external factors beyond the control of the team may affect progress. It is important to be as realistic as possible about what can be achieved in a given time, especially when field work is involved: difficulties associated with transport, weather, availability of materials and so forth can all cause delays.

Figure 2.3 Key elements of a design brief (Source: Author).

2.2 THE POWER OF PROTOTYPING AND ROLE OF DEMONSTRATORS

A key part of UCD/product development is prototyping, which has two main roles, each of which add considerable value to the final product. Firstly, it makes the concept visible or tangible and this helps to explore the concept and explain it to customers and other stakeholders. By making the product concept real enough so that the intended customer can touch, see, simulate and experience it, fresh insights are gained into what is and is not attractive about the idea and what needs to be improved. Secondly, prototyping allows key aspects of functionality and customer appeal to be tested and proven. As we will see, prototyping adds value not just in terms of getting closer and closer to what customers actually want, but also in terms of demonstrating what is technically possible and catalysing the commercial development of the product.

Prototyping can range from the physical – whether early-stage simple mock-ups using off-the-shelf components or more sophisticated simulations of a final product for an in-home test – to the virtual, using computer-aided design software to bring ideas to life and adapt and refine them. And these can be combined in mixed prototyping which has the potential for short learning cycles (Elverum *et al.*, 2016). There are different strategies for prototyping and some thought should be given at the outset to how it will be used and the key questions to be answered. This will help to keep costs down, particularly at the later stages when there may seem to be a need for increasingly sophisticated prototypes (Elverum *et al.*, 2016). Prototyping can be used at different points and in different ways throughout the design and product development process. The example in Text Box 2.2 highlights how it helps to focus on the key features of a new product that matter to customers, and how they engage with it physically and emotionally, essential factors for generating demand and customer satisfaction.

Prototyping is also critical in understanding and proving the functionality of a new product: it can address questions such as how well does the product perform, how

---

Text Box 2.2: The Art of Prototyping

The WaterSHED team in Cambodia have developed a number of user-led innovations in hygiene and sanitation. Prototyping has been an integral part of their design process and they have learnt a lot about its value and how best to use it. Geoff Revell, WaterSHED's co-founder, described (Revell, 2017) the challenges of prototyping a product, the HappyTap handwashing station, that would eventually be molded in plastic: because of the high cost of molds the team had to be sure of the key features of the product before they committed. So they considered what the key features of the product were and mocked up two versions of each of those, and gave them to households to try one after the other. This 'head to head' use of prototyping gave them a sense of which version was preferred and they could build that into the overall design.

They also learnt that you need the right kind of prototype for the kind of feedback you are seeking. If you want to learn about aesthetics cardboard mock-ups are not helpful. There was an 'Aha!' moment for Geoff when he realized that his target consumers were seeing the product not in terms of narrow function or form but as a way to change their lifestyle for the better. His over-riding advice is to probe as deeply as you can qualitatively to gain such insights. However he recognizes there will always be a tension between designing based on expressed wishes and designing based on a powerful entrepreneurial vision. Such tensions can of course be powerful drivers of creativity.

satisfied are potential customers with its performance, what are the issues that still need to be resolved? Again it is about choosing the right kind of prototype for the stage you are at and the information you need to move forward. For products involving new technology, once they get beyond lab proof of concept the first big jump is to make a prototype that people can use, typically referred to as a field or application prototype under real world conditions. This allows the innovator not just to test how well the product performs technically but also to explore how the user interacts with it and to get their feedback on what they like and dislike. It is a major step towards the development of a commercial product. The next step thereafter is typically to make a commercial application prototype, which builds in any necessary refinements and improvements and is closer to a finished product in terms of cost, materials and manufacturability. This in turn will go through further refinements to make it into a product suitable for large-scale manufacturing. As new technologies become available, product improvements and new features become possible. The product improvement and development process does not stand still.

When science and technology make things possible that were not previously possible and drive and enable new product concepts, prototyping takes on an additional significance for value creation. In the case of such emerging technologies, and industries based on them, where the applications are not initially fully understood and the route to market is uncertain, it is clear from a wide-ranging review that the process of emergence from the lab to the mass market goes through several well characterized stages as shown in Figure 2.4 below (Phaal *et al.*, 2011). The catalyst for moving from one stage to another is a particular kind of prototype, which the authors call a demonstrator: a physical representation of what is possible that allows the next stage to proceed and the

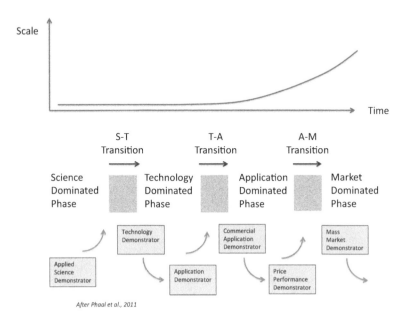

Figure 2.4 Role of demonstrators in moving from science to mass market in technology and industry emergence. (Diagram by author, after Phaal *et al.* 2011).

final outcome to be achieved. The stages and different types of demonstrator and their relationship to growth in scale, are illustrated in Figure 2.4.

The stages are separated by transition zones and getting through these zones is critical to making progress towards eventual mass scale. The first such zone marks the transition between the S (science-dominated) phase into the T (technology-dominated) phase. The end of the S phase and entry into this S–T zone is marked by an *applied science demonstrator* – essentially proof-of-concept, showing the feasibility and practical possibilities for the underlying technical approach. Leaving this zone and entering the T phase is associated with creation of *technology demonstrators* – prototypes that demonstrate that the technology can be integrated to perform the desired functions. During the T phase these prototypes and their components are refined and issues resolved under controlled conditions to the point where it is possible to make one that is sufficiently robust to test outside the laboratory. This is the *application demonstrator*, which is associated with the T–A (application-dominated) transition. Some prior 'field demonstrators' may have been necessary to reach this point depending on the complexity of the operational environment. Application demonstrators bring the concept to life and allow further refinements and improvements to accommodate user feedback and operational challenges in the real world. This can then lead to the creation of a *commercial application demonstrator* which marks the end of the T–A transition and entry into the application phase. This demonstrator is essentially the V_0 product and marks the beginning of market trials and sales. During the application phase the product is likely to go through further iterations in design, performance and price; this leads to the development of a *price–performance market demonstrator* which shows the feasibility of a commercial mass market and marks the start of the A–M (mass-market-dominated) transition. This transition is crucial to achieve scale. It ends with the development of a *mass market demonstrator* which is positioned for substantial growth. For those more familiar with TRL nomenclature, the S, T and A phases approximate to the Research (TRL1–3), Development (TRL4–6) and Deployment (TRL7–9) stages.

This framework is based on a retrospective analysis of many different technologies and industries, ranging from cheese to digital cameras (Phaal *et al.*, 2011). The value capture of the transitional zones is evident and crucial to achieve the full potential of a new concept. There is no reason to believe that new sanitation technologies along the IFSVC should not follow the same pattern. Two such technologies, Zyclone Cube and Tiger Toilet, are compared in Figure 2.5 using this framework.

The Tiger Toilet is an on-site sanitation system based on vermifiltration (Furlong *et al.*, 2014, 2015, 2016). As shown in Figure 2.5 , progress from lab to field demonstrators was quite fast and the results from the household field trials gave real confidence in terms of technical performance and user satisfaction (Furlong *et al.*, 2016). This confidence enabled Bear Valley Ventures and PriMove India to work together to start market testing and development of the first application prototype: to date around 5000 household units have been sold and the technology has also been adapted to enable wastewater and faecal sludge treatment at different scales (TBF, 2020). The applications of this technology are being explored by a number of organisations, and recently the International Worm-Based Sanitation Association (IWBSA) was formed to share, develop and promote best practice: according to their estimates, over 210 000 people in nine countries are benefitting from solutions installed by IWBSA contributors (IWBSA, 2021). Mass-market scale has yet to be achieved, however, and the model in Figure 2.5 suggests that, at least for domestic units, further technical development is necessary to catalyse the jump to this level.

The Zyclone Cube is one of a suite of innovative decentralized waste-treatment systems developed at AIT, Thailand for applications in Asia and beyond (Koottatep

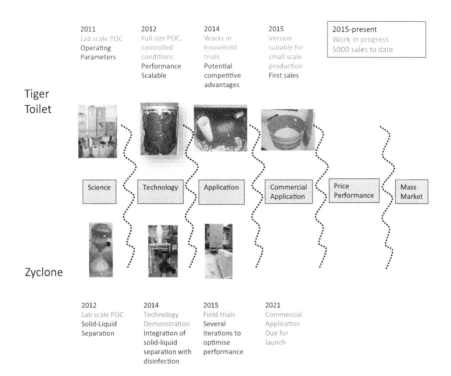

Tiger
Toilet

Zyclone

Figure 2.5 Demonstrators and value capture for emergent sanitation technologies. **Legend:** text in red, nature of demonstrator; text in blue, key value added by demonstrator. Figures for the Tiger Toilet refer to work done by the Sanitation Ventures team at London School of Hygiene and Tropical Medicine (2009–2012); Bear Valley Ventures and PriMove (2012–2015); Bear Valley Ventures, Primove and ITT (2016–2017); and latterly TBF Environmental Solutions Pvt Ltd (2018–present). *Image credits: 2011, 2012 and 2014 images, Dr Claire Furlong; 2015 image, Walter Gibson).* Figures for the Zyclone Cube refer to work done by the NATS project team at the Asian Institute of Technology, Thailand (2011–present) and by SCG chemicals Co. Ltd (2018–present). *Image credits: taken from final report of the innovative DEWAT Technologies Project, supported by Bill & Melinda Gates Foundation and Asian Institute of Technology, November 2019).*

et al., 2018). Following successful initial lab testing of technology prototypes, the team made rapid progress to a field prototype, using a factory setting to test the design with users. The prototype was monitored regularly for technical performance by the team over about a year and feedback was obtained from users. As a result they learnt a huge amount about its strengths and weaknesses, giving the team confidence that it would work under real-life conditions and allowing them to focus on how it could each be improved further. AIT have transferred their intellectual property to SCG Chemicals Co. Ltd who are developing a commercial product, with AIT providing technical support. Sales of 200 units were projected in 2021.

If this model holds for other new product development in sanitation, then it illustrates how important the product development process is for enabling and progressing to scale. Thinking in terms of the next demonstrator in order to capture more value and move from one level of scale to another is crucial to realizing the full economic and impact potential of innovations within the IFSVC. Frameworks such as this can help innovators understand where they are in the process and what they have to do to get to the next stage. As we will see in Chapter 7, scale has been identified as a vital factor in the commercial viability of an innovative value recovery approach, the use of Black Soldier Fly (BSF) larvae to process faecal waste and turn it into high-value products such as lubricants, biodiesel, chitosan and soil conditioners. This began life as a lab experiment in 2011 to prove that BSF larvae could grow and develop on faecal waste (Banks *et al.*, 2013), then went through technology and application demonstrators at increasing scale in South Africa (The BioCycle; see Chapter 7 for more details) and Kenya (Sanergy; see Text Box 2.3 for more details), during which there was considerable value added in terms of operational parameters and commercial insights. This appears then to be another technology approaching the A–M transition.

However important the product development process is, it should be recognized (as we will see in Chapter 7) that the business model and market considerations (e.g., distribution, pricing, demand creation, regulations), partnerships and investment are also crucial to delivering mass scale. Technical progress alone may not be sufficient to convince investors and other stakeholders to get behind a new product. Further, in the case of sanitation because of the social and health impact potential, building evidence for impact and creating a scaling system with all the right partners aligned is also critical (McLean & Gargani, 2019).

2.3 INTEGRATING PRODUCT DEVELOPMENT WITH MARKET AND BUSINESS MODEL DEVELOPMENT

Achieving scale is vital for business sustainability and realizing the full potential value of a new concept. While the product development process plays an important role, it must go hand in hand with exploring and verifying the market potential and working out and proving the business model. Customer feedback from early commercial application demonstrators will guide subsequent development: the price may be too high, or the performance unsatisfactory in some respect. Thus there should be iteration through market trials of improved versions until the product gets close to the price/performance demonstration required to begin the transition to mass scale. Likewise key aspects of the business model need to be explored: cost of customer acquisition, levels of demand, and route to market will all have an influence on the product and vice versa. All of this work adds value because it is only when all three elements come together that a mass-scale demonstrator is feasible and real scale-up and value generation can begin.

Although in sectors such as sanitary hardware, with well established consumer markets dominated by large companies, the business model will typically be well developed and the market well understood, this is not generally the case for businesses trying to operate within the IFSVC in low-to-middle income countries. In such contexts the IFSVC is fragmented and characterized by many small operators (e.g., emptying truck operators), early-stage enterprises piloting innovative products and services, and large companies mainly engaged in collection to resource recovery, as well as cleaning products. Most businesses serve customers at only a limited number of points and relatively few enterprises as yet span the whole chain (Mason *et al.*, 2015): the total current size of the IFSVC is as yet a small fraction of its full potential.

How can that value be realized? It is unlikely that the many small operators can grow their revenues substantially: they operate direct to householders with simple technology and it will be difficult for them to grow their markets. For enterprises pioneering a new product or service of the types cited above, increasing value creation means creating not just the product but the business model as well. In some cases even the market itself has to be proven and developed. Individually some of the markets along the IFSVC are potentially very large, especially with over two billion people still without access to toilets. But perhaps the greatest prize is for more and more businesses to operate either alone or in partnerships along the whole length of the IFSVC, with value being added at every stage. This generates value beyond the sale of toilet products and services by selling products derived from the waste itself, but there must be incentives all the way along for the chain to work. As an illustration of how the business model is factored into design, new technology development and value creation, and how the IFSVC can be incentivized, consider Sanergy, one of the few enterprises to span the whole IFSVC (see Text Box 2.3 below; Walske & Tyson, 2016).

Even with these innovations and enhanced value creation, cost recovery is still a challenge for container-based sanitation services like Sanergy (Russel *et al.*, 2019). In addition to cost savings, for example through scale, there is scope to explore further innovations at the end of the value chain to increase revenue. A number of options have been reviewed and compared in terms of potential market value by Diener *et al.* (2014). These authors highlighted the need to consider local market conditions in selecting technologies for value recovery, as demand can vary considerably from one location to another. This need to match technologies to markets is also typical of the process by which innovations reach scale and is often a key part of the commercialization process. As we will see in Chapter 7, market experiments are underway with a number of different reuse products and business models.

Text Box 2.3: Sanergy – Business Model and Relationship to Design and Value Creation

From the outset the founders of Sanergy sought a business model that was integrated and incentivized across the whole sanitation value chain from design and production through to the final marketplace (Walske & Tyson, 2016). Working in the slums of Nairobi, Kenya, their solution involves a network of off-grid, waterless toilets operated by franchisees under the Fresh Life brand, and collection and treatment of waste by Sanergy into fertilizer and insect-based animal feed. The toilets are designed and produced by Sanergy, incorporating features which provide incentives for users, operators and Sanergy themselves. Toilets are leased by the operators who keep all the pay-per-use fees. Sanergy have made it easier for new operators to make the monthly service payment by making financing available. Because Sanergy's own revenues come from the sale of biomass recovery products to farmers, it is vital to grow this network to ensure the supply of raw material: making sure toilets are used consistently is thus important and toilets are monitored regularly to make sure standards of cleanliness and hygiene are high. Sanergy have introduced new technology at the final biomass transformation step to increase revenue potential by using BSF larvae to convert waste into protein for animal feed. Sales of the product Kuza Pro (which offers a 30% increase in yield for farmers) commenced in 2019; the market is very large, estimated to be $520M in Kenya alone (according to personal communication with Sanergy in 2021). So this added value could be a significant driver of business growth and growth in sanitation provision, the source of the feedstock for animal feed production.

2.4 KEY CONSTRAINTS AND CHALLENGES

Every individual enterprise in the IFSVC will have its own challenges in terms of product design and development. Many of these relate to scaling good ideas to the level required to make a real impact on the problem at a global level. Some are typical of the constraints and challenges faced during product development in any sector, such as competitor activity, finding suitable technology, and getting costs down. However, there are also some particular constraints and challenges which affect the future development and functioning of the IFSVC as a whole. These include:

2.4.1 Constraints
- Regulatory factors – certain uses of products of bioconversion, for example insect-larvae-based products for animal feed, are not permitted under current legislation in some countries and regions. The first such product to gain approval was Sanergy's KuzaPro animal feed in Kenya in 2019.
- Lack of resources for start-up enterprises – not just funding but lack of access to key specialist skills and facilities. Many of the innovations adding value to the IFSVC are being brought to the market by SMEs, which often lack such resources and find them difficult to access.
- The intrinsic difficulty of working in low-income settings – the need to obtain solid evidence at each stage of the process in low-income settings can affect the rate of progress.

2.4.2 Challenges
- Finding novel, higher-value bioconversion products for which there is market demand to provide greater economic incentives, encourage product development and build partnerships along the chain. There is a need for more options creating more revenue per kilogram of waste than biogas and fertilizer.
- Scaling innovative technologies so that their full commercial and impact potential can be realized. This demands not just technical advances but also significant investment and, in all likelihood, major partnerships (see Chapter 7) to undertake trials at scale and provide the evidence required to justify wider rollout.
- Introducing fresh perspectives – technical thinking has dominated the sanitation sector historically, but innovation in and integration with design and business model thinking is important for full value creation.
- Developing effective partnership models – this is vital across the value chain to provide better products and services at all stages and greater incentives to make it function more effectively.

2.5 NEW HORIZONS

Some of these challenges are more amenable than others. For example, it may take many years and many studies to convince every regulator that it is safe for humans to eat animals fed with meal derived from insect larvae fed on human waste. The use of insect larvae is however a major step forward in terms of value creation compared with fertilizer and biogas and points the way to what can be achieved. The potential of bioconversion is huge however and discovery of new high-value bioconversion products will drive even greater opportunities. According to a UNESCO report 'Future research and innovation trends in the field of wastewater will probably focus on resource recovery to reinvent the economics of the treatment and disposal of wastewater and sludge' (WWAP, 2017). The term 'wastewater' here is used broadly and this statement can also be taken to apply to faecal sludge processing. This report cites several promising new technologies under development for energy, nutrient and high-value product recovery, such as microbial fuel cells, advanced methods for recovering N and P, and environmentally friendly microalgae for production of a range of high-value products including bioplastics and cosmetic ingredients. A combination of careful economic analysis of possible markets, identification of appropriate business models and biotechnological advances will be needed to identify the most viable options for any given situation: the context in terms of markets, regulations and consumer perceptions will also be important. The tools for such analyses are available. As an example, one such recent thought is to 'rewire' anaerobic digestion (AD) away from methane production to the production of short-chain fatty acids, which have a high value as intermediates in many different chemical industries (Text Box 2.4).

Text Box 2.4: Rewiring AD

The potential for resource recovery using advanced AD technologies to drive improved sanitation was put forward by Chandran (2014), who identified chemicals such as short-chain carboxylic acids (e.g., acetic, lactic, propionic, and butyric acids) and methanol as more economically attractive products that could be produced by fermentation of human waste. This approach has multiple potential social and environmental benefits: it reduces pressure on natural resources such as fossil fuels, from which such chemicals are normally derived; it saves energy in processing waste by conventional aerobic technology; and it generates revenues which can be used to offset the cost of sanitation provision.

The ability to arrest anaerobic digestion of wet organic waste after the acidogenesis stage, resulting in the production of short-chain carboxylic acids (from C_2 to C_4), is now well established (Bhatt *et al.*, 2020). Such acids have a wide range of possible uses and are high-value feedstocks for products including polymers, pharmaceuticals and cosmetics. While challenges remain to optimize production and separation of short-chain acids, the analysis by Bhatt *et al.* suggests that it has the potential to be cost-competitive even at small plant scales and a decentralized approach can be adopted due to temporal and compositional variability across different regions.

In developed countries the commercial application of this approach could change the current paradigm of wastewater treatment from being cost-intensive to revenue-generating and have considerable environmental and sustainability benefits through reduced use of fossil fuels and biomass reuse. In developing countries, the potential remains to be explored: a key question for future research is whether the revenue generation potential is large enough and can be effectively harnessed through an appropriate business model to drive affordable, safe and low environmental impact sanitation for everyone. While there are still technical challenges to be overcome, there is a strong economic and environmental case for this approach and it represents an exciting new opportunity to be explored in the context of the vision for the IFSVC.

2.6 Take action

(I)   Identify businesses/enterprises and other players involved in the design and development of sanitation products.

(II) Draw an illustration that indicates an overview of a value chain map for sanitation product design and development in your country, and also show the global linkages.

2.7 Journal entry

Consider any existing sanitation product and write an essay on the design and development from conceptualization to creation.

2.8 Reflection

Customer insights are vital for conceptualizing, designing, and developing sanitation products for the market; through an informal survey within your locality, explore users' preferences for particular products' design, functionality and performance.

2.9 Guiding questions

(I)   What is a sanitation product design and development value chain?
(II)   What are the implications of value creation being central to product design?
(III)   Why is it important to focus on what the customer wants as opposed to what they need in product design and development?
(IV)   What is the role of context in product design and development and what does it entail?
(V)   What is the full implication of not understanding customer preferences in product design and developments and what is the benefit of spending time to gain customer insight in sanitation product design and development?
(VI)   'Experiencing what your customer experiences everyday can be powerful and influential in terms of the final outcome.' How true is this statement?
(VII)   Why is prototyping necessary in product design and development?
(VIII) What role did the Bill and Melinda Gates Foundation 'inventing toilet' challenge contribute to the development of innovative sanitation product design and development? What is the current status of those designs?
(IX)   What are the possible constraints and challenges of enterprises in IFSVC with regards to product design and development?

2.6 TOOLS

There are several toolkits on user-centred design available to help teams get started on this process and navigate some of the processes discussed in this chapter. These include:

- Gates Foundation (https://www.ucdinsanitation.com/)
- IDEO (https://www.designkit.org/)
- Stanford Design School (https://dschool.stanford.edu/resources/design–thinking–bootleg)

REFERENCES

Albert J., Luoto J. and Devine D. (2010). End–user preferences for and performance of competing POU water treatment technologies among the rural poor of Kenya. *Environmental Science & Technology*, **44**(12), 4426–4432, https://doi.org/10.1021/es1000566

Auerbach D. (2015). Sustainable sanitation provision in urban slums – the Sanergy way. *Sustainable Sanitation Practice*, **24**, 4–8.

Banks I. J., Gibson W. T. and Cameron M. M. (2013). Growth rates of black soldier fly larvae fed on fresh human faeces and their implication for improving sanitation. *Tropical Medicine & International Health*, **19**, 14–22, https://doi.org/10.1111/tmi.12228

Bhatt A. H., Ren Z. and Tao L. (2020). Value proposition of untapped wet wastes: carboxylic acid production through anaerobic digestion. *iScience*, **23**, 101221, https://doi.org/10.1016/j. isci.2020.101221

Borkowski J. and Perez E. (2019). A Market-Based, Pro-Poor Approach to Rural Sanitation. Global Communities Case Study. Global Communities, Silver Spring, MD, USA, 8pp.

Chambers R. and Myers J. (2016). Norms, knowledge and usage. Frontiers of CLTS: Innovations and Insights. IDS, Brighton, UK, Issue 7, pp. 1–22.

Chandran K. (2014). Technologies and framework for resource recovery. In: Water Reclamation and Sustainability, S. Ahuja (ed.), Elsevier, Oxford, UK, Ch. 17, pp. 415–430. https://doi. org/10.1016/B978-0-12-411645-0.00017-1 (accessed February 2021)

Coffey D., Gupta A., Hathi P., Spears D., Srivastav N. and Yvas S. (2017). Understanding open defecation in rural India: untouchability,pollution and latrine pits. *Economic and Political Weekly*, **52**, 59–66.

Diener S., Semiyaga S., Niwagaba C., Muspratt A. M., Gning J. B., Mbeguere M., Ennin J. E., Zurbrugg C. and Strande L. (2014). A value proposition: resource recovery from faecal sludge – can it be the driver for improved sanitation? *Resources, Conservation and Recycling*, **88**, 32–38, https://doi.org/10.1016/j.resconrec.2014.04.005

Economic Times (2020). Washroom-on-wheels: how a Pune firm is turning buses into women's toilets. Economic Times, 21 February 2020. Times Internet, Gurugram, Haryana, India. https:// economictimes.indiatimes.com/news/politics-and-nation/washroom-on-wheels-how-a-pune-firm-is-turning-buses-into-womens-toilets/buses-or-womens-toilets/slideshow/74239692.cms (accessed 2021)

Elverum C. W., Welo T. and Tronvoll S. (2016). Prototyping in new product development: strategy considerations. *Procedia CIRP*, **50**, 117–122, https://doi.org/10.1016/j.procir.2016.05.010

Furlong C., Templeton M. R. and Gibson W. T. (2014). Processing of human faeces by wet vermifiltration for improved on-site sanitation. *Journal of Water, Sanitation and Hygiene for Development*, **4**, 231–239, https://doi.org/10.2166/washdev.2014.107

Furlong C., Gibson W. T., Templeton M. R., Taillade M., Kassam F., Crabb G., Goodsell R., McQuilkin J., Oak A., Thakar G., Kodgire M. and Patankar R. (2015). The development of an onsite sanitation system based on vermifiltration: the Tiger Toilet. *Journal of Water, Sanitation and Hygiene for Development*, **5**, 608–613, https://doi.org/10.2166/washdev.2015.167

Furlong C., Gibson W. T., Oak A., Thakar G., Kodgire M. and Patankar R. (2016). Technical and user evaluation of a novel, worm-based, on-site sanitation system in rural India. *Waterlines*, **35**, 148–162, https://doi.org/10.3362/1756-3488.2016.013

Gupta A., Khalid N., Deshpande D., Hathi P., Kapur A., Srivastav N., Yvas S., Spears D. and Coffey D. (2020). Revisiting open defecation: evidence from a panel survey in rural north India, 2014–18. *Economic and Political Weekly*, **55**, 55–63.

Hennigs J., Ravndal K. T., Blose T., Toolaram A., Sindall R. C., Barrington D., Collins M., Engineer B., Kolios A. T., McAdam E., Parker A., Williams L. and Tyrrel S. (2019). Field testing of a prototype mechanical dry toilet flush. *Science of the Total Environment*, **668**, 419–431, https://doi.org/10.1016/j.scitotenv.2019.02.220

Hutton G., Haller L. and Bartram J. (2007). Global cost–benefit analysis of water supply and sanitation interventions. *Journal of Water and Health*, **5**, 481–502, https://doi.org/10.2166/wh.2007.009

iDE (2021). Every family deserves a toilet. [online]. https://www.ideglobal.org/story/sanitation (accessed 2021)

Ideo (n.d.). An Enterprise That's Committed to Putting a Toilet in Every Home. Ideo.org, San Francisco, New York, Nairobi. [online] https://www.ideo.org/project/clean-team (accessed 2020)

Ishiyama D. (2017). You can't smell a latrine from your desk. [video] YouTube, https://youtu.be/byS5MGFxu5Y (accessed 2020)

IWBSA (2021). https://iwbsa.org/impact (accessed 2021)

Kisker J. and Drabble S. (2017). The Pula App: a customer acquisition and tracking tool for vacuum tanker businesses. WSUP Practice Note. November 2017. www.wsup.com/content/

uploads/2017/11/PN031-ENGLISH-The-PULA-app-a-customer-acquisition-and-tracking-tool-for-vacuum-tanker-businesses-1.pdf (accessed 2020)

Koottatep T., Chapagain S. K., Polprasert C., Panuvatvanich A. and Anh K.-H. (2018). Sanitation situations in selected Southeast Asian countries and application of innovative technologies. *Environment, Development and Sustainability*, **20**, 495–506, https://doi.org/10.1007/s10668-016-9892-6

LIXIL (2021). https://www.sato.lixil.com/ (accessed 2021)

MacArthur J., Riggs F. C. and Chowdhury R. (2015). Disruptive design in sanitation marketing: lessons from product and process innovations in Bangladesh. Proceedings of 38th WEDC International Conference, Loughborough UK. https://wedc-knowledge.lboro.ac.uk/resources/conference/38/MacArthur–2176.pdf (accessed 2020)

Mason N., Matoso M. and Smith W. (2015). Private sector and water supply, hygiene and sanitation, ODI Report, Overseas Development Institute, London, UK, 74pp.

McLean R. and Gargani J. (2019). Scaling Impact: Innovation for the Public Good. Routledge & IDRC, Ottawa, Canada. ISBN 9781138605558.

Mulumba J. N., Nothomb C., Potter A. and Snel M. (2014). Striking the balance: what is the role of the public sector in sanitation as a service and a business? *Waterlines*, **33**, 195–210, https://doi.org/10.3362/1756-3488.2014.021

Ojomo E., Elliott M., Goodyear L., Forson M. and Bartram J. (2015). Sustainability and scale-up of household water treatment and safe storage practices: enablers and barriers to effective implementation. *International Journal of Hygiene and Environmental Health*, **218**, 704–713, https://doi.org/10.1016/j.ijheh.2015.03.002

Pedi D., Jenkins M., Aun H., McLennan L. and Revell G. (2011). The 'hands–off' sanitation marketing model: emerging lessons from rural Cambodia. In: The Future of Water, Sanitation and Hygiene in Low-Income Countries – Innovation, Adaptation and Engagement in A Changing World, R. J. Shaw (ed.), Proceedings of 35th WEDC International Conference, Water Engineering and Development Centre, Loughborough University, Loughborough, UK, pp. 6–8, July 2011. https://wedc-knowledge.lboro.ac.uk/index.html

Pedi D., Kov P. and Smets S. (2012). Sanitation marketing lessons from Cambodia: a market-based approach to delivering sanitation. Water and Sanitation Program: Field Note. https://www.wsp.org/sites/wsp.org/files/publications/WSP-Sanitation-marketing-lessons-Cambodia-Market-based-delivering-Sanitation.pdf (accessed 2020)

Phaal R., O'Sullivan E., Routley M., Ford S. and Probert D. (2011). A framework for mapping industrial emergence. *Technological Forecasting and Social Change*, **78**, 217–230, https://doi.org/10.1016/j.techfore.2010.06.018

Porter M. (2001). The value chain and competitive advantage. In: Understanding Business: Processes, D. Barnes (ed.), Routledge & Open University, London, UK, Ch. 5, pp. 50–66.

Quicksand (2011a). Potty Project: understanding sanitation experiences of the urban poor. http://quicksand.co.in/work/the-potty-project (accessed 2020)

Quicksand (2011b). Potty Project Blog. https://pottyprojectindia.tumblr.com/search/insights (accessed 2020)

Quicksand Sammaan Brief (n.d.). http://quicksand.co.in/media/downloads/project-sammaan/ProjectSammaan_ProjectBrief.pdf (accessed 2020)

Revell G. (2017). How real should a prototype feel? [video] YouTube. https://youtu.be/Rdoq2k5lqc4 (accessed 2020)

Russel K., Hughes K., Roach M., Auerbach D., Foote A., Kramer S. and Briceno R. (2019). Taking container-based sanitation to scale: opportunities and challenges. *Frontiers in Environmental Science*, **7**, 00190, https://doi.org/10.3389/fenvs.2019.00190

Sanergy (2021). http://www.sanergy.com/ (accessed 2020)

Surfaces Reporter (2020). Sammaan: a sustainable improved sanitation model for urban slums/BMC/BMGF/JPAL/QDS. Surfaces Reporter, 28 November 2020. https://www.surfacesreporter.com/articles/72186/sammaan-a-sustainable-improved-sanitation-model-for-urban-slums-bmc-bmgf-jpal-qds (accessed 2020)

TBF (2020). http://www.tbfenvironmental.in/ (accessed 2020)

Walske J. M. and Tyson L. D. (2016). Sanergy: Tackling Sanitation in Kenyan Slums. Berkeley Haas Case Series, Haas School of Business, Berkeley, CA, USA. https://cases.haas.berkeley.edu/2016/07/sanergy/ (accessed 2020)

WWAP (UN World Water Assessment Programme) (2017). The United Nations World Water Development Report 2017. Wastewater: The Untapped Resource. UNESCO, Paris, 198pp. https://unesdoc.unesco.org/ark:/48223/pf0000247153/PDF/247153eng.pdf.multi (accessed 2020)

doi: 10.2166/9781789061840_0037

Chapter 3

Product/equipment manufacturing

Peter Emmanuel Cookey, Thammarat Koottatep and Chongrak Polprasert

---

Chapter objectives

The aim of this chapter is to help the reader understand the sanitation manufacturing value chain in the IFSVC and the complex range of activities implemented by various actors such as primary producers, processors, traders, and service providers, globally, nationally and locally to bring a sanitation product through production to its sale.

3.1 INTRODUCTION

Sanitation manufactured products and equipment (SMPE) have been traditionally focused on households and in some cases educational and healthcare institutions, but in reality these products and equipment are ubiquitous and our survival are intricately linked with them. This simple perspective fails to capture the complex inter-relationship, information and materials flow between the various players and systems in the sanitation/hygiene manufacturing/production value chain and the final consumers/users (SMVC/SPVC) (Srai & Shi, 2008). In other words, sanitation products and equipment are found in homes, schools, public facilities, government, public and private buildings, religious facilities, markets, shopping malls, airports, bus/train stations and transport facilities, cinemas/theatres, stadia, event centres, hotel and tourism facilities, parks, playgrounds, recreation or community centres, hospitals and healthcare facilities, prisons and correctional facilities as well as at constructions sites and outdoor event areas. Sanitation/hygiene product enterprises (SHPE) are involved with software, hardware and consumables across the integrated functional sanitation value chain (IFSVC) and originate through series of value-added research and development, up to production, distribution, installation and after-sales services (Koottatep *et al.*, 2019a). They include any sanitation/hygiene good (tangible or intangible), service or idea that is produced by labour or effort and/or a result of an act or process that can be offered to a market and/or end-users to satisfy a want or need (Wikipedia, 2019; Wiktionary, 2019). Equipment refers to apparatus, gear, hardware, kit and/or materials that are used in operations or activities that relate to sanitation and/or wastewater, sewage, faecal sludge, and so on.

For instance, sanitation equipment manufacturers produce end-to-end equipment and technologies from non-network systems such as composting toilets, ventilated improved pit latrines and suchlike, to basic network components such as pipes, to complex sewage/ wastewater treatment plants.

Some of these products/equipment and/or technologies are sector-specific, whereas others have wider uses, for example, construction, measurement, plastics, chemical and mechanical applications and the transport of liquid in general (PWC, 2012). These products and equipment are manufactured with rigorously quality control and assurance formulated standards that ensure uniformity, replicability, scalability, reliability, specificity and functionality (Gasiorowski-Denis, 2018; Koottatep *et al.*, 2019a; Lazarte, 2016). They, along with the related services result from value-added processes and transformation of goods, especially via manufacturing processes of industries/ enterprises (UNIDO, 2011). It is important to understand that value addition is the work of industry/enterprise in the sector done in-house and not that which others are paid to perform (Meckstroth, 2016). Another interesting point to note is that sanitation products, equipment and services are not restricted to toilets/latrines but include other consumables that enhance the use of sanitation infrastructures and facilities, (e.g., hygienic products) as well as the total wellbeing of humanity (Table 3.1).

Sanitation products/equipment manufacturing value-added activities are directly concerned with the change of form or dimensions of input materials to products and services (Singh, 2006). It covers the entire life cycle of products from the beginning of life (BOL), which extends from product design to actual production; to middle-of-life (MOL) which extends to product use, after-sales service and maintenance; as well as end-of-life (EOL), which then extends to product reuse with refurbishing, reuse of components with disassembly and refurbishing, material reclamation without disassembly, material reclamation with disassembly, and final disposal (Rolstadaas *et al.*, 2008). In addition, supporting activities such as transportation and handling or storage of parts – even though they are not directly concerned with the changing of form and/or dimensions of the part produced – are critical components of production (Porter, 1985).

Thus the Sanitation Manufacturing Value Chain (SMVC) looks at the complex range of activities implemented by various actors (primary producers, processors, traders, service providers) globally, nationally and locally to bring a raw material through production to the sale of the final product. SMVC starts from how raw materials move along production processes with other enterprises engaging in trading, assembling, processing activities, and so on. (M4P, 2008). In a nutshell, it covers raw material processing, intermediate production, manufactured parts assembly, sales and service, and also addresses the challenges of technological innovations. SMVC, then, is the continued addition of value that occurs while products pass from one enterprise in the chain to the next, gradually increasing its degree of transformation (UNIDO, 2011). The key function is to offer customers/end-users a range of sanitation/hygiene products that match their preferences and budgets by encouraging private/public sector investments in all the stages of the SMVC, particularly to enhance the viability and participation of local entrepreneurs and positioning them appropriately in the global sanitation value chain (USAID, 2018).

The nature of industries/enterprises participating in SMVC are fragmented, active and contribute in different ways to the sanitation global value chain (SGVC), specifically, sanitaryware, treatment plants and hygiene products (Meckstroth, 2016). The focus here is not necessarily on activities implemented by a single enterprise, but rather all the backward and forward linkages up to when the product is delivered to the final customers and/or end-users (M4P, 2008). The main actors of SMVC are enterprises and

Table 3.1 Some sanitation and hygiene manufactured products.

Classification of Products	Product Types
Cleaning and Disinfecting Supplies	Cleaning solutions
	Disinfectants and sanitizers
	Towels and wipes
Cleaning Tools	Brooms, dust pans and accessories
	Carts and mop buckets
	Cleaning brushes and pads
	Mops and accessories
	Squeegees
	Vacuums, sweepers and accessories
Restroom Products	Bathroom cleaners and clog removers
	Fragrance dispensers
	Hand cleaners, sanitizers and soap
	Toilet and urinal products
Towels and Tissues	Dispensers
	Tissues
	Towels and wipes
Trash and Recycling	Trash and recycling containers
	Trash liners and trash bags
Basins	Wall-hung basins
	Over countertop basins
	In countertop basins
	Under countertop basins
	Vanity basins
	Semi-recessed basins
	Totem basins
	Pedestals
Semi-pedestals	Basin complements
Faucets	Basin faucets
	Bidet faucets
	Bath faucets
	Shower faucets
	Shower programme
	Kitchen faucets
	Laundry sink faucets
	Flush valves for toilets
Flush valves for urinals	Faucets complements
Furniture	Base units
	Auxiliary units
Furniture complements	Countertops

(*Continued*)

Table 3.1 Some sanitation and hygiene manufactured products (*Continued*).

Classification of Products	Product Types
Toilets	In-tank toilets
	Close-coupled toilets
	Wall-hung toilets
	Single floor standing toilets
	One-piece toilets
	Toilet cisterns
	Toilet seats and covers
	Toilet mechanisms
	Squatting pans
	Container-based toilets
	Portable toilets
	Mobile toilets
	Smart toilets
Urinals	Standard urinals
	Electronic urinals
	Urinal divisions
Baths	Rectangular baths
	Angular baths
	Other shaped baths
	Bath panels
	Bath complements
Shower trays	Rectangular shower trays
	Corner shower trays
	Square shower trays
	Shower trays complements
	In-drain systems
Installation systems	Operating plates
	Systems for toilet
	Systems for urinal
	Systems for bidet
	Systems for basin
	Installation systems accessories
Containment	Packaged septic tanks
	Containers latrines/toilets
Sewage and faecal sludge emptying equipment	Suction sewage truck
	Sewerage truck
	Jet cleaner – high-pressure water cleaner
	Grit sweeper & clean cuum
Small scale pit latrines equipment	Vacuum trucks, pumping systems, mechanical augers, and so on.,

(Continued)

Table 3.1 Some sanitation and hygiene manufactured products (Continued).

Classification of Products	Product Types
Products and materials used in plumbing	Metallic and non-metallic materials used in pipework
	Earthenware pipes
	Sanitary fixtures
	Concrete products
Treatment plants	Sewage Treatment Plants (STPs)
	Effluent Treatment Plants (ETPs)
	Activated Sludge Plants (ASPs)
	Common and Combined Effluent Treatment Plants (CEPTs)
	Faecal sludge treatment plants
Resource recovery and reuse equipment	Several technologies deplored for resource recovery and conversion of sanitation biomass

businesses involved in product and service research and development, design, supply of raw materials and equipment parts, production/manufacturing, distribution/route to market, buyers and after sales services. They are supported by a range of technical, business and financial service providers. In the value chain, the various business activities in the different segments become connected and to some degree coordinated. Simply put, SMVC encompasses all the activities and interactions required in the creation of a sanitation product or service, from primary production to transformation, commercialization and end-users and/or customers (UNIDO, 2011). Global industries involved in mass production of sanitation/hygiene products and semi-finished products (foreign value-added) which serve as input-materials used by domestic industries for further product development (Meckstroth, 2016) are part of the chain as well as small-scale producers (S-SP), distributors, retailers, and service providers for the functionality of on-site sanitation systems. In many cases S-SP operate in the informal sector, are not registered and are often off the radar of the municipalities and financial institutions, and therefore lack scale. Nevertheless, they are the main providers of sanitation/hygiene products to low-income customers (Narayanan et al., 2011; Nothomb et al., 2014; Valfrey-Visser & Schaub-Jones, 2009).

3.2 SANITATION MANUFACTURED PRODUCTS/EQUIPMENT VALUE CHAIN

Manufacturing processes lie near the centre of a complex value chain composed of an upstream chain that gathers materials and services and a downstream sales chain that moves goods to market and sells as well as services manufactured goods (Meckstroth, 2016). This involves making products/equipment from raw materials by various industrial equipment, machines and processes (Singh, 2006). The emphasis is on value-added activities that transform raw materials into finished goods by using different means of production such as human capital, knowledge and machinery/technologies (UNIDO, 2011). The value chain examines the value for each and every activity as the product/service moves through its lifecycle (Acharyulu et al., 2015). Sanitation manufacturing enterprises and the ancillary businesses can use value chain analysis to examine all of their activities and to see how they are connected systematically, and how business inputs are changed into outputs and the how to further develop the sector towards achieving SDG6 (Acharyulu *et al.*, 2015). SMVC is a network of independent activities/ processes that produces sanitation goods/services and at the same time creates value for the enterprise owners. As the product moves from one player in the chain to another, it is assumed to gain value (Hellin & Meijer, 2006). Thus, value is the amount the customers and/or end-users are willing to pay for the products or services that a firm provides. Every value activity faces costs such as raw materials and other purchased goods and services for 'purchased inputs', human resources (direct and indirect labour) and technology to transform raw materials through other ranges of activities required to bring a product or service from conception through the different phases of production, distribution to end-users into finished goods and final disposal (McGee, 2015; Porter, 1985; Zamora, 2016).

These chains of interrelated activities/processes of production and support activities include order processing, product design and manufacturing of tools/apparatus, dies, moulds, jigs, fixtures and gauges, selection of material, planning, managing and maintaining control of production processes and reliable quality of processed product at different locations in chains and/or networks of enterprises with proper coordination. This systematic cooperation and integration of the whole network of a manufacturing system leads to economical production and effective marketing of designed sanitation products/equipment in the minimum possible time (Singh, 2006). SMVC is not essentially a one enterprise/firm/actor activity but consists of inputs from several value chain actors with several components interacting together in a dynamic-integrated manner, which takes inputs and delivers manufactured products to the customers and/or end-users (Singh, 2006). It operates more on the concept of global value chains (GVCs) based on the fact that the main sanitation (and related hygiene) products, equipment and services (sanitaryware, hygienic cleaning and disinfectant products, wastewater, sewage and/or faecal sludge treatment plants and their ancillary infrastructures, etc) and their trade and consumption span several countries and are increasingly carried out by various entities and/or networks of industries/enterprises located in different countries (Research and Markets, 2021; Sturgeon *et al.*, 2012; Sturgeon & Memedović, 2011)

The implication is that while advanced economies are outsourcing and offshoring sanitation production facilities, the developing economies are participating more in the export markets (Jones *et al.*, 2019), even though most of the raw materials required for these goods are sourced from the developing countries. Therefore, the existing SMVC is made up of buyer-driven and producer-driven value chains. Buyer-driven chains where large retailers, merchandisers and trading companies play a central role in establishing production networks usually arise in developing (exporting) countries, while the producer-driven value chain is dominated by large transnational corporations that play a key role in managing the production network (Abecassis-Moedas, 2006). There are also local product development and manufacturing, whereby partly manufactured products and raw materials from other countries and locations are finished locally.

Regardless of what drives the chain, value additions should reflect through the natural sequence of operations from stage to stage. Value addition implies both value creation and value capture, and every activity performed requires an investment in resources and each link in the chain is expected to add value (Chivaka, 2007). In the same vein, a chain actor's ability to compete and succeed depends on its position along the industry chain and how much value it is able to create and capture (Zamora, 2016). GVCs linking firms, workers and consumers around the world could provide a stepping stone for firms and workers in developing countries to integrate into the global sanitation economy (Gereffi & Fernandez-Stark, 2011). For many developing countries with poor access to safely managed sanitation, their ability to effectively insert themselves into GVC, SMVC or upgrade their local sanitation value chain is a vital requirement to achieve SDG6.

3.3 SANITATION MANUFACTURED PRODUCTS VALUE CHAIN (SMPVC) MAPPING

SMPVC mapping provides a visual representation by identifying products end markets, business operations (functions), chain operators and their linkages as well as the chain supporters. Value chain mapping is the most essential method and the core of any value chain analysis (Springer-Heinzer, 2018). As opposed to other stages of the IFSVC, manufactured products are composed of inputs from different sources. After all, a water closet is more than processed clay. The critical point is that many components and services are needed to make sanitaryware and other sanitation-related products. There is no single dominant input that could be used to characterize the SMPVC (Springer-Heinzer, 2018). SMPVC mapping is aptly illustrated by Porter's value chain model which identifies a number of primary and support manufacturing activities (Porter, 1985). In this regard, the SMPVC has six primary activities and four supporting activities as presented below and shown in Figure 3.1.

The SMPVC primary activities are as follows:

(I)   research and development;
(II) design (concept design, specification, detailed design and production design);
(III) sourcing of inputs and supplies (raw materials and semi-processed goods);
(IV) production processes (which could include four main value-added products: manufactured sanitation and allied products; prefabricated concrete/plastic products; *in situ* constructed products; and disruptive innovative products) with over-arching quality control and standardization;
(V)   distribution/route to markets and end-users; and
(VI) after-sales services (in addition to other services provided, e.g. assembling, installation, operations and treatment system equipment owners).

The SMVC's supporting activities are: (i) infrastructure; (ii) human resources; (iii) technology; and (iv) procurement (see Figure 3.1).

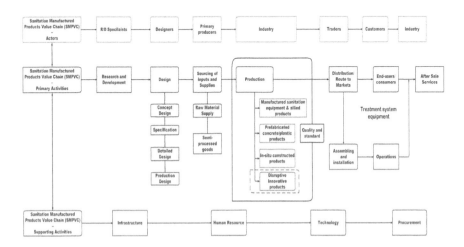

Figure 3.1 Generic sanitation manufactured products value chain map (Sources: Authors).

3.3.1 SMVC primary activities
3.3.1.1 Research and development
This activity deals with discovering new scientific information and technological developments that can be used to design and create innovative products as well as improve existing manufactured products and services. As a stage in SMPVC its focus is on conceptualizing and developing new and/or enhancing already existing sanitation and related products and services. The research aspect occurs when a company's R&D team or a research institute/team in an educational organization tests the viability of a potential sanitation product/technology. Once research produces these inventions/innovations, then development progresses to transforming the discovered science into a useful product that the company can market and sell to customers, either to end-users or as input materials for secondary products (Tarver, 2020). R&D provides manufacturing with the opportunity to explore products/services for customers through careful market analysis, including the measurement of customer trends. Manufacturers can then use these reports to make peculiar products available to their customers (Tarver, 2020). On the other hand, enterprises can invest in R&D when their product lines are becoming outdated to gain and maintain competitive edge (Tarver, 2020). Several new sanitation systems (NSS) products have been developed within the last two decades in order to improve the sanitary conditions of 4.5 billion people without access to safely managed sanitation. The most recent and prominent funder for sanitation R&D is the Bill and Melinda Gates Foundation (BMGF), which initiated the 'Reinvent the Toilet Challenge' (RTTC) in 2011 that supported new research and development approaches for toilet technologies that safely and effectively managed human waste. The R&D of NSS marks a paradigm shift from end-of-pipe wastewater management systems to resource-oriented sanitation systems (Girardet & Mendonça, 2009; Koottatep *et al.*, 2021; Schuetze & Santiago-Fandiño, 2014; Tilley *et al.*, 2008). However, the major challenge remains as translating some of these new sanitation technologies from the pilot phase to real products in the marketplace for the large population that needs them.

3.3.1.2 Design
This describes the process of imagining, creating and iterating products, devices, objects, and services that solve users' sanitation problems and/or address their specific needs in a given market (IDSA, 2020). It is the collection of business activities associated with all phases of product engineering, including research and development (Wognum *et al.*, 2002; Zhu *et al.*, 2011) as well as specification, concept design, detailed and production design (Hartley *et al.*, 1997). All manufactured products must first past the design stage before they are mass produced in the factory floor and/or before they are constructed/fabricated as well as installed, especially for the end-users of the non-sewered sanitation systems commonly found in developing countries. The objective is to develop products, devices, objects, and services with designs that meet both functional and performance requirements as well as being produced at a reasonable cost with minimum technical problems at the highest possible quality in the shortest possible time (Groover, 2010). In other words, its focus is not only on the appearance of a product, but also on how it functions, is manufactured and ultimately the value and experience it provides for the users (IDSA, 2020). Design with respect to SMVC/SPVC is aimed at optimizing the functions of products, devices, objects, and services and minimizing the total production costs and/or achieving other settled targets (Baglieri & Secchi, 2007; Lee & Gilleard, 2002; McIvor *et al.*, 2000, 2006; McIvor & Humphreys, 2004; Zhu *et al.*, 2011). Product design passes through four major stages, namely:

(1) *concept design* – where product architecture contributes key ideas/concepts/critical components and establishes interfaces between products' subsystem;
(2) *specification design* – which focuses on avoiding ambiguity and information distortion, identifying early changes and key component requirements;

(3) *detailed design* – which deals with selection of proprietary parts and component tolerance design, prototype testing and demonstration, design for manufacturability, materials selection and process design; and

(4) *production design* – which looks at tooling design for manufacturability, quality control and assurance as well as raw materials consideration. (Zhu *et al.*, 2011).

Designing for improved service approaches in low-resource settings, especially for products that focus on basic needs like sanitation/hygiene and related concerns, differ from standard industrial design processes and are sometimes referred to as 'design for the bottom of the pyramid' or 'design for the developing world' (Lucena & Schneider, 2008) or 'social sector human-centred design' (Mitcham & Munoz, 2010; Sharpe *et al.*, 2019). These approaches often frame the end-beneficiary as customer, even when they are not, and retail consumers are the actual end-users. The end-user is simultaneously the subject of monitoring and recipient of the data – for feedback between product designers and other production stakeholders (Sharpe *et al.*, 2019).

Thus, the designers lead the process of product design, in order for a product to be well designed for reliability and scalability, the expertise of the designers must be multifaceted. Also, the designers must have deep knowledge of manufacturing processes in order to enable any degree of scale or even a successful pilot project. Designing products without manufacturing knowledge will result in suboptimal parts. Therefore, designers are responsible for the sustainability and safety of their designs (Mattson & Wood, 2014). It is, therefore, necessary for the designers to have knowledge about the product lifecycles and the impacts of design decisions (Sharpe *et al.*, 2019).

Furthermore, design considerations for manufacturing and assembling sanitation products should provide closer interaction and better communication between designers and manufacturing personnel. This will help smooth the transition into production, shortening the product's transition to the market by enabling fewer components in the final product, easier assembly, lower costs of production, higher product quality and greater customer satisfaction (Bakerjian & Mitchell, 1992; Chang & Melkanoff, 2005; Groover, 2010). In the end, product quality will depend on design and leaner operations (Lamming, 1993), in particular as design taps into other external relationship such as supplies, and so on (Zhu *et al.*, 2011).

3.3.1.3 *Sourcing of inputs and supplies*

Inputs and supplies constitute the materials that enterprises/firms use in the process of producing final goods and services for customers. They come in the form of raw materials, upgraded raw materials and/or semi-processed goods (Florén *et al.*, 2013, 2019; UNIDO, 2011). This can also describe the activities that go into acquiring, purchasing, refining, developing and delivering sufficient amounts of raw materials of sufficient quality to ensure that the strategic and operational objectives of the firm are achieved. Efficient sourcing of raw materials and procurement of inputs can help firms reduce costs and become more competitive, as well as lower risks with fewer inputs and defects (Florén *et al.*, 2013, 2019; UNIDO, 2011). Clearly, input materials put heavy constraints on output products and if inputs are not available, the production process can be negatively affected (UNIDO, 2011). Given the importance of raw materials as the critical input and the increasing challenges related to raw materials supply, a systematic and effective approach to the management of raw materials is critical to any firm in the SMVC/SPVC (Florén *et al.*, 2013, 2019). The way manufacturing concerns deal with the raw material challenges affects both short-term operations as well as long-term opportunities. From a short-term perspective, when supplies of raw materials are smooth, operations may progress favourably, but if raw material supply is interrupted the impact on businesses is often immediate and severe (Florén *et al.*, 2013, 2019; UNIDO, 2011). On a long-term perspective, changes initiated in-house or forced by other input supply actors in the value chain could have both constraining and restraining effects on the value proposition(s) of the sanitation product to be manufactured. An example of the constraining effect on the value proposition is a change in raw material quality delivered to the market (Florén *et al.*, 2013, 2019). Therefore, sourcing practices and input supply are important not only to the sanitation product manufacturer but also to those who provide them. For example, manufacturers of sanitaryware products can improve product quality, increase production and raise profitability if they maintain transparent and reliable contractual relationships with buyers (UNIDO, 2011).

3.3.1.4 Production
This covers any of the methods used in industry to create goods and services from various resources. It is aimed at converting raw materials into finished goods with the application of different types of tools, equipment, and machinery. It covers the capacity of firms in the value chain to match with their physical installations, machines, equipment and space for production and their ability to extend those in the short run as well as the quality of the final product and the time that is required to finish this product. The objective of production is to satisfy the demand for such finished goods and services. All production systems are 'transformative processes' – processes that transform raw materials into useful goods and services. The transformation process typically uses common resources such as labour, capital (for machinery and equipment, materials, etc), and space (land, building, etc.) which economists call factor of production while to effect a change. Production managers refers to them as men, machines, methods, materials and money. Generally, there are three basic types of production systems:

(I)   *A project' (one-shot) system or job production* comprises a single, one-of-a-kind product whereby resources are brought together only once to work upon a single job and complete it before proceeding to the next similar or different job. It requires fixed type of layout for developing same product (Groover, 2010; Singh, 2006).

(II) *Batch production* uses general-purpose equipment and methods to produce small quantities (low in number, say 200 to 800) of output (goods and services) with specifications that vary greatly from one batch to the next. Whenever the production batch is over, the same manufacturing facility is used for production of other batch product or items (Groover, 2010; Singh, 2006).

(III) *Continuous production (mass production)* in which items to be processed flow through a series of steps and/or operations; it involves production of a large number of identical products (say more than 5000) that need a production-line plant layout. It is highly rigid and involves automation and a large amount of investment in special-purpose machines to increase production (Groover, 2010; Singh, 2006).

A production system includes both automated and manually operated systems. The distinction between the two categories is not always clear because many manufacturing systems consist of both automated and manual work elements, for example, a machine tool that operates on a semi-automatic processing cycle but also must be loaded and unloaded each cycle by a human worker (Groover, 2010; Singh, 2006).

3.3.1.4.1 Manufactured sanitation equipment and allied products (MSEaP)
MSEaP are industrially manufactured finished sanitation and related products and/or parts material for the production of other secondary products such as sanitaryware, personal, home, healthcare, and industrial hygiene and sanitizers. Sanitaryware products are sanitary appliances found in installations, such as toilets and bathrooms. In the narrowest sense, it could include water closet bowls (WC), cisterns, bidets, urinals and washbasins (and sometimes sewerage pipes that have traditionally been manufactured

from porcelain (a ceramic material made from clay, sometimes described as 'vitreous china' when coated with enamel), see Figure 3.2.

Sanitary appliances are now made from a wide range of materials including metal, acrylic, glass, plastic and so on. In addition, they also now include a wider range of appliances that might be found in sanitary installations such as baths, showers, bins, incinerators, macerators, sinks, bidets and drinking fountains. Some such appliances are not even connected to water and wastewater systems, for example a composting toilet or drain or waterless urinal (Designing Building Wiki, 2021).

Toilets are critical user-interface components and essential plumbing fixtures that are installed everywhere – in residences, businesses, commercial facilities and others. They come in a variety of styles and types such as one- and two-piece units, floor-mounted and wall-mounted toilets, tank and tankless styles, top flush, side flush, or rear flush, round bowl or elongated bowl and so on. There are also toilets for specialized applications, for example, installations in aircraft, boats, and other vehicles such as buses and trains. Besides stationary toilets, there are markets for portable toilets that serve temporary or short-term needs such as construction sites, movie sets, campgrounds, marathons, or natural zones (Figure 3.3).

Table 3.2 shows the list of top manufacturers of toilets used in the USA (listed alphabetically). The table shows the company name, location, and estimated annual revenue, where data was available. Annual revenue amounts are shown in US dollars. Revenue values denoted in foreign currencies were converted to US dollars using exchange rates as of 8 April 2020 (Thomas Publishing Company, 2021).

Figure 3.2 Ceramic sanitaryware casting industrial assembly. (Source: Fortuna, A., Fortuna, D. M., and Martini, E. (2017) An industrial approach to ceramics: sanitaryware. *Plinius,* **43**, 138–145).

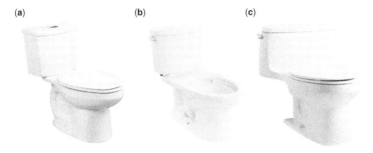

Figure 3.3 Ceramic toilet types and their manufacturer (a) American Standard H2Option; (b) Toto Drake II; (c) Kohler Santa Rosa (Thomas Publishing Company, 2021).

Table 3.2 Top global value chain manufacturers of water closet toilets (Thomas Publishing Company, 2021).

Company Name	City, State Location	Estimated Annual Revenue
American Standard	Piscataway, NJ	$1.5 billion
Delta Faucet	Indianapolis, IN	$570 million
Duravit	Hornberg, Germany	$494.2 million
Kohler	Kohler, WI	$7 billion
Mansfield Plumbing	Perrysville, OH	$51.8 million
Niagara Corp.	Flower Mound, TX	$75 million
PROFLO® (Ferguson)	Newport News, VA	$18.4 billion
Saniflo	Edison, NJ	$48.6 million
Sloan Valve Company	Franklin Park, IL	$2.9 million
Toto	Kitakyushu, Japan	$5.39 billion
Zurn Industries	Milwaukee, WI	$131.5 million

Company Summaries

(I)   *American Standard,* based in Piscataway, NJ, is a major manufacturer of bathroom and kitchen fixtures including sinks, faucets, and toilets for both residential and commercial applications.

(II)   *Delta Faucet* is a US-based provider of sink faucets, shower fixtures, kitchen faucets, toilets, and bathroom accessories. They are headquartered in Indianapolis, IN.

(III)   *Duravit,* headquartered in Hornberg, Germany, is a global manufacturer of sinks, shower trays, toilets, bathtubs, and bathroom accessories.

(IV)   *Kohler,* based in Kohler, WI, is a leading manufacturer of kitchen and bath products including toilets, sinks, faucets, and decorative products.

(V)   *Mansfield Plumbing,* of Perrysville, OH, is a manufacturer of lavatories, bidets, toilets, urinals, and other bathroom products for residential and commercial applications.

(VI)   *Niagara Corp.* is a manufacturer of products designed to conserve water resources, including toilets, showerheads, and water aerators. They are headquartered in Flower Mound, TX.

(VII)   *PROFLO®* (Ferguson) is a wholesale supplier and distributor of commercial and residential plumbing supplies and owns the PROFLO® brand of toilets. They are headquartered in Newport News, VA.

(VIII) *Saniflo,* located in Edison, NJ, is a manufacturer of toilets and macerating and grinding pump systems for both residential and commercial applications.

(IX)   *Sloan Valve Company,* headquartered in Franklin Park, IL, is a manufacturer of bathroom fixtures, flushometers, water closets, sinks, faucets, hand dryers, and other bathroom and plumbing accessories.

(X)   *Toto* is a global manufacturer of toilets, lavatories, faucets, bidets, toilet seats, and showers for residential and commercial use. They are headquartered in Kitakyushu, Japan.

(XI)   *Zurn Industries,* headquartered in Milwaukee, WI, is a manufacturer and supplier of plumbing fixtures, water control and safety products, building drainage products, and grease, oil, and sediment separation solutions for commercial and residential applications.

On the other hand, a dry toilet forms a contrast with the wet toilet (water closet) presented above because it operates without flushwater. The dry toilet may be a raised pedestal on which the user can sit, or a squat pan over which the user squats. In both cases, excreta (urine and faeces) fall through a drop hole (Tilley *et al.*, 2014). A classic example of dry toilets is the composting toilet (Figure 3.4) that uses a predominantly aerobic processing system to treat excreta using no water or only a small volume of flush water, via compositing or managed aerobic decomposition (Kubba, 2017).

Several types of manufactured sanitation-related products for hygiene purposes exist in the market and some good examples are:

(I) Disposable super-absorbent hygiene products that promote cleanliness by containing bodily fluid and excreta often used for infants and adults in institutional settings and allowing the waste to be disposed of in appropriate ways. In addition to providing skin protection benefits, disposable super-absorbent diaper products provide better containment of excreta and reduce the environmental spread of pathogenic organisms. This category also includes female hygiene products for menstrual flow management.

(II) Some healthcare/hygiene products, usually also available over the counter, are normally used for to prevent infection and transmission of diseases, provide hygiene, and enhance care in the hospital ward and operating room.

(III) Personal hygiene products such as bath soaps and gels, shampoos/conditioners, toothpaste/mouthwash, and their dispensers and so on, are classified according to the health risks they may present.

(IV) Cleaning, hygiene and disinfectant products are critical to help combat and/or prevent the spread of pathogens to humans and animals and are used in households or in professional and institutional environments such as hospitals, homes for the elderly and food factories, and so on.

These products result from value-added activities that collect integrated equipment and human resources to convert semi-finished products or raw materials into finished products (Groover, 2010; Singh, 2006). Sanitation manufacturing basically implies making goods or articles and providing services to meet the sanitation needs of mankind through value creations that apply support systems used by firms to solve inherent

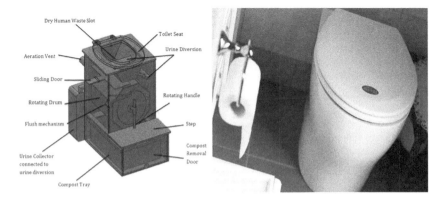

Figure 3.4 Schematic of a composting toilet with urine diversion (Source: Wikipedia contributors, 31 December 2020).

problems with production, supply and delivery to the end-users in order to ensure quality (Groover, 2010). The enterprises involved in the production and supply processes here make up part of the SMPVC.

3.3.1.4.2 Prefabricated concrete/plastic products

This is the use of pre-cast concrete and/or plastic components for off-site fabrication of toilet/latrine products, which can be turnkey and/or component parts for the construction of toilet/latrine systems. These products range from those pre-cast by local masons who use concrete to construct toilet products for user-interfaces such as concrete slabs for latrines/ toilets, pedestals, pans or urinals and biogas receptacles as well as containment systems (septic tanks). More sophisticated prefab toilet components are produced on factory floors and assembled at construction sites (Rahman *et al.*, 2013). The fully assembled prefab toilets are designed, coordinated and tested in advance under controlled factory production environments. Prefabrication technology that is conceptualized at early design stage provides significant advantages over conventional *in situ* (cast-in-place) systems such as reduction in construction time, less skilled labour requirements and construction cost savings as well as improved service delivery (Rahman *et al.*, 2013).

Prefabrication of concrete and/or plastic latrine/toilet slabs is becoming a popular way to encourage private sector investment in the provision of value-added sanitation infrastructure, which entails but is not limited to new products development, procurement of inputs materials, production, marketing, distribution, finance and customer service that can support faster toilet construction, especially during emergencies (CAWST, 2014; Harvey, 2007; Holm *et al.*, 2018; Mchenga & Holm, 2019; WEDC, 2012). Value is typically added beyond material/information aggregation by fabricating some key toilet substructures and interface components (e.g. concrete pit rings, slabs with integrated pan) to provide customer value through ready-to-install packages. In addition, services related to the substructure – such as delivery or installation of the substructure and/or materials for the superstructure – are optional add-ons (USAID, 2018). The most commonly prefabricated latrine slab design available on the market in most developing countries, particularly in sub-Saharan Africa, are pre-cast slabs that come shaped either as a flat square or a circular dome (Mchenga & Holm, 2019; WEDC, 2012). The slabs in most cases have additional holes to accommodate the vent pipes and covered with fly-proof screens in the case of ventilated improved pit latrines (Mchenga & Holm, 2019), see Figure 3.5a and b. Also, the use of pre-cast slabs provides a market for those engaged in small-scale emptying and/or conveyance transport equipment to provide services for onsite sanitation systems.

Research and development on innovative offsite prefabrication for toilet construction in countries like Malaysia and Singapore could help change people's perceptions so that toilet are considered as not just a space for basic sanitation but also a space of comfort and luxury. Off-site produced toilets are seen as a revolutionary approach to toilet construction that will replace the conventional labour-intensive and time-consuming *in situ* toilet. Prefabricated toilet systems are commonly used in shipbuilding, aircraft industries and even in buildings design in some developed countries (Rahman *et al.*, 2013). The two most common prefab toilets available in the market are:

(I)   lightweight panel systems that can be assembled and disassembled on site; and
(II) pre-assembled systems whereby a complex box with fittings and accessories is assembled in the factory and delivered to site for installation.

Both systems can be further categorized based on the material in which it is used. The floor materials could be pre-cast concrete or moulded fiberglass, while the wall materials can be fibre-cement board, ferro-cement board or pre-cast concrete panels (Rahman *et al.* 2013), see Figure 3.6.

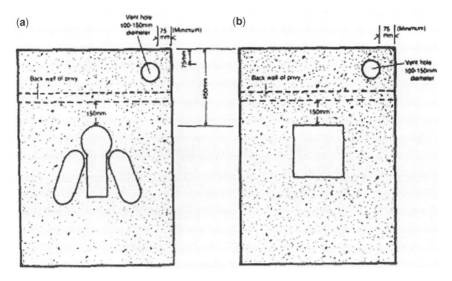

Figure 3.5 (a) Top view of circular and square pre-cast toilet/latrine slabs (Source: http://www.clean-water-for-laymen.com). (b) Pre-cast concrete and galvanized metal pour-flush bowls for squatting slabs (Source: http://www.clean-water-for-laymen.com).

Figure 3.6 Illustration of pre-cast concrete toilet and HDPE plastic portable flush toilet (Source: Toppla Toilet, China). (a) pre-cast concrete toilet. (b) HDPE Plastic portable flush toilet, Toppla Toilet, China.

(a)   (b)

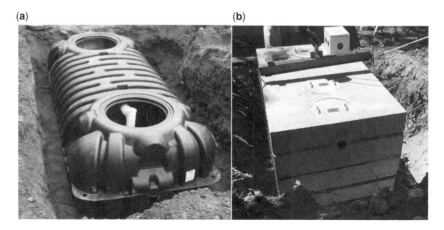

Figure 3.7 Prefabricated plastic and precast concrete septic tanks (Source: https://theconstructor. org). (a) Prefabricated plastic septic tank; (b) Precast concrete septic tank.

The manufacturing of industrial prefabricated concrete/plastic septic tanks has been found to be more cost-effective and durable than the conventional *in situ* tanks designed and installed to receive partially treated raw domestic sewage/faecal sludge and wastewater (USEPA, 2018). The prefabricated structures last longer, take half the effort to install and lower the overall septic tank cost. If properly designed and manufactured, they do not leak like the conventional ones (Figure 3.7).

Consequently, all enterprises and businesses involved in the production of these artefacts are lined up on the value chain to improve the quality of precast sanitation products and services (Cutler & Frank, 2010). Studies have also shown that pre-cast concrete septic tanks are durable and can last for several decades, are very resistant to cracking and are generally watertight throughout their lifespan – an advantage they have over the conventional septic tanks (Rahman *et al.* 2013). Manufacturers and suppliers of these products and their component parts also form part of the SMPVC.

3.3.1.4.3 In-situ constructed products

There are also businesses/enterprises and/or private individuals involved in the SMPVC that provide *in situ* construction products and services to toilet/latrine users, especially local masons. The mason as a production enterprise provides low-income households with access to safely managed sanitation products. They also provide more affordable and desirable products to traditionally un-served consumers. The masons engage in both the supply and value chain and work within the networks of producers, receive materials from several suppliers, and then deliver final products to the end-users (USAID, 2018), see Figures 3.8 and 3.9.

This is the most common way sanitation infrastructure and/or facilities are provided in most developing countries via the construction of substructures and superstructures. While the substructure provides safe disposal or reuse of human wastes, the superstructure is meant to provide privacy of the toilets. Designing an *in situ* toilet/latrine involves site selection, calculating the size of the vault, and determining the labour, materials, and tools needed for construction. *In situ* reinforced concrete (IRC) is still the preferred method for construction for toilets, especially the whole toilet supporting structural elements such as columns, beams and slabs (Rahman *et al.*, 2013).

Figure 3.8 In-situ block work septic tank by local mason (Source: https://niwa.co.nz).

Figure 3.9 Example of *in situ* toilet/latrine construction by local mason: five stance lined pit latrine at Kisomoro Primary School, Uganda (Photo by Rutenta Allan, Source: https://rsr.akvo.org/fr/project/3890/update/19237).

3.3.1.5 Distribution/route to market
is the process of storing and distributing the finished products for sale (Porter, 1985) as well as the place where the products on the value chain will be sold without further transformation (M4P, 2008; UNIDO, 2011). The distribution route to market include planning and dispatch, distribution management, transportation, warehousing; promotion management; domestic sales such as products sold to traders, wholesalers and retailers or directly to the customers (Acharyulu *et al.*, 2015); all of which make up a major part of the SMVC. The primary objective here is to ensure that the products reach the clients/customer in an efficient and effective manner, thereby ensuring satisfied customers and increased sales growth (M4P, 2008; UNIDO, 2011). The manufacturing firms in the value chain need to consider end-market demands, not only to determine

how best to sell products, but also to understand the nature and quality of the products that they will be able to sell in the future as well as the barriers that can prevent them from entering markets and selling their products. Such barriers include trade regulations, standards and export restrictions, as well as the market power of competitors (M4P, 2008; UNIDO, 2011). Other major challenges in the sanitation products marketing is that most businesses in the value chain do not interact directly with the end-consumers of their products and in most cases the 'sanitation professionals' do not consider the main producers of sanitation and related products and the building/construction industries as part of the main actors that can accelerate the race towards SDG 6. Often there are a range of intermediaries, exporters, importers, wholesale distributors, retailers, services providers and brokers involved in marketing and trade. In other cases large manufacturers may deal with retailers directly, but rarely with the consumers (M4P, 2008; UNIDO, 2011).

3.3.1.6 End-users/customers
Manufactured products value chain is inherently dependent on the satisfaction it provides to end-users/consumers (Ellis *et al.*, 2019). Understanding the consumer's defined value in satisfaction and the process in which it is attained can aid in increased value creation through process optimization and product development (Ellis *et al.*, 2019). Thus, sanitation and related commodity producers should pay more attention to the preferences and needs of individual customers/end-users and serve them differently according to their importance and value through products, information and transaction customization (tailoring of products and services to customers' individual needs) (Cho & Lau, 2014; Syam *et al.*, 2005). Studies should be conducted from consumer perspectives in the quest to provide more value to the end-users of these products (Dekker, 2003; Zokaei & Simons, 2006). The consumer should be viewed as the starting point of the entire process because the producers and suppliers are rewarded not only for providing a product but for the performance of the activities in providing the product. Demand drives the market and thus production, processing, and market approaches should focus on consumer needs (Christopher, 1998). The end-users demonstrate appreciation of value for a product by the willingness to pay and it is essential to note that the product is being purchased to derive a more direct value or satisfaction. Consumers usually perceive value more differently than the products actual monetary value thus need to assess value beyond price by evaluating end-user consumption chain (Ellis *et al.*, 2019). It is increasingly apparent that end-users are key to the success of sanitation and its related products value chain, especially when new well-designed sanitation technological products fail because they do not fit the standards of comfort the end-users uphold or were incompatible with cultural beliefs or religious codes (Koottatep *et al.*, 2019b). Domestic end-users are vital to co-producers of change in sanitation if they are taken seriously as system users and invited to rethink sanitation practice (Hegger *et al.*, 2007; Spaargaren *et al.*, 2007; van Vliet *et al.*, 2011).

3.3.1.7 After sale services
are set of activities taking place after purchase of the product, devoted to supporting customers in the usage and disposal stage of that product (Rolstadaas *et al.*, 2008; Saccani *et al.*, 2007). These services may include but are not limited to assembly, installation and operations activities designed to provide support engineering and construction services for the installation of treatment and user-interface equipment ranging from simple networks to highly complex treatment facilities (PWC, 2012). Also, aftermarket business and after-sales service processes play an integral role in many sanitation manufacturing and service concerns of the container-based sanitation enterprises and other innovative

smart sanitation products in the market as well as in wastewater, sewage and faecal sludge treatment plants. These equipment producers/providers have used the expertise they acquired in a particular technology to lay claim to act as integrators, installers, and operators often in the sphere of treatment facility activities. In recent times, the application of technologies like desalination has enabled a number of production and construction companies to develop operator-type skills (PWC, 2012). After-market which often refers to downstream value chain businesses is four to five times larger than the original sanitation equipment businesses (Cohen *et al.*, 2006). After sale services can be in form of:

(I)   Field and online support services performed directly at customers' sites or via online connections; onsite fault elimination, remote monitoring, management of call desks and helpdesks; and on-call services, reliability solutions;

(II) Repair services that are performed along the sanitation service chain (repair and calibration); and

(III) Logistics services to support and/or optimise customers service processes such as spare parts management, supply of instruments and tools (Rolstadaas *et al.*, 2008; Saccani *et al.*, 2007).

Also, after-sale services can be a way to encourage people to buy the product in the first place and could influence future sales.

3.3.2 SMPVC supporting activities

SMPVC supporting activities are services and/or activities that are not directly involved in the conversion process but support the primary/main activities in their function. They allow the proper operations of the primary activities. The supporting services are categorized into four and they include:

(I)   Infrastructure that consists of many services to facilitate production such as general management, planning, finance, legal, and external affairs;

(II) Human resource provides skilled personnel, premises and plant information technology and systems. It also deals with recruitment, hiring and training, developing, rewarding and sanctioning people in the organization;

(III) Technology defines product characteristics and is concerned with the equipment, hardware, software, technical skills used by a firm to transform inputs to outputs and supporting limited activities of businesses, such as accounting, and so on; and

(IV) Procurement manages supplier relationships and is concerned with the acquisition of inputs or resources for smooth production processes (Acharyulu *et al.*, 2015; Porter, 1985).

The goal is to ensure that the primary manufacturing value chain is effective and efficient to the extent that if it is done well, the production enterprises can increase the value-added and margin of the value output over cost in input (McLeod, 2012).

3.4 PRODUCT QUALITY AND STANDARDS

Product quality is an important aspect in value chain development because the quality of a marketable good is not just about product features but also the processes that occur in the value chain (Springer-Heinzer, 2018). Apart from the intrinsic aspects of product quality such as the materials used and the processing quality, the characteristic of business processes count: resource efficiency, the technologies used, conditions of employment and other factors all contribute to the quality of the product. There are two quality aspects of manufactured products:

(I) Product features which depict the characteristics of the product that result from design. These are the functional and aesthetic features of the product intended to appeal and provide satisfaction to the customer. The sum of a product feature is usually referred to as 'grade' (Evans & Lindsay, 2005; Groover, 2010; Juran & Gryna, 1993); and

(II) Free from deficiencies which means that the product does what it is supposed to do (within the limitations of the design feature), that is absent of defects and out-of-tolerance conditions, and that no parts are missing.

This aspect of quality includes components and sub-assemblies of the products as well as the product itself (Evans & Lindsay, 2005; Groover, 2010; Juran & Gryna, 1993). Quality benchmarks of products can be grouped into four major parts:

(I) Legal requirements regulating the minimum level of product safety;

(II) Industry-specific technical norms and quality grades facilitating contracts;

(III) Quality criteria defined by individual enterprises to position a product in the market; and

(IV) Sustainability standards on a wide variety of issues of social and political interests.

The first two points constitute the basic rules of any kind of business activities. Every enterprise first has to comply with the current laws and regulations, both in the country of production and in the country where the product is to be sold. Technical norms and grades are necessary to facilitate business linkages (Springer-Heinzer, 2018). Product simplification and standardization is required to achieve a higher efficiency in production, better quality and reduced production cost. Simplification is a process of determining limited number of grades, types and sizes of components or products or parts in order to achieve better quality control, minimize waste, simplify production and, thus, reduce cost of production. By eliminating unnecessary varieties, sizes and designs, simplification leads to manufacturing of identical components or products for ease of interchangeability and maintenance purposes during parts assembling.

Standardization techniques include the determination of the optimal manufacturing processes, identifying the best possible engineering materials, and allied techniques for products and services as well as adhering to very strict and better standards (Groover, 2010; Singh, 2006). Thus, definite standards are set up for a specified product with respect to its quality requirement, equipment and machinery, labour, material, processes and the cost of production (Singh, 2006). In other words, standards are a set of rules describing product and process quality as well as documents established by consensus and approved by a recognized body to provide for common and repeated use, rules, guidelines or characteristics for activities or their results, aimed at achieving optimum degree of order in a given context (Springer-Heinzer, 2018). The players involved in all these aspects also feed the SMVC to ensure that the sanitation products and services (that these products provide) meet standard and acceptable quality.

3.5 DISRUPTIVE INNOVATION

Lack of safely managed sanitation in developing countries and aging sanitation infrastructure as well as budget constraints in developed countries are increasing pressure on governments to find new solutions to old and new problems. New suites of technologies could improve sanitation and provide viable alternatives to traditional toilets, septic tanks and sewerage systems. Some disruptive innovative ways to remove pathogens from human waste includes among others:

(I) Wet oxidation (where materials suspended in water are broken down using oxygen);
(II) Dry combustion (where human waste is converted to charcoal briquette instead of flushing it away with water);
(III) Electrochemical processing (that makes use of metal oxides);
(IV) The uniquely solar septic tank (modified conventional septic tank with a solar-heated water system to create higher temperature than ambient inside the septic tank) (AIT, 2015 – Figure 3.10).

The blue diversion toilet (essentially a urine diverting dry toilet improved with a separate water cycle (blue) diversion for hand-washing, anal cleansing, menstrual hygiene and flushing of the front compartment) is being developed as a way to make sanitation sustainable, safe, accessible, and affordable (Tobias *et al.*, 2017 – Figure 3.11).

Other prototypes tested are the non-fluid pedestal incorporated mechanical flush toilet system, which is activated by moving the toilet lid and then a gear connects the lid to a rotating bowl that turns downward as the lid is closed. A swipe situated inside the pedestal is connected to the bowl-lid-gear-system. As the bowl rotates, the swipe moves downwards, clearing remaining faeces out of the bowl. This mechanism acts as a barrier for visual and olfactory irritation of the user (Figure 3.12).

The effort to unlock the potential of these solutions is in its earliest stages, and many technologies are still under development and this market is estimated to be worth around $6bn annually by 2030. Many of the new sanitation technologies have emerged from the 'Reinvent the Toilet Challenge', an initiative of the Bill & Melinda Gates

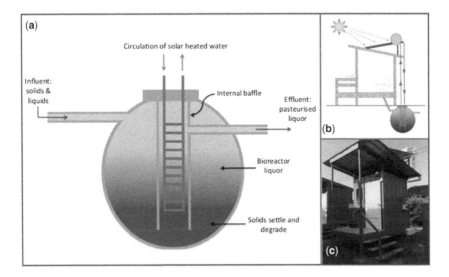

Figure 3.10 Schematic (a) illustrates the principle of the solar heating applied to the SST in contrast to the CT which operates at ambient temperature and without an internal baffle. Schematic (b) illustrates the installation of the SST at the field test site, showing the buried septic tank and the solar collection unit on the roof of the served toilet block seen in the photograph. (From Connelly *et al.* (2019) Under CCA 4.0 license, © 2019 by the authors).

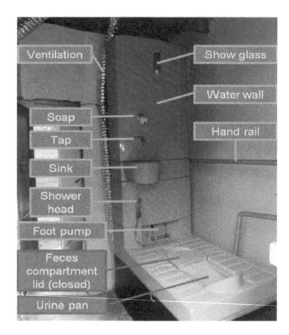

Figure 3.11 The working model of the Blue Diversion Toilet as used in the Kampala field test. The water wall contains the water tanks and bio-reactor. The metallic tube (labelled ventilation) is part of the active ventilation of the faeces compartment to prevent odour. (From Tobias *et al.* (2017) under CCA 4.0 license, © 2016 by the authors.)

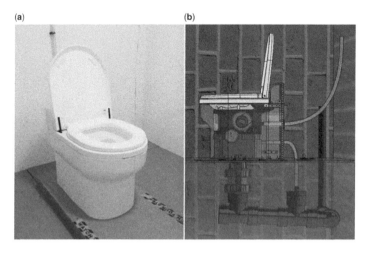

Figure 3.12 (a) Prototype pedestal with mechanical waterless flush, installed in a dedicated toilet room adjacent to the laboratories of the Pollution Research Group at the University of KwaZulu-Natal; (b) schematic of the installation: the pedestal is connected to the sewer mains and has a ventilation pipe from inside the unit; the gear system is shown to be on the side of the pedestal, underneath the cover. (From Hennigs *et al.* (2019) under the CCA 4.0 license, © 2019 by the authors).

Foundation (BMGF) launched in 2011, which funded innovators developing alternative sanitation technological solutions for the urban poor living on less than $2 a day. BMGF called for high user comfort, zero emissions to the environment, on-site solutions for resource recovery, and low costs of $0.05 per person per day. The costs are comparable to the lifecycle costs of community-based sanitation solutions with simple anaerobic technologies reported for Indonesia which stand at $0.05 per person per day (Tobias *et al.*, 2017).

Also, many of the systems are being designed to operate off the grid without connections to water and sewer systems or electrical lines. This means that sanitation solutions could be installed in parts of the world lacking access to power supply and other infrastructure. With modular, portable, easy-to-install formats, they could allow for increase in use as populations expand and make it possible to extend safe sanitation to remote locations where sewage or septic tanks might not be feasible. These promising innovative sanitation technologies can be deployed in both developing and developed countries, and the size of the opportunity varies depending on the setting. In developed countries that are building on current sanitation infrastructure, a potential market exists of about $1.7bn annually, while emerging economies with high needs, market readiness, and national policies prioritizing improvement in sanitation offer a $3.2bn opportunity.

3.6 CASE OF SOME EXAMPLES OF SANITATION AND RELATED PRODUCTS

3.6.1 Sanitaryware products

The product categories of sanitaryware are highly diversified as typical products include bidets, pedestals, sinks, showers, tanks, flush toilets and wash basins and the weights of these products range from 7.6 kg to 39.5 kg per piece (Lv *et al.*, 2019). The world ceramic sanitaryware production is estimated to grow from $32.1bn in 2020 to $44.6bn by 2025 (Research and Markets, 2021) (Figure 3.13). Toilet sinks/water closets (*water closet is defined as the cistern, bowl and plastic/metal attachments used to connect to the plumbing supply*) is projected to account for the largest share of the overall ceramic sanitary ware market in terms of value between 2020 and 2025.

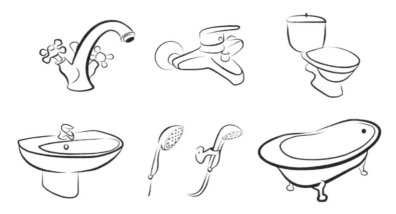

Figure 3.13 Illustration of ceramic sanitaryware products (Source: https://www.pinterest.com).

This is because of the awareness and activities created during the era of the MDGs and now the SDGs have led to increased demand for safely managed sanitation in most developing countries (Research and Markets, 2021) and could be responsible for the most significant growth of the sanitaryware production market being seen in Africa (+300%), the middle East (+181%) and South America (+163.7%). The latter region is driven by Brazil as the world's second largest producer country (24 million pieces) while China in particular is the world's largest producer and has almost doubled its volumes to reach an estimated 120 million pieces (MECS-Acimac Research Department, 2017). According to European data, sanitaryware production is estimated to be about 350 million pieces per year and total sales close to €18 billion in 2016 (Silvestri *et al.*, 2020). Also, a country like India is taking initiatives under its Swachh Bharat Abhiyan program to build public toilets in rural areas (*see section 8.4.2*) and increase their production capacity of ceramic sanitaryware products.

On the other hand, the uses and patronage of foreign manufactured ceramic water closet are high in Nigeria, and supply is met through importation despite an abundance of ceramic raw materials locally. Research shows that Nigeria ranks 13th among the world's consumers of ceramic products, mostly ceramic water closets (Elakhame *et al.*, 2020; Research and Markets, 2021). The inadequate development of the ceramic sanitaryware value chain in Nigeria is due to the lack of product development raw with the available ceramic raw materials and support industry as well as an absence of skilled manpower such as ceramic designers, engineers, scientists and/or technologies (Elakhame *et al.*, 2020). With a population of over 200 million, Nigeria would be a huge market for locally manufactured sanitation products, which would then be affordable.

Manufacturing processes for sanitaryware are similar to common ceramic products. These sanitation products help meet the increasing demand for safely managed sanitation facilities such as better toilets and improved sanitary ware products. The valued-added activities are a set of processes required to transform a clay slip produced by a mix of raw materials and water then stored in tanks for subsequent slip cast and this can occur in separate moulds or through the employment of pressure casting machines. After the casting process has been completed, pieces are dried allowing transportation and further processing. Ware surface are glazed by spraying and fired in a specific kiln. The colour and vibrancy are given to underline shape by an additional process. The desired colouring is obtained by using pigments (metal oxides) in combination with the glaze. Finally finished sanitaryware are tested, packed and dispatched to a storage facility (Lv *et al.*, 2019; Silvestri *et al.*, 2020; Womack & Jones, 1996. See Box 3.1).

Box 3.1: Life cycle processes of sanitary ware production (Lv *et al.*, 2019)

(I) **Raw material extraction**, raw materials are extracted from the earth via mining.

(II) **Raw material transportation**, raw materials are then transported to the plant using big trucks.

(III) **Body preparation**, slurry for body is prepared by grinding the raw materials with water in ball mills.

(IV) **Glaze preparation**, glaze is also prepared by wet grinding.

(V) **Mould preparation**, mould is made from Plaster of Paris.

(VI) **Casting**, the slurry is poured into the mould to form sanitary ware body.

(VII) **Drying**, the casted body is dried using hot air flow.

(VIII) **Glazing**, the glaze is sprayed onto the dried sanitary ware body.

(IX)   **Firing**, the sanitary ware is then sent to kiln for firing at a high temperature.

(X)   **Packing**, the sanitary ware is packed and stored in warehouse.

(XI)   **Waste treatment**, wastewater, solid wastes and air pollutants (i.e., CO_2, SO_2 and NO_X) from the production processes are treated, and part of the wastewater and solid wastes are recycled and reused.

(XII)   **Delivery**, the products are distributed to domestic and overseas markets.

3.6.2 Portable toilet products

Prefabrication of toilet comes in form of portable toilets commonly used at construction sites, outdoor parking lots, and other environments where indoor plumbing is inaccessible, and at large outdoor gatherings such as concerts, fairs and recreational events (Advameg Inc., 2021). The main raw material components of the facility is light-weight sheet plastic, such as polyethylene, which forms the actual toilet unit as well as the cabana in which it is contained. A pump and holding tank form the portable sewage system. These items are fastened with assortment of screws, nails, bolts and hinges. The facility is also equipped with a chemical supply container and inlet tube (Advameg Inc., 2021) (Figure 3.14).

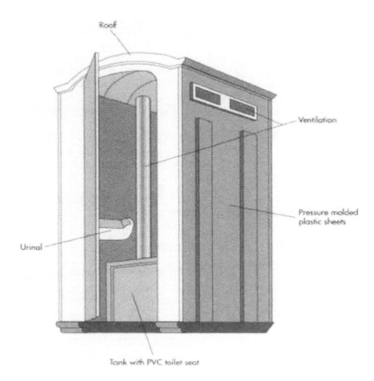

Figure 3.14 Portable toilet for outdoor events (Advameg Inc., 2021).

According to Advameg Inc. (2021), the value chain manufacturing process for the portable toilet unit comprises of the following steps:

(I)   The toilet unit is formed into a box-like structure and secured with nuts, bolts, and rivets with a rigid, lightweight sheet plastic. The top sheet contains an opening for placement of the toilet tank. The top sheet may not be secured with these permanent fixtures, allowing for its easy of removal for tank cleaning. A lock is placed over the top sheet to prevent unauthorized removal.

(II)   The actual toilet tank, which is placed in this unit, is made of the same material and shaped with a flat, corrugated front wall and rounded rear wall. The upper edge of the toilet tank is formed as a peripheral flange that extends outward and downward.

(III)   The toilet tank is fitted with a cover of two flat semi-circular plastic sheets. The lower sheet has a peripheral edge lip that extends downward, the upper sheet has a front lip that extends downward, and the rear lip extends upward and outward to latch onto the peripheral flange of the toilet bowl. Both sheets are fitted with a central toilet opening.

(IV)   A conventional toilet seat made of plastic is placed over the toilet bowl and connected to the assembly with hinges.

(V)   The seat is fitted with a pin, which pushes upward against a metal wear plate, which is secured to the bottom surface of the seat. The pin extends downward through the cover and a bracket. Under the bracket, a coil spring is placed around the pin. The upper end of the coil engages a washer fastened to the pin so that the seat maintains an upright position when not in use. (Note: Not all portable toilets are flushable. Those that are not do not contain this or the following two steps in the manufacturing process, but merely contain chemicals in the holding tank).

(VI)   A piston is placed underneath the lower end of the pin, and a mechanical bellows-type pump is placed beneath the piston. The pump contains a spray opening and is connected to an inlet tube which is, in turn, connected to a chemical supply container. When the seat is raised, the piston will activate the pump.

(VII)   The toilet opening is fitted with a pair of flat, plastic doors secured by hinges to bosses fastened to the bottom of the tank. These doors are connected to the toilet seat with metal links so that they are activated when the seat is lowered and raised.

Portable toilets will always be necessary as long as humans continue to congregate in outdoor areas and other sites without plumbing.

3.6.3 Treatment systems' equipment and ancillary products

Treatment plants collects effluent, sewage and/or faecal sludge from domestic, municipal and industrial sources and treat same to a level of purification that enable their reuse in agriculture, industry and even as potable drinking source (Meticulous Market Research, 2020). The producers/manufacturers of treatment equipment of either material and immaterial goods range from technology providers for treatment plants and/or components manufacturers such as provision of material goods to treatment systems for sewered (sewage/wastewater) and non-sewered (faecal sludge) to an engineering service provider or software developer (provision of immaterial goods) (Bombeck *et al.*, 2013). The producers/manufacturers of treatment equipment are more complex than that for basic equipment because there are more fragmented and no equipment represents more than 10% of a given installation's value; besides barriers to entry are powerful (PWC, 2012). Such equipment can be categorised into five major types:

(I)   Sewage Treatment Plants (STPs);
(II) Effluent Treatment Plants (ETPs);
(III) Activated Sludge Plants (ASPs);
(IV) Combined Effluent Treatment Plants (CEPTs); and
(V) Faecal sludge treatment plants (FSTPs).

In the case of new technology solutions that typically enjoy intellectual property protections, specialist providers can impose high prices and reap healthy profit margins from value addition. This market is expected to grow at a CAGR of 6.5% from 2019 to 2025 to reach $211.3bn by 2025 (Meticulous Market Research, 2020) because of growing focus on sewage treatment in many developed and faecal sludge treatment in developing countries. However, high costs of equipment, operations and disposal obstruct the growth of this market to some extent. In addition, aging infrastructure, excess energy consumption and rising expenditure due to excess sludge production are some of the major challenges in the sanitation treatment systems market (Meticulous Market Research, 2020). The overall sanitation treatment systems market is mainly segmented by product category such as:

(I)   Treatment technologies,
(II) Delivery equipment, and
(III) Treatment chemicals, and instrumentation.

The major global players in this sector range from large international groups (from the United States, Germany, Japan, China, Switzerland, the United Kingdom, Denmark and France) with an extensive range of products to local players with generally narrower product offerings. Also, markets are often regional or even local in nature because of the variety of standards and technology solutions across areas (PWC, 2012).

3.6.4 Tissue paper products

The production of tissue paper includes all paper products used for hygienic and sanitary purposes both at home and in public places. They include toilet paper, kitchen towels, tablecloths, napkins and wipes, but toilet paper dominates the market (Masternak-Janus & Rybaczewska-Błażejowska, 2015). The global average toilet paper consumption per person reaches up to 55 kg and the global tissue paper market is expected to grow at a compound annual growth rate (CAGR) of 6.45% during 2020–2025 (Research and Markets, 2020a, 2020b). Tissue paper production globally is expected to exceed 44 million tonnes in 2021, which is an increase of more than 14 million tonnes over 2010. Among all the tissue paper products, bathroom tissue remains the key tissue category, driving the tissue paper market forward through a combination of necessity and the general westernization of toilet culture. Increased development of organic tissue paper, rising disposable income, and government policies to promote public health are also some of the major factors driving the growth of the market. Also, there is a sudden spike in the demand for tissue papers due to the ongoing spread of the COVID-19 pandemic. The manufacturers of these tissues are producing 20% more than normal levels which might strain the supply chain (Research and Markets, 2020a) Figure 3.15.

The value chain of tissue paper spreads across three main lines: entry level or first price products, branded products, and private label products. The *'entry level or first price products'* are often low-quality raw materials based and sold with minimum packaging

Figure 3.15 End-user using toilet tissue (image: Freepik.com).

(standard packs with few colour images). They are present in hard discounts while in supermarkets they are identified as 'first or best price'; the same ones can be found on hawkers' benches, local markets, suburban bazaars (Galli, 2017). The *'branded products'* are manufactured by big local and multinational companies and are internally developed and very often supported by dedicated advertising campaigns. The best techniques both for the paper production and packaging available in each company are used to design the products. The *'private label products'* which are produced by third party companies for big retailer chains requiring higher product quality comparable with the best ones in the market and used to indicate by consumer loyalty to the retailer brand as they love lower selling price as competitive advantage over branded products (Galli, 2017).

Tissue paper can be manufactured from either virgin pulp and/or recycled waste paper (Masternak-Janus & Rybaczewska-Błażejowska, 2015) and must have the following technical characteristic for customer satisfaction: softness, dry strength, wet strength and perforation efficacy (Galli, 2017). The tissue paper production from virgin materials has two basic units: the stock preparation and the paper making process in the paper machine. The stock preparation consists of the following stages: fibre refining, the removal of impurities, and finally the pulp is fed to a paper machine where it is formed and most of its properties are determined (Masternak-Janus & Rybaczewska-Błażejowska, 2015). In addition, the process of tissue paper manufactured from recycled waste paper is as follows: waste paper storage, repulping of the dry recovered waste paper, mechanical removal of impurities (screening, cleaning) and bleaching. Finally, the pulp is pumped to the storage chests that serve as a buffer between the stock preparation and paper machine (Masternak-Janus & Rybaczewska-Błażejowska, 2015; Michniewicz *et al.*, 2005).

3.6.5 Hand sanitizer

Hand disinfectants are commercially available in various types and forms such as anti-microbial soaps, water-based or alcohol-based hand sanitizer. Different types of delivery systems are also formulated – for instance, rubs, foams, or wipes (Jing *et al.*, 2020).

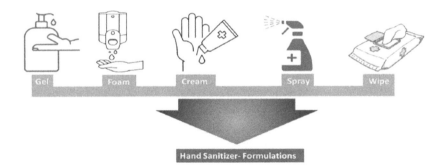

Figure 3.16 Various types of hand sanitizer dosage forms. (From Jing *et al.* (2020) under CCA 4.0 license, © 2020 by the authors).

The World Health Organization (WHO) recommends alcohol-based hand sanitizer (ABHS) in line with the proven advantages of their rapid action and a broad spectrum of microbicidal activity offering protection against bacteria and viruses (Jing *et al.*, 2020). The market is also witnessing an influx of bio-based or organic ingredients in hygiene products in several countries because of the awareness of the ill-effects of chemical-based hand rubs (Jing *et al.*, 2020; Research and Markets, 2020b), see Figure 3.16.

Hand sanitizers have an advantage over conventional hand-washing products as they can be applied directly without water. Also, renowned manufacturing companies such as Henkel Corporation, Unilever, and Procter and Gamble have been offering hand sanitizers in convenient packaging such as sachet and mini bottles, which can easily be carried in bags or a pockets by the consumers (Grand View Research, 2020).

The global hand sanitizer market size is valued at \$2.7bn in 2019 and is expected to grow at a compound annual growth rate (CAGR) estimated by Research and Markets (2020b) of over 17% during the period 2019–2025 while Grand View Research (2020) estimated CAGR of 22.6% from 2020 to 2027. The gel-based hand sanitizer segment dominated the global market in 2019 with a share of more than 49%. Gel sanitizers are usually thin and watery in formulation and therefore provide the convenience of getting spread easily and penetrate into the skin to kill most of the bacteria. The foam-based hand sanitizer, however, is expected to dominate the market with a revenue base of CAGR of 23.1% from 2020 to 2027. The product is gaining prominence in the market owing to its ability to penetrate the skin and stay there for a longer period of time (Grand View Research, 2020).

One of the main factors contributing to the growth of hand sanitizers is the outbreak of the coronavirus (COVID-19) a global pandemic by the World Health Organization (WHO). The outbreak has reinforced the significance of regular hand sanitizing and cleaning practices among consumers and is among the prominent factors driving the market (Grand View Research, 2020; Research and Markets, 2020b). Also, the introduction of fragrance-based hand sanitizers is identified as one of the major factors responsible for the growth of the global hand sanitizer market. This innovation has offered positive dividends and has boosted the market growth. Other contributory factors to the growth of the hand sanitizer market include (Research and Markets, 2020b):

(I)   Increasing influence of internet in shaping end-users purchasing behaviour;
(II) Growing demand of flavoured and organic hand sanitizers;
(III) Growth in promotional activities; and
(IV) Rise in health consciousness among consumers.

3.7 SMPVC – THE FUTURE OF SAFELY MANAGED SANITATION PRODUCTS

SMPVC are at the heart of providing safely managed sanitation and its related products and services to over 60% of the world's population without access to these vital goods and services. Proper diagnosis and understanding of the SMPVC holds the key to unlocking the bottlenecks surrounding access to sanitation products and services in most developing countries. This is based on the fact that most sanitation and other connected products and input materials are products emanate from the GVC that operates outside the shores of most of these countries with inadequate access to safely managed services. Even when they are domestically manufactured within developing countries, the operating businesses and enterprises are operated by trans-national corporation seeking cheaper labour and other resources. In the end, the costs of these products and services are beyond the reach of the vast majority of the population in these developing countries most of whom ear minimal income. For instance, some studies have shown that even when sanitation enterprises are profitable in these countries in the global South, the complexity of the business and the capital required may make its attractiveness lower than other alternatives (USAID, 2018).

The information in this chapter has shown that developing countries without access to safely managed sanitation services are not active participants in the SMPVC. This is why sanitation (USAID, 2018). In essence, it is not that sanitation businesses and enterprises cannot be market-driven or even create a sanitation economy, the barrier is in the absence of local players in the industry-sector most of the time. Considering that all sanitation management and its related products for end-use and service delivery (whether final products, semi-finished products, input materials and non-tangible services) in developing countries are commodified, even in the remotest areas of the least developed countries. These products and services are paid for by consumers as they are needed, which indicates that there is market somewhere. The challenge, then is that most of the materials needed for sanitation management are imported and by the time they get to the market, a large number of the population cannot afford them so they make do with a makeshift options. Now, if there is some domestic active level of participation of enterprises and businesses in the sanitation sector at local and national levels then affordability and acceptability will surely drive up demand and market. When home-grown players participate in the SMPVC, directly and indirectly, it will also serve as an incentive to prioritise sanitation at government and business levels with policies and laws thereby strengthening the SDG6 aspirations in developing countries.

Such a value-added network of integrated production systems will require well thought-out and coordinated value-chain governance that supports active participation of local businesses/enterprises to grow. It also requires facilitating business/enterprise linkages between the GVC and domestic investors/players to solve the challenge of access to markets for input equipment, services and their products. Moving forward will require proper diagnosis of the SMPVC to understand how actors operate and coordinate their businesses to ensure that input materials are transformed, stored, transported and reached, in certain form and quality and finally to the end-consumers. It should also look at the various effects that operations in the chain have on vulnerable groups of the society who live below the poverty level (UNIDO, 2011). The diagnosis should provide a clearer picture of (UNIDO, 2011):

(I)   Who are the actors that participate in sanitation businesses across value chains?

(II)   Are there actors that coordinate activities in the overall SMPVC?

(III) What are the contractual arrangements under which actors buy and sell products?

(IV) How do actors exchange information and learn about solutions to improve products and business performance?

(V) What technical, business and financial services are available to support actors in the chain?

(VI) How much value do actors add to the product in the different steps in the SMPVC, what are their costs and how is this value distributed?

(VII) What are the power relations in the SMPVC and to what extent do they determine how economic gains and risks are distributed among chain actors?

(VIII) What kinds of barriers exist for firms to enter the SMPVC?

(IX) What is the level of competitiveness of firms in the SMPVC?

(X) What bottlenecks exist and what opportunities are available for development (upgrading) of the SMPVC?

(XI) Which policies and institutions constrain/support chain actors and facilitate SMPVC development?

The results of the diagnosis can inform government officials and key stakeholders about the required interventions and which part of the SMPVC need immediate and long-term attention as well as provide insights on how those interventions should be designed (UNIDO, 2011).

3.8 Take Action

(I) Identify the major sanitation, hygiene and related products and equipment used in your area and highlight which one is sewered and non-sewered.

(II) Take an informal survey of sanitation, hygiene and related products and equipment in your city and surrounding area and then visit some of the major players.

3.9 Journal Entry

(I) List different sanitation, hygiene and products and indicate equipment and where they are found and/or used.

(II) Write short notes on innovation for sanitation products and equipment manufacturing and the impacts on the value chain.

3.10 Reflection

(I) What kind of innovation should be considered in the manufacturing of sanitation systems for transportation facilities such as air, land, rail and sea?

(II) How can mobile sanitation systems be enhanced and made attractive for urban centres in developing countries?

3.11 Guiding Questions

(I)   What is a sanitation product and equipment manufacturing value chain?
(II)   Why are sanitation products and equipment ubiquitous? How are they intricately linked to our survival?
(III)   What is the place of value-addition in sanitation product and equipment value chain?
(IV)   Describe the nature of industries/enterprises participating in the sanitation manufacturing value chain?
(V)   Explain these concepts 'buyer-driven' and 'producer-driven' value chain and how they relate to the sanitation manufacturing value chain.
(VI)   With the aid of a well labelled value chain map, explain the primary activities of the sanitation manufactured products value chain.
(VII)   'All production systems are transformative processes', how is this so?
(VIII)   What are product quality standards? Give examples relevant to sanitation
(IX)   What are the four major routes of disruptive innovative technologies?

REFERENCES

Abecassis-Moedas C. (2006). Integrating design and retail in the clothing value chain: an empirical study of the organization of design. *International Journal of Operations and Production Management*, **26**(3/4), 412–428, https://doi.org/10.1108/01443570610650567

Acharyulu S. G., Subbaiah K. V. and Rao K. N. (2015). Value chain model for steel manufacturing sector: a case study. *International Journal of Managing Value and Supply Chains (IJMVSC)*, **6**(4), 45–53. https://doi.org/10.5121/ijmvsc.2015.6404. https://aircconline.com/ijmvsc/V6N4/6415ijmvsc04.pdf (accessed 1 January 2021).

Advameg Inc. (2021). How Products Are Made. http://www.madehow.com/Volume-3/Portable-Toilet.html (accessed 18 January 2021).

AIT (Asian Institute of Technology) (2015). Innovative DEWAT technologies. Reinventing the Toilet – Progress Report 3, submitted to Bill & Melinda Gates Foundation.

Baglieri E. and Secchi R. (2007). Strategic goals and design drivers of the supplier portal: lessons from the ferrari case. *Production Planning and Control*, **18**(7), 538–547, https://doi.org/10.1080/09537280701529647

Bakerjian R. and Mitchell P. (1992). Tool and Manufacturing Engineers Handbook, 4th edn., Vol. **VI**, Design for Manufacturability. Society of Manufacturing Engineers, Dearborn, Michigan.

Bombeck M., Gregarek D., Hilbig J. and Rudolph K.-U. (2013). Integrated Water Resources Management in the "Middle Olifants" River Basin, South Africa (MOSA). Phase II Summary Report, Research Project No. 033048 A-Iz. Institute of Environmental Engineering and Management at the Witten/Herdecke University. http://www.iwrm-southafrica.de/wp-content/uploads/2020/01/2.5_Water_Market_Value_Chain_Approach.pdf (accessed 4 March 2021).

CAWST (Centre for Affordable Water and Sanitation Technology) (2014). Latrine Construction Manual. CAWST, Calgary, Canada.

Chang C-H. and Melkanoff M. A. (2005). NC Machine Programming and Software Design, 3rd edn. Prentice Hall, Inc., Upper Saddle River, New Jersey.

Chivaka R. (2007). Strategic cost management: value chain analysis approach, *Accountancy SA*, 24–27, August.

Cho V. and Lau C. (2014). An integrative framework for customizations on satisfaction: The case of an online jewelry business in China. *Journal of Service Science and Management*, **7**, 165–181. Published Online April 2014 in SciRes. https://www.scirp.org/pdf/JSSM_2014043010000976.pdf (accessed 4 March 2021).

Christopher M. (1998). Logistics and Supply Chain Management. Financial Times, London.

Cohen M. A., Agrawal N. and Agrawal V. (2006). Winning in the Aftermarket (HBR OnPoint Enhanced Edition), Harward Business Review OnPoint Article, May, 2006.

Connelly S., Pussayanavin T., Randle-Boggis R. J., Wicheansan A., Jampathong S., Keating C., Ijaz U. Z., Sloan W. T. and Koottatep T. (2019). Solar septic tank: next generation sequencing reveals effluent microbial community composition as a useful index of system performance. *Water*, **11**(12), 2660. https://doi.org/10.3390/w11122660

Cutler P. P. E. and Frank D. P. E. (2010). Watertight Precast Concrete Septic Tanks. National Precast Concrete Association / Precast Magazines / Archive – 2004–2008 / Watertight Precast Concrete Septic Tanks. https://precast.org/2010/05/watertight-precast-concrete-septic-tanks/ (accessed 18 January 2021).

Dekker H. C. (2003). Value chain analysis in interfirm relationships: a field study. *Management Accounting Research*, **14**(1), 1–23, https://doi.org/10.1016/S1044-5005(02)00067-7

Designing Building Wiki (2021). Sanitaryware. https://www.designingbuildings.co.uk/wiki/Types_of_sanitary_appliances (accessed 14 February 2021).

Elakhame Z. U., Obe Y. J., Shuaib-Babata Y. L., Bankole L. K., Omowunmi O. J., Ambali I. O., Akinsanya O. O., Unachukwu A. J., Adeyemo R. G. and Ikusedun M. (2020). Utilisation of indigenous ceramic raw materials for the production of water closet. *Journal of Minerals and Materials Characterization and Engineering*, **8**(5), 364–337, https://doi.org/10.4236/jmmce.2020.85023

Ellis E., Kwofie E. M. and Ngadi M. (2019). Value beyond price: End user value chain analysis. *Int. J. Food System Dynamics*, **10**(4), 347–360. http://centmapress.ilb.uni-bonn.de/ojs/index.php/fsd/article/view/23/908 (accessed 23 February 2021).

Evans J. R. and Lindsay W. M. (2005). The Management and Control of Quality, 6th edn. Thomson/South-Western College Publishing Company, Mason, Ohio, USA.

Florén H., Frishammar J., Lee C., Ericsson M. and Gustafsson S. (2013). A framework for raw materials management in process industries. Conference: R &D Management Conference At: Manchester, UK, June 2013. https://www.researchgate.net/publication/256079372_A_framework_for_raw_materials_management_in_process_industries (accessed 6 January 2021).

Florén H., Frishammar J., Löf A. and Ericsson M. (2019). Raw materials management in iron and steelmaking firms. *Mineral Economics*, **32**, 39–47. https://doi.org/10.1007/s13563-018-0158-7

Fortuna A., Fortuna D. M. and Martini E. (2017). An industrial approach to ceramics: sanitaryware. *Plinius*, **43**. https://www.socminpet.it/dwl.php?file=files/download/vol-43/fortuna.pdf. (accessed 22 January 2022).

Galli E. (2017). Introduction to process and properties of tissue paper. 138–145. https://www.tecnicelpa.com/files/20171013_EnricoGalli.pdf (accessed 24 February 2021).

Gasiorowski-Denis E. (2018). ISO focus highlights water and sanitation for UN Decade of Action. *ISO Focus*. https://www.iso.org/news/ref2256.html (accessed 22 February 2018).

Gereffi G. and Fernandez-Stark K. (2011). Global Value Chain Analysis: A Primer. Published by the Center on Globalization, Governance & Competitiveness (CGGC) Duke University Durham, North Carolina, USA, https://www.researchgate.net/publication/265892395_Global_Value_Chain_Analysis_A_Primer (accessed 13 February 2021).

Girardet H. and Mendonça M. A. (2009). Renewable World: Energy, Ecology, Equality. Green Books, Cambridge, UK, p. 256.

Grand View Research (2020). Hand Sanitizer Market Size, Share & Trends Analysis Report By Product (Gel, Foam, Liquid), By Distribution Channel (Hypermarket & Supermarket, Drug Store, Specialty Store, Online), By Region, And Segment Forecasts, 2020–2027. https://www.grandviewresearch.com/industry-analysis/hand-sanitizer-market#:~:text=Report%20Overview,22.6%25%20from%202020%20to%202027.&text=The%2077.0%25%20population%20in%20the,and%2062.5%25%20of%20female%20users (accessed 28 February 2021).

Groover M. P. (2010). Fundamentals of Modern Manufacturing: Materials, Processes, and Systems. John Wiley and Sons, New York, USA. ISBN 978-0470-467002.

Hartley J. L., Meredith J. R., McCutcheon D. and Kamath E. R. (1997). Suppliers' contributions to product development: An exploratory study. *International Journal of Engineering Management*, **44**(3), 258–267.

Harvey P. A. (2007). Excreta Disposal in Emergencies: A Field Manual. Water, Engineering and Development Centre, Loughborough University, Loughborough, UK.

Hegger D. L. T., van Vliet J. and van Vliet B. J. M. (2007). Niche management and its contribution to regime change: the case of innovation in sanitation. *Technology Analysis and Strategic Management*, **19**, 729–746, https://doi.org/10.1080/09537320701711215

Hellin J. and Meijer M. (2006). Guidelines for Value Chain Analysis. Food and Agriculture Organization (FAO), UN Agricultural Development Economics Division, Rome.

Hennigs J., Ravndal K. T., Blose T., Toolaram A., Sindall R. C., Barrington D., Collins M., Engineer B., Kolios L. J., McAdam E., Parker A., Williams L. and Tyrrel S. (2019). Field testing of a prototype mechanical dry toilet flush. Science of the Total Environment, **668**, 419–431. https://doi.org/10.1016/j.scitotenv.2019.02.220

Holm R. H., Kamangira A., Tembo M., Kasulo V., Kandaya H., Gijs Van Enk P. and Velzeboer A. (2018). Sanitation service delivery in smaller urban areas (Mzuzu and Karonga, Malawi). Environment & Urbanization. **30**(2): 597–612. https://doi.org/10.1177/ 0956247818766495

IDSA (Industrial Designers Society of America) (2020). What Is Industrial Design? https://www.idsa.org/what-industrial-design (accessed 05 January 2021).

Jing J. L. J., Pei Yi T., Bose R. J. C., McCarthy J. R., Tharmalingam N. and Madheswaran T. (2020). Hand sanitizers: A review on formulation aspects, adverse effects, and regulations. International Journal of Environmental Research and Public Health, **17**, 3326. https://doi.org/10.3390/ijerph17093326

Jones L., Demirkaya M. and Bethmann E. (2019). Global value chain analysis: concepts and approaches. Journal of International Commerce and Economics, 1–29. https://www.usitc.gov/journals/jice_home.htm (accessed 05 January 2021)

Juran J. M. and Gryna F. M. (1993). Quality Planning and Analysis, 3rd edn. McGraw-Hill, New York, USA.

Koottatep T., Cookey P. E. and Polprasert, C. (eds). (2019a). Chapter 6: resource system. In: Regenerative Sanitation: A New Paradigm For Sanitation 4.0. IWA Publishing, London, UK, pp. 209–282. https://doi.org/10.2166/9781780409689_0209

Koottatep T., Cookey P. E. and Polprasert C. (2019b). Chapter 4 social-ecological system. In Regenerative Sanitation: A New Paradigm For Sanitation 4.0. IWA Publishing, London, UK, pp. 107–140. https://doi.org/10.2166/9781780409689_0209

Koottatep T., Pussayanavin T., Khamyai S. and Polprasert C. (2021). Performance of novel constructed wetlands for treating solar septic tank effluent. Science of the Total Environment, **754**, 142447. https://doi.org/10.1016/j.scitotenv.2020.142447

Kubba S. (2017). Water Efficiency and Sanitary Waste in Handbook of Green Building Design and Construction, 2nd edn, The Netherlands. https://www.sciencedirect.com/topics/engineering/composting-toilet/pdf (accessed 05 January 2021)

Lamming R. C. (1993). Beyond Partnership: Strategies for Innovation and Lean Supply. Prentice-Hall, London.

Lazarte M. (2016). ISO's two-step solution to improving sanitation for 2.4 billion people. ISO Focus Magazine. https://www.iso.org/news/2016/11/Ref2145.html (accessed 22 February 2018).

Lee Y. C. and Gilleard J. D. (2002). Collaborative design: a process model for refurbishment. Automation in Construction, **11**(5), 535–544, https://doi.org/10.1016/S0926-5805(01)00064-4

Lucena J. and Schneider J. (2008). Engineers, development, and engineering education: From national to sustainable community development. European Journal of Engineering Education, **33**(3), 3797.

Lv J., Gu F., Zhang W. and Guo J.-F. (2019). Life cycle assessment and life cycle costing of sanitary ware manufacturing: A case study in China. Journal of Cleaner Production, **238**, 117938. https://doi.org/10.1016/j.jclepro.2019.117938

Masternak-Janus A. and Rybaczewska-Błażejowska M. (2015). Life cycle analysis of tissue paper manufacturing from virgin pulp or recycled waste paper. Management and Production Engineering Review, **6**(3), 47–54. https://doi.org/10.1515/mper-2015-0025. https://journals.pan.pl/Content/89638/PDF/5-janus.pdf?handler=pdf (accessed 24 February 2021).

Mattson C. A. and Wood A. E. (2014). Nine principles for design for the developing world as derived from the engineering literature. Journal of Mechanical Design, **136**, 121403, https://doi.org/10.1115/1.4027984

MCGee J. (2015). Value Chain. Volume 12. Strategic Management, Wiley Encyclopedia of Management, 3rd edition. https://doi.org/10.1002/9781118785317.weom120081

Mchenga J. and Holm R. H. (2019). Can a precast pit latrine concrete floor withstand emptying operations? An investigation from Malawi. Journal of Water, Sanitation and Hygiene for Development, **9**(1), 181–186. https://doi.org/10.2166/washdev.2018.096

McIvor R. and Humphreys P. (2004). Early supplier involvement in the design process: lessons from the electronics industry. *Omega*, **32**(4), 179–199, https://doi.org/10.1016/j.omega.2003.09.005

McIvor R., Humphreys P. and Huang G. (2000). Electronic commerce: reengineering the buyer–supplier interface. *Business Process Management Journal*, **6**(2), 122–138, https://doi.org/10.1108/14637150010321295

McIvor R., Humphreys P. and Cadden T. (2006). Supplier involvement in product development in the electronics industry: A case study. *Journal of Engineering and Technology Management*, **23**, 374–397, https://doi.org/10.1016/j.jengtecman.2006.08.006

McLeod G. (2012). Value Chain/Network Improvement. https://static1.squarespace.com/static/55df0347e4b064f219e05fb4/t/5792056ccd0f686a939c5b99/1469187496779/Value-Chain-Network-Improvement.pdf (accessed 23 February 2021).

Meckstroth D. J. (2016). The Manufacturing Value Chain Is Much Bigger Than You Think. MAPI Foundation Arlington, Virginia. https://static1.squarespace.com/static/58862301f7e0ab81 3935c244/t/58c05a2f6a4963ad69ed3734/1489001008886/PA-165_web_0.pdf (accessed 28 February 2021).

MECS-Acimac Research Department (2017). World sanitaryware production and exports. Tile International 1/2017. https://www.ceramicworldweb.it/filealbum/Home/materialicasa/pdf/tile-international/2017/1/070_075%20Statistic%20SANITARI.pdf (accessed 1 January 2021).

Meticulous Market Research (2020). Water and Wastewater Treatment Market Worth $211.3 Billion by 2025, Growing at a CAGR of 6.5% from 2019- Global Market Opportunity Analysis and Industry Forecasts. https://www.globenewswire.com/news-release/2020/06/18/2050151/0/en/Water-and-Wastewater-Treatment-Market-Worth-211-3-Billion-by-2025-Growing-at-a-CAGR-of-6-5-from-2019-Global-Market-Opportunity-Analysis-and-Industry-Forecasts-by-Meticulous-Researc.html#:~:text=The%20water%20and%20wastewater%20treatment,systems%20more%20relevant%20than%20ever (accessed 4 March 2021).

M4P (2008). Making Value Chains Work Better for the Poor: A Toolbook for Practitioners of Value Chain Analysis, Version 3. Making Markets Work Better for the Poor (M4P) Project, UK Department for International Development (DFID). Agricultural Development International: Phnom Penh, Cambodia. https://www.researchgate.net/publication/329574289_Making_Value_Chains_Work_Better_for_the_Poor_A_Toolbook_for_Practitioners_of_Value_Chain_Analysis (accessed 30 December 2020).

Michniewicz M., Janiga M., Sokół A., Żubrzak M., Przybyszewska-Witczak E., Kiszczak B., Jarowski P. and Bartosiak M. (2005). The Best Available Techniques (BAT). Guidelines for the Pulp and Paper Industry [in Polish: Najlepsze dostępne techniki (BAT). Wytyczne dla branży celulozowo-papierniczej], Ministry of the Environment, Warsaw.

Mitcham C. and Munoz D. (2010). *Humanitarian Engineering. Synthesis Lectures on Engineers, Technology and Society*, Morgan & Claypool Publishers, 1–87. ISBN-13: 978-1608451517.

Narayanan R., van Norden H., Gosling L. and Patkar Ab. (2011). Equity and Inclusion in Sanitation and Hygiene in South Asia: A Regional Synthesis Paper. UNICEF; WaterAid; WSSCC.

Nothomb C., Snel M., McHugh K., Narracott A. and Sauer J. (2014). Sanitation as a business: unclogging the blockages. In: Sanitation as a Business: Unclogging the Blockages. IRC, PSI, WSUP, Water For People, Kampala, Uganda, p. 38.

Porter M. E. (1985). Competitive Advantage: Creating and Sustaining Superior Performance. The Free Press, New York, NY, p. 38.

PWC (2012). Water challenges, drivers and solutions. https://www.pwc.com/gx/en/sustainability/publications/assets/pwc-water-challenges-drivers-and-solutions.pdf (accessed 4 March 2021).

Rahman N. A. A., Ahmad S. and Zainordin Z. M. (2013). Perception and awareness of leaking for toilet in pre-cast concrete structure. *Procedia – Social and Behavioral Sciences*, **85**, 61–69. https://doi.org/10.1016/j.sbspro.2013.08.338

Research and Markets (2020a). Tissue Paper Market – Growth, Trends, and Forecast (2020–2025). https://www.businesswire.com/news/home/20200928005369/en/Outlook-on-the-Tissue-Paper-Global-Market-to-2025---Bathroom-Tissue-to-Occupy-the-Maximum-Growth-Demand---ResearchAndMarkets.com#:~:text=With%20the%20global%20average%20consumption,14%20million%20tonnes%20over%202010. (accessed 24 February 2021).

Research and Markets (2020b). Hand Sanitizer Market Size By Functional Ingredients, Product Type, End-Users, Distribution Channels, Supply Chain, Geography, Industry Analysis and Forecast 2020–2025. https://www.researchandmarkets.com/reports/5067358/hand-sanitizer-market-size-by-functional (accessed 28 February 2021).

Research and Markets (2021). $44.6 Billion Worldwide Sanitary Ware Industry to 2025 – Impact of COVID-19 on the Market. https://www.globenewswire.com/news-release/2020/06/18/2049942/0/en/44-6-Billion-Worldwide-Sanitary-Ware-Industry-to-2025-Impact-of-COVID-19-on-the-Market.html (accessed 1 January 2021).

Rolstadaas A., Hvolby H. H. and Falster P. (2008). Review of after-sales service concepts. In: Lean Business Systems and Beyond. IFIP – The International Federation for Information Processing, vol. **257**, T. Koch (ed.), Springer, Boston, MA, pp. 383–391. https://doi.org/10.1007/978-0-387-77249-3_40

Saccani N., Johansson P. and Perona M. (2007). Configuring the after-sales service supply chain: A multiple case study. *International Journal of Production Economics*, 110, 52–69. https://doi.org/10.1016/j.ijpe.2007.02.009

Schuetze T. and Santiago-Fandiño V. (2014). Terra Preta sanitation: A key component for sustainability in the urban environment. *Sustainability*, 6, 7725–7750, https://doi.org/10.3390/su6117725

Sharpe T., Muragijimana C. and Thomas E. (2019). Product design supporting improved water, sanitation, and energy services delivery in low-income settings. *Sustainability*, **11**, 6717. https://doi.org/10.3390/su11236717

Silvestri L., Forcina A., Silvestri C. and Ioppolo G. (2020). Life cycle assessment of sanitaryware production: A case study in Italy. *Journal of Cleaner Production*, **251**, 119708. https://doi.org/10.1016/j.jclepro.2019.119708

Singh R. (2006). Introduction to Basic Manufacturing Processes and Workshop Technology. New Age International Limited Publishers, New Delhi. ISBN (10) 81-224-2316-7

Spaargaren G., Mommaas H., Van Den Burg S., Maas L., Drissen E. and Dagevos H. (2007). More Sustainable Lifestyles and Consumption Patterns: A Theoretical Perspective for the Analysis of Transition Processes within Consumption Domains, Contrast Research Report, TMP project. Wageningen: Environmental Policy Group, Wageningen University; Telos: Tilburg University; RIVM; LEI, Wageningen, Netherlands.

Springer-Heinzer A. (2018). ValueLinks 2.0 – Manual on Sustainable Value Chain Development. Value Chain Solutions, volume 2, pp 306, GTZ. https://beamexchange.org/uploads/filer_public/d3/a4/d3a4882e-eb14-4c30-8f7e-6ba4f51f6ec9/valuelinks-manual-20-vol-2-january-2018_compressed.pdf (accessed 13 January 2021).

Srai J. S. and Shi Y. (2008). *Understanding China's Manufacturing Value Chain: Opportunities for UK Enterprises in China*. University of Cambridge Institute for Manufacturing, UK. https://www.ifm.eng.cam.ac.uk/uploads/Research/CIM/china09.pdf (accessed 29 December 2020).

Sturgeon T. J. and Memedović O. (2011). Mapping Global Value Chains: Intermediate Goods Trade and Structural Change in the World Economy. United Nations Industrial Development Organization (UNIDO). Development Policy and Strategic Research Branch Working Paper 05/2010. https://www.unido.org/api/opentext/documents/download/9928658/unido-file-9928658 (accessed 13 January 2021)

Sturgeon T., Linden G. and Zhang L. (2012). Product-level global value chains: UNCTAD study on improving international trade statistics based on global value chains, Massachusetts Institute of Technology, August 29.

Syam N. B., Ruan R. and Hess J. D. (2005). Customized products: A competitive analysis. *Marketing Science*, 24, 569–584. https://doi.org/10.1287/mksc.1050.0128

Tarver E. (2020) Research and Development (R&D) vs. Product Development. Publiahws in Investopedia. https://www.investopedia.com/ask/answers/042815/what-difference-between-research-and-development-and-product-development.asp#:~:text=The%20difference%20between%20research%20and,existing%20products%20with%20new%20features (accessed 4 January 2021).

Thomas Publishing Company (2021). Top Toilet Manufacturers and Companies in the USA. https://www.thomasnet.com/articles/top-suppliers/toilet-manufacturers-and-companies/ (accessed 14 February 2021).

Tilley E., Lüthi C., Morel A., Zurbrügg C. and Schertenleib R. (2008). Compendium of Sanitation Systems and Technologies. Swiss Federal Institute of Aquatic Science and Technology (EAWAG), Duebendorf, Switzerland, p. 158.

Tilley E., Ulrich L., Luethi C., Reymond P. and Zurbruegg C. (2014). Compendium of Sanitation Systems and Technologies, 2nd Revised edn. Swiss Federal Institute of Aquatic Science and Technology (EAWAG), Duebendorf, Switzerland.

Tobias R., O'Keefe M., Künzle R., Gebauer H., Gründl H., Morgenroth E., Pronk W. and Larsen T. A. (2017). Early testing of new sanitation technology for urban slums: The case of the blue diversion toilet. *Science of the Total Environment*, **576**, 264–272. https://doi.org/10.1016/j. scitotenv.2016.10.057

UNIDO (2011). Industrial Value Chain Diagnostics: An Integrated Tool. United Nations Industrial Development Organization (UNIDO), Vienna, Austria. https://www.unido.org/sites/default/ files/2011-07/IVC_Diagnostic_Tool_0.pdf (accessed 29 December 2020).

USAID (2018). Scaling Market Based Sanitation: Desk review on marketbased rural sanitation development programs. Washington, DC., USAID Water, Sanitation, and Hygiene Partnerships and Learning for Sustainability (WASHPaLS) Project. https://www.issuelab. org/resources/31021/31021.pdf?download=true (accessed 17 February 2021).

USEPA (2018). Septic Systems. https://www.epa.gov/septic/types-septic-systems#conventional (accessed 19 January 2021).

Valfrey-Visser B. and Schaub-Jones D. (2009). Supporting Private Entrepreneurs to Deliver Public Goods: Engaging Sanitation Entrepreneurs. 34th WEDC International Conference, Addis Ababa, Ethiopia.

van Vliet B. J. M., Spaargaren G. and Oosterveer P. (2011). Sanitation under challenge: contributions from the social sciences. *Water Policy*, **13**, 797–809, https://doi.org/10.2166/wp.2011.089

WEDC (Water, Engineering and Development Centre) (2012). Latrine Slabs: an Engineer's Guide. Loughborough University, Loughborough, UK.

Wikipedia (2019). Product (business). Provided by: Wikipedia. Located at: http://en.wikipedia. org/wiki/Product_(business). License: CC BY-SA: Attribution-ShareAlike. https://courses. lumenlearning.com/boundless-marketing/chapter/what-is-a-product/ (accessed 4 March 2021).

Wikipedia contributors (2020, December 31). Composting toilet. In Wikipedia, The Free Encyclopedia. Retrieved 20:02, March 16, 2021, from https://en.wikipedia.org/w/index.php? title=Composting_toilet&oldid=997478251 (accessed 13 January 2021)

Wiktionary (2019). product. Provided by: Wiktionary. Located at: http://en.wiktionary.org/wiki/ product. License: CC BY-SA: Attribution-ShareAlike. https://courses.lumenlearning.com/ boundless-marketing/chapter/what-is-a-product/ (accessed 4 March 2021).

Wognum P., Fisscher O. and Weenink S. (2002). Balanced relationships: management of client-supplier relationships in product development. *Technovation*, **22**(6), 341–351, https://doi. org/10.1016/S0166-4972(01)00031-1

Womack J. and Jones D. (1996). Lean Thinking: Banish Waste and Create Wealth in Your Corporation. Simon & Schuster, New York, NY.

Zamora E. A. (2016). Value chain analysis: a brief review. *Asian Journal of Innovation and Policy*, **5**(2), 116–128. https://doi.org/10.7545/ajip.2016.5.2.116. https://www.researchgate.net/ publication/315688804_Value_Chain_Analysis_A_Brief_Review (accessed 12 February 2021).

Zhu Y., Alard R., You J. and Schönsleben P. (2011). Collaboration in the design-manufacturing chain: a key to improve product quality. In: Supply Chain Management – New Perspectives, S. Renko (ed.), InTech, Croatia. ISBN: 978-953-307-633-1, Available from: https://www.intechopen. com/chapters/18508 (accessed 5 January 2021).

Zokaei K. A. and Simons D. W. (2006). Value chain analysis in consumer focus improvement. A case study of the UK ed meat industry. *The International Journal of Logistics Management*, **17**(2), 141–162.

doi: 10.2166/9781789061840_0075

Chapter 4

Facility integration, installation and construction

Harinaivo A. Andrianisa, Asengo Gerardin Mabia and Peter Emmanuel Cookey

Chapter objectives

The aim of this chapter is to help the reader understand sanitation facility integration, installation and construction (SFIIC) and the interactions and linkages between the enterprises involved delivering installation, integration and construction services within the chain so that the supply of products and/or services are better aligned to the demands of customers. This stage of the IFSVC adds value to manufactured products and/or semi-finished products and related services and presents them in usable forms to customers/end-users.

4.1 INTRODUCTION

The sanitation facility integration, installation and construction value chain (SFIICVC) describes the businesses/enterprises that plan, design, integrate, install and construct sewered and non-sewered sanitation infrastructure. 'Sanitation facility' refers to a space in the built environment where sanitation infrastructure and technologies are installed and used by people and their organizations (Lian, 2019). Value addition is derived from end-users' satisfaction with installed facilities as they (i.e., the end-users) represent the source of the value inherent in the SFIIC value chain. The focus is on the downstream sector of the sanitation industry wherein value creation is only guaranteed through the eyes of the customers – the end-users (European Commission, 2016). Furthermore, a value chain approach is concerned with improving the organization and delivery of installation, integration and construction services within the chain so that the supply of product and/or services are better aligned to the demands of customers (European Commission, 2016). And so, this stage of the IFSVC adds value to manufactured products and/or semi-finished products and presents them in usable forms to customers/end-users. For instance, the end-user value is related to the fulfilment of physico-psychological needs such as privacy, comfort, aesthetic appeal or status from using a toilet and provisions of technologically sound final treatment and reuse facilities. Therefore, the amount of money the final user is prepared and able to pay for this benefit is the amount that the value chain as a whole captures (De Groote & Lefever, 2016).

4.2 SFIIC VALUE CHAIN'S MAIN SERVICES

The SFIIC is a complex process involving a number of businesses and enterprises with an array of activities and actors. In some cases the services are provided by informal actors, operating as freelance skilled manpower, that are in most cases uncoordinated and with conflicting interests – including contractors, installers, architects and suppliers. This coalition of actors is characterized by a complex value chain involving on-site/ off-site construction activities, together with input materials supply and the assembly and integration of manufactured construction materials and products that contribute to the 'upstream' construction supply chain. In contrast with other stages in the IFSVC, the types and amounts of actors and businesses/enterprises involved vary during a single facility installation process. Enterprises and actors are selected depending on parameters such as the project scale, planned number of workers and user preferences, depending on whether the work requires upgrading, retrofitting and rehabilitation of dysfunctional sanitation infrastructures, disused, aged and unimproved sanitation systems, see Figure 4.1 (Koottatep *et al.*, 2019).

Small-scale contractors or installers often act as 'gatekeepers' between suppliers of products and facility owners (De Groote & Lefever, 2016), especially in countries where input materials for the SFIIC value chain activities are wholly imported and in dispersed rural and semi-urban communities owing to limited skilled workforce. In broader perspectives, the SFIIC value chain is not limited to the design and execution of construction works, but also includes the 'user-interface' and eventual rehabilitation or demolition, as well as a range of additional enterprises in the construction value chain, even those connected to the operations and maintenance of sanitation infrastructure (Figure 4.2).

The chain of businesses in this category includes fabrication and installation, logistics and transportation, architecture and estate developers, sanitary engineering, public/ environmental health practitioners, town and urban planners/designers, wastewater treatment plant installation, faecal and sewage sludge treatment plant installation, recovery, recycling and reuse, home and commercial building toilet installation, plumbing and cement works, public and mobile toilet installation, special case toilet design and

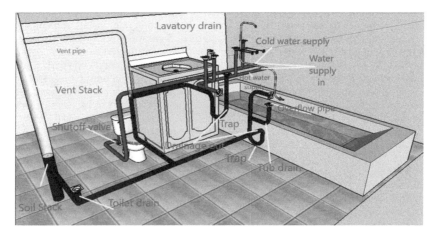

Figure 4.1 Installation and integration of toilet and bathroom plumbing system (Source: https://drpipe.ca).

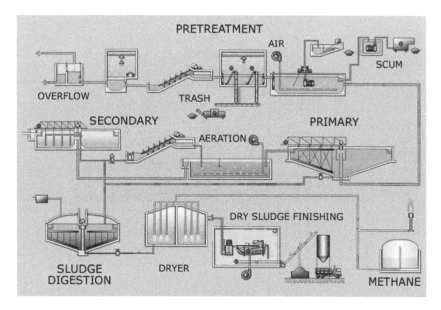

Figure 4.2 Illustration showing the integration of wastewater and sewage treatment. (Source: LibreTexts (2021) *wastewater and sewage treatment*. LibreTexts libraries under CC BY-NC-SA 3.0 license, © 2021 by authors).

installation (luxury airplanes, trains, buses, ships etc), microenterprise services, sales, installation and construction, local masons, metal, woodwork and concrete building contractors, sanitation testing and quality control laboratories, and so on.

The SFIIC enterprises are made up of a skilled and unskilled workforce who are engaged as the need arises by the business owners. The skilled workforce includes professionals like engineers, scientists, planners, technicians and/or artisans like masons, plumbers and so on. The activities of these skilled persons cover providing installation, integration and construction services for new sanitation infrastructure and/or improve existing dysfunctional systems, which may include hard- and soft-asset sanitation solutions at public and private levels, such as faecal sludge treatment plants (Figure 4.3), sewers, septic tanks, latrines/toilets, treatment plants, disposal trucks, and so on. which must take into account all the technical solutions and all economic, social, organizational, institutional and environmental aspects (Koottatep *et al.*, 2019). The unskilled workforce provide assistance to the skilled workers (see Figure 4.4).

Then, professional, skilled and knowledge businesses may include engineers employed by consultants working for agencies with the primary responsibility to provide technical sanitation services. It may also include enterprises majorly in the built environment that are involved in the provision of sanitation services such as architects, planners, civil works and construction engineering firms involved with designs, planning and implementation of sanitation systems and/or services (WHO, 1992). Others are knowledge-providing services firms, which may engage behavioural scientists, anthropologists, health staff, geologists, economists, and others having specialist expertise that are involved in some planning or implementation stage of the sanitation systems, programme and/or services (WHO, 1992).

Figure 4.3 First successful pilot faecal sludge treatment plant (FSTP) installation and integration in India in Devanahalli, Karnataka. The main treatment steps followed in this FSTP are solid-liquid separation, stabilization, dewatering of sludge and pathogen removal. (Source: Consortium for DEWATS Dissemination Society).

The artisan enterprises such as masons, bricklayers, drain-layers, carpenters, plumbers and others are also involved in the provision of sanitation services. These groups of workers often have experience in construction of houses and other buildings in ways that include providing of sanitation infrastructure in such facilities (e.g. water closets and septic tanks in most urban and rural housing projects/programs). They also have special skills required for the construction of toilets/latrines in specific communities. This category also includes local contractors engaged by householders to carry out certain tasks for them such as lining pits or constructing squatting slabs and/or pans for pour flush toilets, ventilated improved pit latrines (VIP) and other kinds of toilets/latrines (WHO, 1992).

The provision of sanitation infrastructure in towns and cities of developed countries is facilitated by the installation, integration and construction of sewered networks, which then require operations and maintenance. In developing countries, however, a high number of non-sewered sanitation infrastructures are supported by faecal sludge treatment plants (Figure 4.4) as their main treatment system (PWC, 2012). Figure 4.5 shows a schematic of the design, construction and installation of the common pit latrines built up with concrete rings (left) and the newly proposed urine diverting dehydrating toilet (UDDT, right) (Uddin *et al.*, 2013).

4.3 THE SFIIC VALUE CHAIN MAPPING

Mapping the SFIIC value chain here aims to identify and highlight the main and supportive businesses/enterprises and all the related components and relationships between them (Springer-Heinze, 2018). The mapping process is a key step that needs to be implemented for proper understanding of the activities in the value chain and to ensure the transformation of the system towards sustainability. For the SFIIC value chain it allows experts to expand their perspectives on the opportunities and risks as well as improving the quality and efficiency of performance within each stage of the value chain. Generally, practitioners use value chain mapping to develop sustainability strategies and materiality assessments of numerous potential environmental, social and governance issues that could affect businesses/enterprises and/or stakeholders within

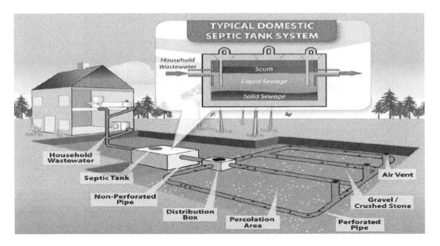

Figure 4.4 Septic system installation and integration with the system environment. (https://medium. com/waste-disposal-hub/how-septic-tanks-work-and-when-to-empty-them-346a4fe4fe6f)   (Source: https://www.angelsepticsconstruction.com)

the value chain (Holweg & Helo, 2014). Value chain maps will systematically evaluate the performance of current business processes to identify bottlenecks, challenges and improvement initiatives. It is carried out as an initial activity to study the needs, impacts and values of each entity in the chain, starting from the plan and design, input material supply, installation, integration, and construction, and then operations and maintenance. There are several reasons why mapping the SFIIC value chain would be a good solution for transforming sanitation businesses towards sustainability:

(I)   Value chain mapping provide a platform for communication and discussion with stakeholders or actors in the sector. This enhances the internal understanding of business opportunities arising from the external environment (customers and users).

(II) Mapping may reveal missing information including needs, desires, impacts, and gaps for each stage and/or entity in the value chain. This could enable organisations to see which stakeholders need to be involved and at what stage of sanitation businesses/enterprises, and/or which stakeholders need to be examined for further disclosure and to identify business activities that need improvement.

(III) Mapping the value chain may extend the perspectives of experts in providing concrete ways of thinking about the external environment of the sanitation installation, integration and construction sub-sector.

The SFIIC value chain, like any other value chain according to Michael Porter's model, includes two categories of activities where value is created: core/primary activities and support activities (Dubey *et al.*, 2020). Core activities or operational functions are directly involved in the provision of services along the chain. They include activities such as: plan and design; input material supply; installation, integration and construction; operations and maintenance; rehabilitation and/or demolition which may generate waste for recycling and/or disposal. Support activities or support functions help operational

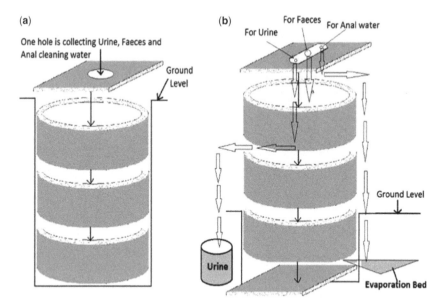

Figure 4.5 Pit latrine and urine diversion dehydration toilet (UDDT). Reprinted from Uddin *et al.* (2013) with permission from IWA Publishing. (a)Pit latrine; (b) Urine diversion dehydration toilet (UDDT).

functions to carry out their missions efficiently. They include research and development; human resources management including recruitment of skill and unskilled manpower, capacity development for staff/workers, finance and management; and infrastructure (which includes governance, planning, quality policy, sustainable development, ethics, etc.); and information systems. Figure 4.6 gives an overview mapping of the SFIIC value chain.

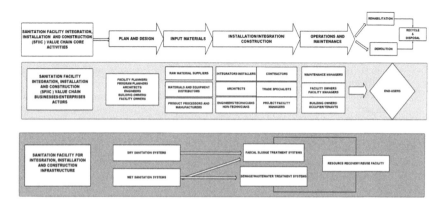

Figure 4.6 Generalized sanitation facility integration, installation and construction (SFIIC) value chain mapping. (Source: authors).

4.3.1 The SFIIC primary activities and actors

SFIIC value chain enterprises/businesses could be grouped into four main activities: planning and design, input materials, installation/integration/construction, and operations and maintenance; and they are made up of producers, suppliers, skill/service providers and contractors (Deshmukh *et al.*, 2014). SFIIC value chain activities are required to bring products and/or services from conception, through to the intermediary phases of production of sanitation infrastructural products and components which are used in construction, installation and delivery of safely managed sanitation facilities, to the final user. It can also be considered as all the different activities that are needed to deliver safely managed sanitation systems within the larger building and construction industry (Deshmukh *et al.*, 2014).

4.3.1.1 Planning and designing services

Planning and designing requires identifying activities and resources that support integration, installation and construction of physical realities (Hendrickson & Au, 2008) to enable the best use of available and limited resources. Facility planning addresses issues around location, layout, design and machinery/plant layout. The location refers to its placement with respect to the end-users and how it interfaces as well as integrates with other facilities. Design components consist of facility systems, the layout and the handling systems (Korhan, 2017) that creates the outlook of a new facility and/or modifications/improvements of existing facility, usually represented by detailed plans, specifications and construction.

Construction is the actual implementation of the plan and design envisioned by architects, planners and engineers. In both design and construction, numerous operational tasks must be performed with a variety of precedence and other relationships among the different tasks (Hendrickson & Au, 2008). For ordinary projects of moderate size and complexity, the owner often employs a planner and designer (an architectural/ engineering firm) which prepares detailed plans and specifications for the constructor (a general contractor). The designer also acts on behalf of the owner to oversee the project implementation during construction. The general contractor is responsible for the construction itself even though the work may actually be undertaken by a number of specialty subcontractors. (Hendrickson & Au, 2008).

At the planning phases the scoping and feasibility stages happen during designing and consist of schematic, development and tender documentation, especially in large sanitation project implementation. The scoping is done to outline the extent of the projects and what is needed in terms of sanitation systems and approaches (Lian, 2019). Planning and designing tools vary from checklists and cookbook-type approaches to highly advanced mathematical modelling approaches (Korhan, 2017). The plan and design could be for city-wide facilities improvement, or public infrastructures such as airports, hospitals and so on, or for provision of simple sanitation facilities at household and community levels. In addition to the above, the objectives could also be to rehabilitate and maintain existing infrastructure and services, construct more sanitation facilities and services and provide institutional support for the long-term sustainability of infrastructure (ADB, 2006). It is also important to note that without planning and design, it will be difficult to ensure that resources are used where they are most required and in a way that coordinates the actions of different stakeholders (Tayler *et al.*, 2000).

During planning and design stages of sanitation facilities development and/or improvement projects, the owner usually pays a fee to architectural/engineering (A/E) firms for most small, medium and large projects in urban and semi-urban areas; but in the rural areas, village technicians like plumbers and masons as well as other unskilled labour are contracted. In addition to planning and designing the facility, A/E firms also to some degree supervise the construction as stipulated by the owner (Hendrickson & Au, 2008).

4.3.1.2 Input materials supply services

The activities and businesses involved in the supply of input hardware and software for facilities' integration, installation and construction services vary and are selected depending on parameters such as the project scale, planned work and end-users' preferences (De Groote & Lefever, 2016). The input materials value chain service providers generally include: importers/wholesalers, retailers, contractors, concrete materials producers, masons and the end-users (USAID, 2015). They may also include: other raw materials suppliers, materials and equipment distributors, and hiring, as well as intermediate product processes and manufacturing. The basic materials are toilets and bathroom products, components of treatment plants, systems and infrastructure. In addition, other distributors supply roofing, guttering and insulation products for the super-structure.

Input materials providers could be split into two groups: (i) natural resources materials contractors who supply limestone, gravel, stone, sand, and wood (wood, lumber, wood panelling and mill-work products fall into this category); and (ii) manufactured resources such as bricks, cement, concrete, pipes, taps, toilets, washbasins, washbasin devices, slabs, aluminium, and so on. Actors involved in enterprises in group (i) are the owners of stone quarries, subcontractors supplying sand, gravel, and so on., while for group (ii) the actors are brick kiln companies, pre-manufacturers (craftsmen) of aluminium, steel and slabs, as well as the companies distributing plumbing and sanitary materials (wholesalers and retailers). The plumbing system is comprised of the entire piping networks, fixtures and appliances used for sanitation services consisting of wash basins, water closets, urinals, traps, soil waste pipes, vent pipes, septic tanks, and so on.

The main input materials for latrine construction and installations such as a ventilated improved pit latrine (VIP), pour-flush latrine, urine diversion ecosan toilets and so on, consist of a hole in the ground, a concrete slab or floor with a small hole, and a shelter which is the latrine superstructure that provides privacy and protection for the user. The procurement processes focus on the supply of materials, equipment and labour for the integration, installation and construction services. The businesses in this process are the ones involved in material and equipment supply, the manufacturers who sell turn-key equipment like treatment plants, wholesalers, retailers, labourers, contractors and subcontractors (Deshmukh *et al.*, 2014). The wholesalers could be a dealership that is an authorized agent of a manufacturer and receives direct supply from the manufacturer either within the country and/or outside the country (global value chain). Also, transporters are involved in haulage, especially with large and heavy materials that require special handling. The retailers purchase goods from wholesalers or other retailers (USAID, 2015). The labour market in this group includes both formal and informal operators, the latter being employed by small suppliers.

4.3.1.3 Installation/integration/construction services

The is the actual construction/implementation phase of the SFIIC value chain and should include all of the components and documentation for the project actualization. The actors involved in this phase will be responsible for requests for information, change order management, conflict resolution, inspection, submittal reviews, adhering to schedules and coordinating timely payments. Oversight in this area is critical because it has significant impact on the project's total cost (WBDG, 2017). Once the contract is awarded to a main contractor according to the design described and expressed in the contract drawings and documents, the consultants' team monitors the project's cost, quality and schedule to ensure that the project is delivered within budget, meets the quality requirement and is completed on time (Lian, 2019). If there are any changes to the design, this will be addressed through the variations procedures outlined in the contract and the contractor can claim for additional costs and time. This means that construction and installation can be part of a larger project (e.g., building) or a stand-alone (e.g., FSM treatment plants).

The construction and installation sites include a mix of small and large sites. Small sites mainly refer to residential complexes and commercial arcades, while large sites refer to schools, airports, hospitals, public markets, administrative buildings, and so on. The economic actors are construction companies, installation companies (plumbing and sanitary) and other registered small and medium-sized construction and installation companies. This segment is regulated by municipalities and the state. Generally, the labour market in this group includes both formal and informal workers, the latter being predominant (bricklayers, plumbers, architects, etc.). The general aim of this stage is to improve quality access to sanitation facilities such as toilets, latrines and urinals, handwashing facilities, bath structures, small and large-scale wastewater treatment systems and faecal sludge treatment plants for urban and rural housing structures, educational facilities, healthcare infrastructure, public and private office complexes, markets and malls, public packs, public transportation systems, and so on (Deshmukh *et al.*, 2014; Lian, 2019). The actual integration, installation and construction services are provided by architects, designers, builders, installers, contractors, engineers, masons, plumbers, and so on (Deshmukh *et al.*, 2014).

4.3.1.4 *Operations and maintenance (O/M) services*
Operations and maintenance (O/M) in the SFIIC value chain refers to the decisions and actions regarding the control and upkeep of sanitation facilities and equipment such as user-interface devices, storage, collection systems, emptying/conveyance/transport systems, treatment systems, and disposal as well as recovery and reuse systems, in an effective manner by various technical personnel, which includes routine functions (GoI, 2005). Maintenance refers to the art of keeping the structures, plants, machinery and equipment and other facilities in optimum working order and includes preventive maintenance or corrective maintenance of mechanical adjustments, repairs and planned maintenance (GoI, 2005). These include, but are not limited to the following:

(I) action focused on scheduling, procedures, and work/systems control and optimization; and
(II) performance of routine, preventive, predictive, scheduled and unscheduled actions aimed at preventing equipment failure or decline with the goal of increasing efficiency, reliability and safety (Sullivan *et al.*, 2002).

Regular maintenance preserves the system and ensures sustainability because studies have shown that preventive and predictive maintenance programmes reduce the cost of maintenance and energy consumption. Also, a properly maintained facility also produces less waste, less environmental impact and improves end-users' morale and perception (Lian, 2019). Therefore, O/M encompasses a broad spectrum of services, competencies, processes, and tools required to assure that the sanitation facilities will perform the functions for which they were designed and constructed. It typically includes the day-to-day activities necessary for the built-facility structure, its systems and equipment, and capacity to perform their intended functions (WBDG, 2017). Regular upgrades and rehabilitation of facilities are also needed. These include improvements to maintain service standards for the end-users, safety and energy upgrades, and reduction of environmental impacts of sanitation facilities and infrastructure as well as major overhauls of the entire systems (Lian, 2019).

To deliver operations and maintenance services firms required varied activities that could range from complex to very simple depending on the amount and type of sanitation equipment to be maintained. Operations plans focus primarily on the many processes and approaches associated with the management of facilities that set the expectations, standards, and precedents for broad facility management. On the other hand, maintenance plans are sometimes part of the operations plan and focus specifically on the upkeep of the facilities – from janitorial services, to housekeeping, plumbing, electrical and so on. Maintenance plans offer both preventive and reactive guidance for facility maintenance, including schedules for routine service and action plans for vital system failures (Schwartz, 2020). In specific terms, quality operations and maintenance of sewerage systems consist of the optimum use of labour, equipment and materials to keep the system in good condition so that it can accomplish efficiently the intended purpose of sewage collection. But, for non-sewered sanitation systems, maintenance varies depending on its volume and treatment method. If the volume is small, in many cases the owner controls the facility voluntarily. If the volume is medium or large, or if the facility employs an advanced faecal sludge treatment method, a separate maintenance firm controls the facility (GoI, 2005).

4.3.1.5 End-Users

Taking end-users perspective is vital for effective and efficient SFIIC value chain service delivery (De Groote & Lefever, 2016). End-users are closely related to the final utilization of the sanitation facilities and proper understanding of them will support designs and the provision of infrastructure that meet their expectations and assured willingness to pay for service (Cardone *et al.*, 2018; De Groote & Lefever, 2016; Koottatep *et al.*, 2019). If facilities' installation, integration and construction are below what the users are willing to tolerate, the services and assets will fall into decline, and are eventually abandoned toilet blocks that nobody uses for sanitation. The failure of most sanitation infrastructure is the service providers' false assumption of usage, willingness and ability to pay, and level of efforts to maintain facilities. These misconceptions lead to incompatibilities between technologies and the values, beliefs, and experiences of the users (Cardone *et al.*, 2018). The end-users are responsible for final use of the facility and will determine the comfort, performance and quality of the product and/or service, thus the SFIIC value-addition is the derived-satisfaction of the end-users. The challenge is that end-user behaviour is most often unpredictable and precarious and can change unexpectedly over time depending on existing situations and expectations. Innovations at this phase of the IFSVC/SFIIC should be connected to enhancing the end-use functionalities of comfort and safety, which could be a strong demand-pull for the more than four billion would-be end-users that currently lack access to safely managed sanitation (De Groote & Lefever, 2016).

4.4 BUSINESS MODEL, FINANCIAL MECHANISM AND SUSTAINABILITY

Business model refers to the drivers, processes and resources for the way the enterprise in the value chain creates and captures value within a market network of producers, suppliers and end-users (Vorley *et al.*, 2009). The business model concept is linked to business strategy (the process of business model design) and business operations (the implementation of a company's business model in organisational structures and systems). In the SFIIC value chain, buyers or customers or end-users are households and public ventures and other investments such as hospitals, schools, markets, restaurants, universities, companies, and so on,. The sellers are engineers, masons, plumbers, craftsmen, hardware or plumbing companies and construction companies or enterprises who market their services and products to these buyers along the chain. The relationship between these two stakeholders, the buyer and the seller, can be described by different types of links (White, 2009):

(I)   The instant market, where the operators (masons, craftsmen, plumbers, hardware stores, etc.) come to sell their services (construction, installation or integration) or products (washbasins, toilets, slabs, superstructures, etc.) and where prices fluctuate; this is the most risky in terms of setting the market price.

(II) A construction or installation contract in return for a grant from the supporting organization. In this case, prices are set by negotiations between the supporting organization and the operators offering the service or product, who are recruited, selected and trained technically by the supporting organization.

(III) A long-term, often informal, relationship characterized by trust or interdependency.

(IV) A company that has achieved full vertical integration.

When services, product sales and marketing depend on an instantaneous market with fluctuating prices and demands, financiers are uncomfortable; they prefer a contractual or partnership structure in a value chain where market risks can be better controlled. This is their comfort zone (White, 2009).

Osterwalder breaks business models up into their constituent elements that create costs and value (Osterwalder *et al.*, 2005). This template shows the importance of market differentiation (building a 'value proposition') and cost management to the success of any business model. In this SFIIC value chain, market differentiation is built on consumer desires, innovative technology, high services quality, and, sometimes, lower prices that are communicated to end-users through own brands. It follows that the partner network the supply chain input materials, suppliers and its coordination is an important source of competitive advantage. The model is highly sensitive to any addition of costs and risks, and it is around this apex that the question of market inclusivity ultimately revolves (Osterwalder *et al.*, 2005).

Now a sanitation company is a commercial entity that provides goods and services to a client who pays for them, generating financial income for the owner of the business/ firm, as the different elements of the sanitation enterprise interacts with each other and also acts on some of the barriers to greater client and contractor participation. The design of the sanitation enterprise is an iterative process. Ideally, the process starts with the selection of the target market, which will determine the other elements of the sanitation enterprise, namely the product system, the sales and marketing activities and the delivery model. However, interventions sometimes need to work with existing sanitation enterprises that may already have chosen these elements. The selection of the target market transcends purely socio-economic dimensions to include factors such as:

(I)   the number of customers who show an interest in building, upgrading or replacing their toilets/latrine facilities.

(II) willingness and ability to purchase toilets/latrines through savings and/or cash equivalents (e.g., credit, partial subsidy).

(III) ease of access for local products and service providers.

(IV) extent of innovation required (e.g. new products, new financing solutions) to activate demand.

For instance, in Mozambique, a low-cost sanitation project targeted the city of Maputo, where 90% of clients had access to unimproved pit latrines (Brandberg, 1997).

The project, therefore, focused on designing a low-cost cement slab as an add-on to improve the interface of existing toilets rather than developing a completely new product (Brandberg, 1997). Also, in Peru, 3.4 million households did not have access to improved toilets in 2014 even though most of these households were willing to purchase a toilet/latrine including superstructures. Thus, the Water and Sanitation Program (WSP) was able to identify 500 000 households as potential targets for the local private sector that consist of the major suppliers of components and materials for toilet/latrine construction, including the superstructure (Balcazar *et al.*, 2015).

The challenges to most sanitation business models arise due to lack of product standardization, especially in rural areas with low purchasing power where some local sanitation enterprises may over-specify substructures and user interfaces since their consumers lack appropriate understanding, while also offering poor production process, which can lead to increase in costs. For example, by tradition, manufacturers of cement concrete rings in Cambodia often increase manufactured concrete casing rings with a thickness of almost double what was required; this considerably increased the cost of materials and the product cost price (Pedi *et al.*, 2012). Standardization and improved material optimization could be achieved through the following strategies:

(I) Product re-engineering to reduce input materials or to incorporate less expensive alternatives while maintaining durability and quality;

(II) Efficient and effective production techniques and processes;

(III) Standardization of certain components, such as prefabricated sub-structure blocks, as was done in the 3Si and WaterAid projects in India and Nigeria respectively;

(IV) Obtaining a grant: to reduce the net cost to clients through the incorporation of subsidy programme or making the product eligible for a subsidy programmes;

(V) Simplified product components that make it easier for clients to install, which could also lower installation costs. Client installation is made possible by offering ready-to-install toilets or by providing instructions on the supply of materials and the construction processes.

A variety of marketing methods can be used in the SFIIC businesses including mass marketing or one-way communication channels (e.g., billboards, TV or radio commercials, social networks), toilet demonstrations, branding and interpersonal communication. Word of mouth by satisfied customers can also play a crucial role in convincing community members who do not yet have an opinion on improved and comfortable safely managed toilet/latrine systems.

4.5 CASE STUDIES

4.5.1 Sewered SFIIC value chain for six local government areas in USA

The sewered infrastructure SFIIC value chain incorporates and illustrates the complex set of activities and actors needed to coordinate the integration, installation and construction of sewer infrastructure (Daly *et al.*, 2015). SFIIC value for sewered infrastructure is made up of four major stages: (i) design and planning, (ii) materials and components, (iii) construction and installation, and (iv) maintenance and monitoring (see Figure 4.7). This case study covers the SFIIC value chain set of activities, actors, and policies across pre-construction, construction, and post-construction phases of sewered infrastructure in six local government authorities within the United States (Cleveland, OH; Louisville, KY; Omaha, NE; Philadelphia, PA; San Francisco, CA; Seattle, WA). The major actors in each segment of the value chain can be described across pre pre-construction, construction, and post-construction phases (Daly *et al.*, 2015).

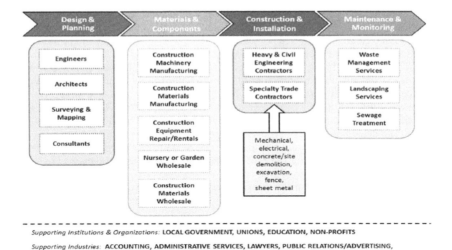

Figure 4.7 Sewer Infrastructure Value Chain (SIVC). (Source: Daly *et al.* (2015) © 2015 Center on Globalization, Governance & Competitiveness, Duke University).

The *designing and planning* phase of sewer infrastructure involves engineers, architects, surveyors (both traditional and geophysical), and environmental and management consultants. The firms in this pre-construction phase of the SFIIC value chain are the lead firms in the value chain, as they act as the prime contractors for many of the other phases of the chain (Daly *et al.*, 2015). The *materials and components* segments of the value chain serves as the inputs into the construction and installation phase. The materials and components suppliers include construction machinery manufacturing, construction material wholesalers, construction equipment repair/rentals, nursery and garden suppliers, wholesalers, and construction materials manufacturing (Daly *et al.*, 2015). The *construction and installation* segment of the value chain is comprised of two groups: (i) businesses whose primary activity is the construction of entire engineering projects; and (ii) companies that specialize in specific trades such as mechanical, electrical, concrete, excavation, fencing, sheet metal, site preparation and other contracting opportunities (Daly *et al.*, 2015).

Maintenance and monitoring on sewer infrastructure projects is either performed by local government or by businesses whose primary focus is construction and installation services. However, there are companies that specialize in sewer or catch basin cleaning, sewer system maintenance or landscape maintenance. The *supporting industries* services are often outsourced because they are outside the core expertise of the sewer infrastructure lead and subcontractors. The professionals often used for these supporting services include, for example, accounting, administrative services, lawyers, public relations, publishing, photography, real estate, security guard services and trucking (Daly *et al.*, 2015). In sewer infrastructure projects, multinational corporation (MNCs) in design and planning, and construction and installation segments of the value chain had the most lucrative contracts, acting as lead firms in the sense they capture about 65–80% of the total value and controlled much of the rest of the chain by dictating which contracts were distributed to smaller subcontractors (Daly *et al.*, 2015).

4.5.2 Wastewater design-build-operate SFIIC value chain, Thailand

This case study is of a firm that provides end-to-end supervision and construction of wastewater treatment plants and the manufacturing and sales of treatment products and chemicals. It is headquartered in Japan (Hassani & Wirjo, 2015). The firm establishes new wastewater treatment plants and undertakes the modification and enhancement of already existing plants. It was responsible for the building of a chemical manufacturing and centralized wastewater treatment facility in Chonburi Province which became operational in 2015 designed to treat wastewater from clients that do not have their own treatment facilities or that prefer to treat them in a centralized facility. The firm's business model is a proprietary biological–chemical treatment which has been developed in-house and used in virtually all of the firm's recommended treatment procedures, which is capable of reducing the cost of its client's wastewater treatment process to between 1/3 and 1/25 of the original cost.

The SFIIC value chain for this firm begins when the firm receives a request from its client to provide wastewater design-build-operate (DBO) services and ends when the firm operates the treatment plant on behalf of its client, which includes services such as monitoring and maintenance and disposal of treatment by-products (Hassani & Wirjo, 2015), see Figure 4.8. A total of at least 118 services can be identified in this value chain and grouped according to the various stages within the firm's SFIIC value chain as shown in Figure 4.9 (Hassani & Wirjo, 2015):

(I)   Services provided during the design stage,
(II)   Pre-building services,
(III) Services provided during the building stage,
(IV) Services during the operation stage, and
(V)   Back-office services.

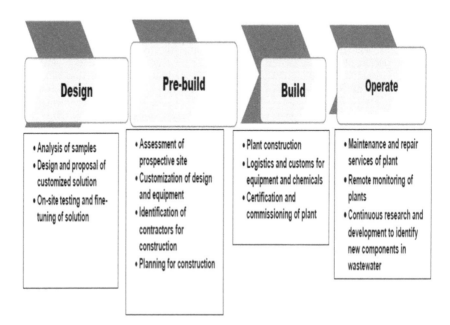

Figure 4.8 Wastewater design-build-operate SFIIC value chain. (Source: Hassani and Wirjo (2015) Under CCA 3.0 license, © 2015 by the authors)

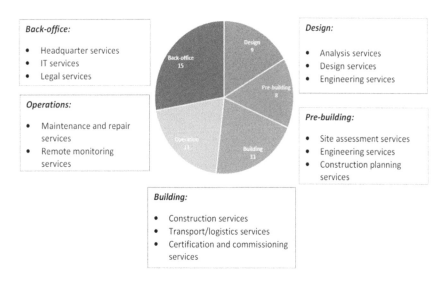

Back-office:

- Headquarter services
- IT services
- Legal services

Operations:

- Maintenance and repair services
- Remote monitoring services

Design:

- Analysis services
- Design services
- Engineering services

Pre-building:

- Site assessment services
- Engineering services
- Construction planning services

Building:

- Construction services
- Transport/logistics services
- Certification and commissioning services

Figure 4.9 Breakdown of services by stage and examples of key services. (Source: Hassani and Wirjo (2015) Under CCA 3.0 license, © 2015 by the authors)

The important role of services in the value chain can also see in the number of staff involved. The complement of five staff in the firm's Thailand subsidiary is generally providing services to clients and none are directly involved in the chemical and equipment manufacturing processes. Among the services activities provided are engineering services throughout the entire SFIIC value chain, remote monitoring services of installed equipment, as well as sales and marketing services for its products-cum-services offering.

4.5.3 Non-sewered SFIIC value chain, Ouagadougou (Burkina Faso)

The value chain for sanitation infrastructure integration, installation and construction in Ouagadougou could be subdivided into 4 different strata/classes as follows.

(I) *Traditional infrastructure services* dominated by households living in low-standard housing with services provided by the household members themselves (e.g. husband or cousin). Some use local masons to build the toilets/latrines, and these are toilets with inadequate sub-structures (unprotected pits) and superstructures;

(II) *End-users' services for improved dry pit toilets/latrines* and others' improved facilities, which is dominated by households living in medium housing and some public places (schools, markets, railway stations, restaurants, etc.). These technologies were designed and installed by National Office of Water and Sanitation (Office National de l'Eau et de l'Assainissement – ONEA) and other skilled and unskilled workers engaged by the private sector. Contractors and quarry owners provide raw materials such as sand, gravel, stone, and so on., which are then used in the facility construction or in manufacturing of prefabricated elements. Artisans are also involved in the prefabrication of slabs, bricks, superstructures and vendors provide other building materials and sanitary facilities. The work execution is done by masons and construction companies. Maintenance during the 'use' phase is carried out by plumbers and local masons;

(III) *End-users services for flush toilets,* mobile toilets, septic tanks, intelligent toilets, as well as integrated systems (including showers, wash hand basins, water closets, etc.). This market is dominated by households in high-standard housing and public places (schools, restaurants, hotels, bars, airport, etc.). The preparation of plans and designs is the business of architects, civil engineers and design offices, which sometimes do it at the same time when designing house plans. Raw materials are supplied by subcontractors, quarry owners, and so on., and pre-fabricated elements by craftsmen (individual contractors or private company). The building materials and sanitary plumbing supply is done by plumbing companies (individual contractors or private wholesalers and retailers). The latter offer different brands of materials depending on the quality and customer expectations; and

(IV) *Construction and installation of major sanitation infrastructure works* such as the construction of faecal sludge treatment plants, sewage treatment plants and the construction of sewerage networks, and so on. These services are managed and regulated by the government via its decentralized organization, which is ONEA. The plan's preparation, design and execution are provided by international consulting firms and construction companies. During the construction, integration and installation phase, local masons and plumbers are hired as day labourers or permanent workers until the end of the works. Raw materials and manufactured products are supplied by subcontractors, sales companies, craftsmen, and so on.

All these companies and individual operators are supported by numerous national and international sector actors such as ONEA, Action Contre la Faim, GIZ, WaterAid, national, regional and international banks, Plan International Burkina, Kynarou and other NGOs. These organizations provide technical and capacity building support as well as expert assistance to the implementation of SFIIC value chain. They also make new technologies available to the actors in order to contribute to the creation of added value. Others intervene through financial support by subsidizing the business of small and large SFIIC entrepreneurs. Above all, the supporters and enablers of the SFIIC value chain act as facilitators of the chain's activities. The ministries in charge of sanitation, trade, urban planning and others facilitate the implementation of activities through regulations and development of major sanitation and hygiene programmes that ensure the achievement of the SDG6 targets (Figure 4.10).

4.5.3.1 Economic analysis of SFIIC value chain in Ouagadougou

The SFIIC determination of added value in Ouagadougou was derived from monthly average data from each category of sector operators in the value chain. It is important to point out that the determination of the actual volume of activities per month in the study area was very difficult and so we opted to use the number of contracts won by each of the operator monthly. This is because for the craftsman, the contract may be to supply prefabricated elements of a diversified nature such as slabs, bricks, doors, claustras, and so on. The same situation is applicable to the mason whose contract represents a call to build a concrete pit or a complete latrine. The average fee for such service rendered takes into account the unit price of input materials, manhours and the skill level of workers engaged. The added values generated (in US dollars) by these operators on a monthly and annual basis are given in Table 4.1.

The values calculated reflect those generated by all SFIIC value chain operators in Ouagadougou. The number of operators taken into account remains approximate because it was very difficult to find the correct number working in the informal sector due to the

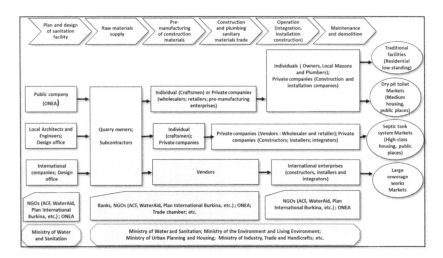

Figure 4.10 Overview of SFIIC value chain mapping in Ouagadougou (Burkina Faso) (Source: Hassani and Wirjo, 2015).

absence of a database. The number of masons and craftsmen used in this case study were derived from those trained by ONEA for sanitation facilities' construction and the supply of prefabricated materials for toilets/latrines. These operators are made up of a total of 1 009 trained masons and 900 craftsmen throughout the national territory, and Ouagadougou accounts for half of these actors (505 masons and 450 craftsmen). Other actors include 200 civil engineers and 217 architects registered and/or affiliated to the Order of Burkina's Architects and the Order of Civil Engineers. More than twenty million US dollars are generated each year by the SFIIC value chain operators in Ouagadougou. The artisans and vendors of sanitary plumbing materials are the ones who create the most value-added products/services in the chain with 17.62% and 15.98% respectively, which represent the total value addition created along the chain in one year, as shown in Figure 4.11.

4.5.3.2 Environmental analysis of the SFIIC value chain in Ouagadougou
The environmental analysis places the value chain into an ecosystem context to identify negative environmental impacts of the value chain activities on the environment as well as vice versa, the impact of the natural resources scarcity and climate change on business operations. A clear view on the environmental problems along the value chain also is the basis for identifying new business opportunities that come with the need to transform businesses to build a regenerative sanitation (Koottatep *et al.*, 2019; Springer-Heinze, 2018). Some of the impacts of SFIIC activities on the environment include the following:

(I) *Impacts on resource depletion:* The construction and installation activities of sanitation infrastructures extract natural resources such as wood for the manufacture of doors, superstructure, and so on. Other input materials are sand, gravel, stone, and so on. The over-consumption of these resources due to a high demand for latrines constitutes the pressure the value chain exerts on the environment. In DR Congo, millions of tonnes of wood are consumed per year and apart from energy use, construction is the second largest wood-consuming

Table 4.1 Value added generated by the Ouagadougou SFIIC value chain operators.

Operators	Quantity/Month Per Operator	Unit	Price Per Unit (XOF)	Number of Operators	Value Generated/Month (XOF)	Other Inputs and Services/Month (XOF)	Intermediate Products/Month (XOF)	Value Added/Month (XOF)	Value Added/Month (USD)	Value Added/Year (USD)
Masons	6	Contract	35 000	505	10 60 50 000	55 000	80 000	105 915 000	190 647	22 87 764
Plumbers	12	Contract	18 000	150	3 24 00 000	60 000	0	32 340 000	58 212	6 98 544
Craftsmen	5	Contract	75 000	450	16 87 50 000	250 000	1 200 000	167 300 000	301 140	36 13 680
Engineers	3	Contract	200 000	200	12 00 00 000	50 000	0	119 950 000	215 910	25 90 920
Architects	2	Contract	225 000	217	9 76 50 000	40 000	0	97 610 000	175 698	21 08 376
Vendors	12	Integrated system	90 000	144	15 55 20 000	400 000	3 400 000	151 720 000	273 096	32 77 152
Subcontractors	2	Contract	200 000	105	4 20 00 000	60000	0	41 940 000	75492	9 05 904
Design office	4	Contract	350 000	55	7 70 00 000	120 000	0	76 880 000	138 384	16 60 608
Construction enterprise	2	Contract	300 000	160	9 60 00 000	350 000	375 000	95 275 000	171 495	20 57 940
Maintenance and installation company	35	Contract	15 000	115	6 03 75 000	80 000	0	60 295 000	108 531	13 02 372
Total					95 57 45 000			949 225 000	1 708 605	2 05 03 260

Note: XOF – The West African CFA franc is the currency of eight independent states in West Africa: Benin, Burkina Faso, Côte d'Ivoire, Guinea-Bissau, Mali, Niger, Senegal and Togo (1 West African CFA franc equals 0.0018 United States Dollar). USD – United States Dollar.

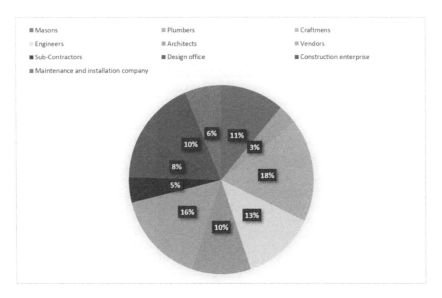

Figure 4.11 Weighting of value added by Ouagadougou SFIIC value chain operator. (Source: Authors)

activity in the country. The SFIIC value chain activities contribute to the depletion of natural resources such as:

- supply of input materials for construction, which is provided by quarry owners and subcontractors;
- prefabrication of components for toilets/latrines construction by craftsmen and prefabrication enterprises; and
- construction of sanitary facilities by masons and construction companies.

(II) *Impacts on ecosystem quality:* Aquatic and terrestrial ecosystems are not immune to SFIIC value chain activities because of its contribution to the destruction of terrestrial habitats through latrine and septic tank construction. It is, therefore, important to emphasize that failure to comply with established standards for pit depths will ultimately lead to the pollution of aquatic ecosystems. In Ouagadougou's precarious neighbourhoods, toilets are built by local masons and, for reasons of sustainability, they dig very deep beyond the limits required by regulators. Some justify these practices as ignorance of the dimensions established by regulations and others by the requirements of households (customers). Also, construction, integration, installation and maintenance activities generate wastes of various kinds that are not adequately managed and this leads to land and aquatic ecosystem degradation. Waste engine oil from sewage trucks used along the value chain constitutes a particular category of waste that impacts on ecosystems quality.

(III) *Impacts on human health:* According to WHO, air pollution in urban areas increases the risk of acute (pneumonia) and chronic (lung cancer) respiratory diseases as well as cardiovascular diseases. SFIIC activities, such as raw materials supply and construction activities generate dust that is dangerous to the worker and also to those in the vicinity of the construction sites.

(IV) *Impacts on climate change:* The high mobility of the machines and trucks used in the logistics of the value chain input materials contribute to emissions of greenhouse gases (although, there is no study on this issue). Quantitative studies should be carried out to quantify the greenhouse gases (GHGs) emitted by the SFIIC value chain.

(V) *Environmental impacts on SFIIC value chain:* Environmental factors such as topography, hydrogeology, and soil conditions can also influence the choice of available sanitation technology options and can increase the cost of toilets. Sandy, loose or unstable soil increases the risk of pit collapse. On the other hand, stony or rocky soil makes it difficult to construct and install sub-surface structures of the sanitation systems. Hostile environments require different products, which may involve designing radically new products instead of trying to adapt existing models that are used in stable or conventional environments. The case of the Easy Latrine product is a tangible example of the impact of the environment on the value chain. This model is not suitable for areas that are prone to flooding or where the water table is high. The company has developed two product options for flood-prone areas, as they are home to 88% of customers who live in harsh environments. Product costs in these areas have surpassed the $200 mark, compared to $35–50 in the rest of Cambodia, making them unaffordable for over a third of the country's population (Wei *et al.*, 2014).

4.5.3.3 *Social analysis of SFIIC value chain in Ouagadougou*
Social analysis of value chains starts with getting to know the vulnerable groups in and around the value chain – the people below the poverty line, women and young. Each of these people have different disadvantages in economic status and face discrimination for different reasons (Springer-Heinze, 2018). Therefore, social aspects such as the type of contract, remuneration (or wage), method and period of payment are taken into account in order to identify the most disadvantaged category (or categories) of actor(s) in the SFIIC value chain. These social aspects, thus, make it possible to understand the level of job security within the value chain, because a sustainable business requires stable employment for the actors involved. In Ouagadougou, the two main actors in the SFIIC value chain are the informal and formal actors. The social strata of actors in this stage of value chain are influenced by their educational, skill level, and types of activities as shown in Table 4.2.

Actors working in the informal sector (such as masons, craftsmen and local plumbers) may benefit from signing contracts and/or verbal arrangements for the services they provide throughout the SFIIC value chain, and the remunerations received are far below the country's minimum wage for people with no level of education, which is $109. Also, they are overexploited at construction and installation sites, as they are paid on a daily basis by the construction or installation companies (the contractors). On the other hand, those involved in the construction and installation of sanitation infrastructure in the formal sector enjoy a certain level of stability. Even though the job opportunities are not so regular, the wages are acceptable because they exceed the minimum wage and contracts are entered into between parties concerned. Some operators such as engineers and architects even receive their remunerations from the banks. In Uganda, for example, builders are recruited by contractors for on-site work (construction of latrines and septic tanks) by signing a paper contract regardless of the duration of the activity (daily, weekly or monthly) (USAID, 2015). The same procedure is adopted in Ouagadougou by ONEA in the recruiting process of local artisans and builders for the construction project of latrines for the benefit of vulnerable households.

The SFIIC value chain is often linked to the construction value chain in large urban centres and is often characterized by a low female participation. The nature of activities carried out at this stage of the IFSVC does seem to not allow for the active involvement

Table 4.2 Social treatment of the SFIIC actors in Ouagadougou.

	Activities	Contract Types	Enterprise Actors	Wage	Payment Method	Wage Settlement Period
Formal sector of the SFIIC value chain	Plan and design	Written contracts	Engineers; Architects; Design office; and so on.	Over $181.24	In cash; In bank; Mobile money	Weekly; Monthly
	Input materials supply	Written contracts and verbal agreements	Subcontractors; Quarry owners; Craftsmen; and so on.	Over $108.75	In cash; In bank; Mobile money	Daily; Weekly; Monthly
	Sanitation facility construction on-site	Written contracts and Verbal agreements to complete tasks spanning over weeks / months	Construction company; trained and local mason; bricklayer; and so on.	Over $181.24	In cash; In bank; Mobile money	Daily; Weekly; Monthly
	Sanitation facility installation	Verbal agreements to complete tasks spanning over weeks / months	Installation company; trained and local plumber; and so on.	Over $27.18	In cash; In bank; Mobile money	Daily; Weekly; Monthly
	Sanitation facility maintenance (user interface)	Verbal agreements to complete tasks spanning over weeks / months	Maintenance company; trained and local plumber or mason	$3.63 to $27.18	In cash; In bank; Mobile money	Daily; Weekly; Monthly
Informal sector of the SFIIC value chain	Raw materials supply	Verbal agreements or verbal contracts	Craftsmen; quarry owners; and so on.	$36.26	In cash	Daily; Weekly or Monthly
	Sanitation facility construction on-site	Verbal agreements or verbal contracts	Local mason; bricklayer; and so on.	$3.63	In cash	Daily
	Sanitation facility installation	Verbal agreements or verbal contracts	Local plumber	$5.45 to $36.26	In cash	Daily
	Sanitation facility maintenance (user interface)	Verbal agreements or verbal contracts	Local plumber and bricklayer	$3.63 to $18.13	In cash	Daily

of women. For instance, there are no female masons and plumbers in the city of Ouagadougou, although females are involved in the sales of plumbing and sanitaryware and in the provision of certain intellectual and administrative services such as funds management within companies, planning and design, and so on.

4.5.3.4 SFIIC value chain actors and enterprises in Ouagadougou

The SFIIC value chain, like the construction industry in large urban centres, is strongly dominated by informal actors. As the market for sanitation facilities' construction and installation is a non-regular market, the majority of actors have other secondary activities to support them. However, SFIIC value chain activities in Ouagadougou are carried out by three categories of actors, including operational actors, support actors and enablers.

Operational actors are those who are directly involved in the creation of value addition throughout the chain. Depending on each step in the chain and the activities that take place from the planning and designing of the sanitation infrastructure up to installation of user interfaces, there are many operational actors involved. Some work individually as service providers in the informal sector or as enterprises and others in association with service providers and products suppliers that not only meet the wishes of clients and users, but also generate value addition in the chain. The operational actors of the SFIIC value chain in Ouagadougou can be grouped into (see Figure 4.12):

(I)   sanitary facilities' constructors;
(II) sanitary facilities' integrators and installers;
(III) the suppliers of input materials and building and plumbing sanitary materials (distributors).

In addition to the operational actors, there are also the support and enabling actors. The support actors facilitate the operational actors in value addition throughout the chain. In the Ouagadougou SFIIC value chain, support actors intervene in various ways. These include financial and technical capacity building support for operational actors. For instance, the ecological sanitation and Ventilated-Improved Pit Latrines (VIP) projects executed by ONEA and ACF were good examples of support activities to actors. Some of the entities and organisations acting as supporters of the SFIIC are ONEA, Action Contre la Faim (ACF), WaterAid, Commune of Ouagadougou, the Burkina

Figure 4.12 Trade of plumbing sanitary materials in Ouagadougou. (Source: Photo by Authors)

International Plan and others. They intervene in the chain by providing the following services:

(I) mobilisation of financial resources from international and regional development banks and make same available for the promotion of sanitation businesses;

(II) strengthening the educational and technical capacities of actors (masons, craftsmen, architects, etc.) in construction, installation and integration of sanitation infrastructures;

(III) technologicainnovation of sanitation facilities in order to improve the comfort of users and boost the creation of added values by the operators.

The enablers, on the other hand, act indirectly on the SFIIC value chain through the instrument of policies and laws that regulate the construction sector and enable actors to create value along the chain. These include government and state entities, as well as international organisations that support the development of sanitation policies, laws and regulations. The Ministry of Water, Sanitation and Hydro-Agricultural Development, the Ministry of the Environment and Living Environment, the Ministry of Urban Planning and Housing, the Ministry of Industry, Trade and Handicrafts, and so on., are some of the government agencies that act as value chain enablers.

4.5.3.5 *End-users of sanitation facilities*

In Ouagadougou, sanitation facilities are used at household levels and public places (schools, hospitals, places of worship, markets, tourism and hospitality, etc.). Users benefit from the value created along the chain. Knowing their desires and requirements is a major asset for operators, as it enables them to offer services and products that meet their expectations (which constitutes a p-value in the chain). The choice of sanitation technology depends on particular households based on comfort, aesthetic aspects, water saving, access to sanitation, and so on. These reasons vary according to social rank and housing standards. Households occupying high-standard dwellings have a penchant for comfort, high technology and aesthetics, as their purchasing power enables them to build the toilets they desire and to buy the sanitary appliances of their choice. Figure 4.13 illustrates a typical toilet for high-class households in Ouagadougou.

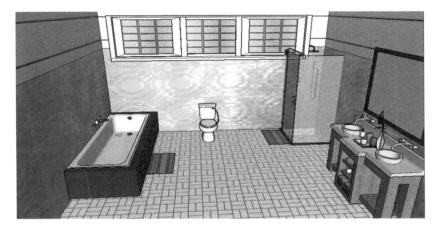

Figure 4.13 Integrated system toilet in Ouagadougou high standard housing. (Source: Authors)

(a) Flushing toilet in middle class housing

(b) Traditional toilet in low-class housing

Figure 4.14 Toilets in middle and low-standard housing. (a) Flushing toilet in middle class housing; (b) Traditional toilet in low-class housing. (Source: Photo by Authors)

Those in middle-class housing are sometimes forced to build and buy only for reasons of access to sanitation and functionality. For this purpose, the cheaper technologies are those used by these households despite the fact that they do not meet their desires. Toilets are less spacious and pose serious problems of mobility and ease of use. Figure 4.14 shows typical toilets in middle and lower-class dwellings in Ouagadougou.

Services are provided by SFIIC operators (masons, plumbers and distributors). Households occupying low-standard dwellings use traditional toilets. On the other hand, some households do not wish to use these technologies because they do not represent their perceived status, but due to a low purchasing power, they are forced to make use of these toilets/latrines. The sanitation facilities found in public squares in Ouagadougou are mostly built by the government. The construction contracts are carried out by the engineering, consultancy and construction companies. The choice of technologies varies according to parameters such as the nature of the activity on the site, the level of affluence, access to sanitation, comfort, and so on. The sanitation facilities' end-users are sometimes disconnected from the service providers and as a result services provided do not always match the expectations and desires of the clients.

4.6 CONCLUSION

The SFIIC value chain consists of enterprises involved in installation and construction of sanitation infrastructure from the users' interface, conveyance, treatment plants and other related sanitation facilities like bathroom plumbing systems and so on, specifically in accordance with customer preferences and requirements (IFC, 2018). The SFIIC cannot be separated from the overall building and construction industry, which is complex and involves a myriad of actors (De Groote & Lefever, 2016). A look into the building, construction and infrastructure sectors' 'value chain' will provide a full picture of the economic sector by identifying all the enterprises and businesses involved in providing value in the SFIIC stage of the IFSVC. This value chain stretches from enterprises that supply sanitation facility construction and installation materials to the delivery of completed and integrated infrastructures that meet the need of users' specifications such as individual households, tourism and hospitality, healthcare, educational, transportation, commercial and manufacturing concerns and/or utilities. It also covers a range of knowledge-intensive services provided by private enterprises and public knowledge organizations, including architectural and engineering consultancy services.

More broadly, taking a 'life-cycle' view, which is not limited to the design and execution of integration, installation and construction works, but also the 'user phase' and eventual rehabilitation or demolition, a range of additional actors in the construction value chain comes into play, including those connected to the operation and maintenance of a buildings as well as their sanitation appurtenances during their lifetime (European Commission, 2016). The major challenge with previous concepts of the sanitation value chain was looking at sanitation as a stand-alone system that is delivered in isolation of other sectors like building and construction. Meanwhile, in the real sense major SFIIC service providers of sanitation facilities and devices are also in the building, construction and infrastructure sector. In this sector, they provide wide-ranging sanitation services, like cesspool fabrication and installation, septic tanks, storm drains, as well as sewage and faecal sludge treatment structures. In a nutshell, the SFIIC VC is composed of specific variation within a fixed framework of distinct stages – plan, design, integrate, install and construct as well as their interactions and the value-addition that bring sanitation projects to fruition. Each of these is comprised of its own internal stages, processes, stakeholders and enterprises/businesses, actors (formal and/or informal) and business models involved in bringing sanitation products and infrastructure to end-users' point of utilization (IFC, 2018).

4.7 Take Action

(I)   Visit major hospitality, educational, recreation and transportation facilities and identify the sanitation systems available and the processes of integration, installation and construction.

(II) Analyse the performance functionality, efficiency and effectiveness of the above identified infrastructure and the value-added to the end-users.

4.8 Journal Entry

(I)   Draw an overview value chain map showing actors and enterprises with the SFIIC value chain of your area.

(II) What are the main activities in SFIIC value chain that are most relevant in your area and identify the workforce at formal and informal levels?

4.9 Reflection

Consider the economic, environmental and social impacts of the SFIIC value chain in your area.

4.10 Guiding Questions

(I)   What is a facility integration, installation and construction value chain?
(II) What are the main activities in the facility integration, installation and construction value chain?
(III) Describe the relationships between the buyers and sellers in the facility integration, installation and construction value chain
(IV) What is a sanitation company and what are the socio-economic factors required to select target markets in the facility integration, installation and construction value chain?
(V)   What strategies do enterprises in this value chain require for standardization and improved materials optimization?
(VI) What are the major stages of facility integration, installation and construction value chain for sewered infrastructure?
(VII) What are the four main classes of facility integration, installation and construction value chain for non-sewered sanitation in Ouagadougou?

REFERENCES

ADB (2006). Model Terms of Reference: Planning Urban Sanitation and Wastewater Management Improvements. https://www.adb.org/sites/default/files/institutional-document/33030/files/planning-urban-sanitation-tor.pdf (accessed 28 March 2021)

Balcazar C., Baskovich M. R. and Málaga I. (2015). Saneamiento–Un Negocio Redondo. WSP, World Bank Publications, Washington DC, USA.

Brandberg B. (1997). Latrine Building: A Handbook for Implementation of the SanPlat System. Intermediate Technology Publications, London, UK.

Cardone R., Schrecongost A. and Gilsdorf R. (2018). Shared and Public Toilets: Championing Delivery Models That Work. World Bank Publications, Washington, DC, USA. https://openknowledge.worldbank.org/bitstream/handle/10986/30296/W18035.pdf?sequence=4&isAllowed=y (accessed 06 April 2021)

Daly J., Brun L. and Guinn A. (2015). Targeting Inclusive Development: A Value Chain Approach to Sewer Infrastructure Investment. Center on Globalization, Governance & Competitiveness, Duke University, North Carolina, USA. https://globalvaluechains.org/publication/targeting-inclusive-development-value-chain-approach-sewer-infrastructure-investment (accessed 09/05/2021)

De Groote M. and Lefever M. (2016). Driving Transformational Change in the Construction Value Chain. Buildings Performance Institute Europe (BPIE). January 2016. http://bpie.eu/wp-content/uploads/2016/01/DrivingTransformationalChangeCVC2016.pdf

Deshmukh P. S., More A. B. and Chavan S. A. (2014). Supply chain management in residential construction sector. *International Journal of Engineering and Advanced Technology (IJEAT)*, 4(2), 87–90.

Dubey S. K., Singh R., Singh S. P., Mishra A. and Singh N. V. (2020). A brief study of value chain and supply chain. In: Agriculture Development and Economic Transformation in Global Scenario. Mahima Publications, Varanasi, India, pp. 177–183.

European Commission (2016). The European Construction Value Chain: Performance, Challenges and Role in the GVC. European Commission, DG GROW, Austrian Institute for Economic Research (WIFO), Brussels, Rotterdam, Vienna.

GoI (Government of India) (2005). Manual on Operation and Maintenance of Water Supply System. Produced by the Ministry of Housing and Urban Affairs, Goverment of India. http://cpheeo.gov.in/cms/manual-on-operation--and-maintenance-of-water-supply-system-2005.php (accessed 02/April/2021)

Hassani A. and Wirjo A. (2015). Wastewater treatment services in GVCs. In: Services in Global Value Chains: Manufacturing-Related Services, Vienna. pp. 121–147. https://www.apec.

org/publications/2015/11/services-in-global-value-chains-manufacturing-related-services (accessed 09/May/2021)

Hendrickson C. and Au T. (2008). Project Management for Construction: Fundamental Concepts for Owners, Engineers, Architects and Builders. Prentice Hall, New York, ISBN 0-13-731266-0. https://www.cmu.edu/cee/projects/PMbook/index.html

Holweg M. and Helo P. (2014). Defining value chain architectures: linking strategic value creation to operational supply chain design. *International Journal of Production Economics*, **147**, 230–238, https://doi.org/10.1016/j.ijpe.2013.06.015

IFC (2018). Construction Industry Value Chain: How Companies Are Using Carbon Pricing to Address Climate Risk and Find New Opportunities'. https://www.ifc.org/wps/wcm/connect/9d25fe8c-64c5-450a-ad01

Koottatep T., Cookey P. E. and Polprasart, C. (eds) (2019). Resource system. In Regenerative Sanitation: A New Paradigm For Sanitation 4.0. IWA Publishing, London, UK, pp. 209–282, https://doi.org/10.2166/9781780409689_0209

Korhan O. (2017). Facilities Planning and Design: A Lecture Note. Eastern Mediterranean University Department of Industrial Engineering. https://ie302.cankaya.edu.tr/uploads/files/file/LectureNotes/IENG441%20Facilities%20Planning%20and%20Design%20-%20Lecture%20Notes.pdf (accessed 27/March/2021)

Lian J. (2019). Facilities Planning and Design: An introduction for Facility Planners, Facility Project Managers and Facility Managers. National University of Singapore, Singapore, https://doi.org/10.1142/11227

LibreTexts (2021). Wastewater and Sewage Treatment. Retrieved 14 March 2021, from https://chem.libretexts.org/@go/page/12423, Licensed under CC BY-SA. https://chem.libretexts.org/Bookshelves/Introductory_Chemistry/Chemistry_for_Changing_Times_(Hill_and_McCreary)/14%3A_Water/14.08%3A_Wastewater_Treatment (accessed 23/January/2022)

Osterwalder A., Pigneur Y. and Tucci C. L. (2005). Clarifying business models: origins, present, and future of the concept. *Communications of the Association for Information Systems*, **16**(1), 1–25.

Pedi D., Kov P. and Smets S. (2012). Sanitation Marketing Lessons From Cambodia: A Market-Based Approach to Delivering Sanitation. WSP, WSP, World Bank Publications, Washington DC, USA.

PWC (2012). Water challenges, drivers and solutions. https://www.pwc.com/gx/en/sustainability/publications/assets/pwc-water-challenges-drivers-and-solutions.pdf (accessed 04/03/2021)

Schwartz K. (2020). Operation and Maintenance Plans for Buildings. https://spaceiq.com/blog/operation-and-maintenance-plans-for-buildings/ (accessed 02/april/2021)

Springer-Heinze A. (2018). ValueLinks 2.0: Value Chain Analysis, Strategy and Implementation. Vol. 1.

Sullivan G. P., Pugh R., Melendez A. P. and Hunt W. D. (2002). Operations & Maintenance Best Practices: A Guide to Achieving *Operational Efficiency*. Prepared byPacific Northwest National Laboratory Under Contract DE-AC06-76RL01831for the Federal Energy Management Program U.S. Department of Energy. https://www.pnnl.gov/main/publications/external/technical_reports/PNNL-13890.pdf (accessed 02/04/2021)

Tayler K., Colin J. and Parkinson J. (2000). Strategic Planning for Municipal Sanitation: A Guide. GHK Research and Training in Association with Water, Engineering and Development Centre (WEDC), London, UK. https://www.ircwash.org/sites/default/files/302.5-00ST-16187.pdf (accessed 27/March/2021)

Uddin S. M., Ronteltap M. and van Lier J. B. (2013). Assessment of urine diverting dehydrating toilets as a flood resilient and affordable sanitation technology in context of Bangladesh. *Journal of Water, Sanitation and Hygiene for Development*, **3**(2), 87–95, https://doi.org/10.2166/washdev.2013.113

USAID (2015). Assessment & Improvement of Sanitation and Hand Washing Supply Chain in Burera, Musanze, Nyabihu & Rubavu Districts in Rwanda. Part A: Main Assessment Report. https://www.usaid.gov/sites/default/files/documents/1860/FINAL-REPORT-Assessment-sanitation-hardware.pdf (accessed 30/March/2021)

Vorley B., Lundy M. and MacGregor J. (2009). Business models that are inclusive of small farmers. In: Agro-Industries for Development, C. A. da Silva, D. Baker, A. W. Shepherd, C. Jenane and S. Miranda-da-Cruz (eds.), CABI, Wallingford, pp. 186–222, https://doi.org/10.1079/9781845935764.0186

WBDG (Whole Building Design Guide) (2017). Project Planning, Delivery and Control. https://www.wbdg.org/project-management (31/March/2021)

Wei Y., Baker T., Roberts M., Taylor S. and Toe V. (2014). Sanitation Marketing ScaleUp (SMSU 1.0) – End of Project Report. iDE.

White A. (2009). From Comfort Zone to Performance Management, White & MacLean Publishing, Geneva, 20pp.

WHO (1992). A Guide to the Development of On-Site Sanitation. WHO, Geneva. https://www.who.int/docstore/water_sanitation_health/onsitesan/ch05.htm

doi: 10.2166/9781789061840_0103

Chapter 5

Sanitation services

Harinaivo A. Andrianisa, Mahugnon Samuel Ahossouhe and Peter Emmanuel Cookey

---

Chapter objectives

The aim of this chapter is to describe the sanitation services value chains (SSVC) activities, services, benefits, and satisfaction that are offered for sale, and/or provided to enhance access to safely managed facilities for safe disposal of human waste (faeces and urine), as well as ensuring adequate hygienic conditions within and around sanitation infrastructure. Services differ from products (facilities) in many ways, but are inseparable (simultaneously delivery and consumption), intangible (can only be experienced), perishable (can't be stored) and heterogenous (variable in performance of the same service).

5.1 INTRODUCTION

Globally, millions of large and small private enterprises and public ventures including a combination of the two are involved in the provision of valuable sanitation services to millions of households and communities. Within both groups exist many 'sanitation entrepreneurs' who seek to grow their businesses in a sustainable manner by providing value addition services to their customers. Sanitation services are described as activities, benefits, or satisfaction which are offered for sale, or are provided, in conjunction with the sale of goods to enhance access to safely managed facilities for safe disposal of human waste (faeces and urine), as well as ensuring adequate hygienic conditions within and around the sanitation infrastructure (Hamilton, 2004). Sanitation services differ from products (facilities) in many ways but are inseparable (simultaneously delivery and consumption), intangible (can only be experienced), perishable (can't be stored) and heterogenous (variable in performance of the same service) (Bhadwal, 2015). When the service has been completely rendered to the end-user, this particular service becomes irreversible and the service provider must deliver the service at the exact time of service consumption.

The service is not manifested in a physical object that is independent of the provider and also service end-users are not inseparable from service delivery (Wikipedia contributors, 2021). For example, the end-user must sit in the toilet/latrine and the emptier must be

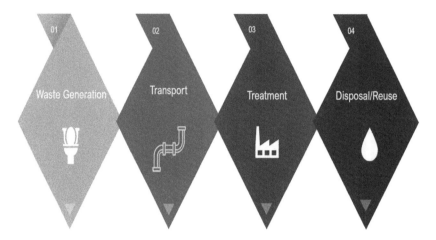

Figure 5.1 Conventional sanitation service chain. (Source: Frost & Sullivan (2021). Under CCA 4.0 license, © 2021 by authors.)

called upon to deliver their services of dislodging of the septic tank. Therefore, services include all sanitation value chain (SVC) activities where output are not a physical product or construction, and usually consumed when produced, and delivered as intangible value-added activities to end-users of the sanitation service chain (SSC) – for example emptying of toilets/latrines (Hamilton, 2004). Furthermore, sanitation services entail the payment of money for the provision of a technology or waste-related service, and overlooking the end-users' perspectives could lead to the development of inappropriate technologies and services (O'Keefe *et al.*, 2015). The sanitation services chain can be conceptualized in three different ways:

(I)   Conventional sanitation services as shown in Figure 5.1, depicting sewered sanitation service provision with excreta flushed away using a fully waterborne sanitation system for transport via networks of sewage pipes connecting each toilet to a main link leading to a treatment facility; treated sewage results in sewage sludge and effluents that are then disposed of (Frost & Sullivan, 2021).

(II) Containerised sanitation services, as shown in Figure 5.2, which collect excreta in containerized systems that are basically dry systems. Urine diversion mechanisms are often installed to reduce the volume and weight of the faecal container; once the containers reach capacity, they are replaced with empty clean containers. Containerized faecal matter is collected from generation sites such as residential homes and schools, and the faecal matter is then transported to treatment facilities where resource recovery can take various forms to derive products such as compost, biogas, animal feed and briquettes (Frost & Sullivan, 2021).

(III) On-site and/or non-sewered sanitation services, as shown in Figure 5.3, provide management in which generated faecal matter is collected and contained in tanks below ground. The level of onsite treatment depends on the technology, which could range from pit latrines to septic tanks. Once the tanks have reached capacity, they are emptied using pumps to remove the faecal matter and load it to a vehicle for transport to a treatment facility where it undergoes further treatment and resource recovery, which can take various forms (Frost & Sullivan, 2021; TNUSSP, 2018).

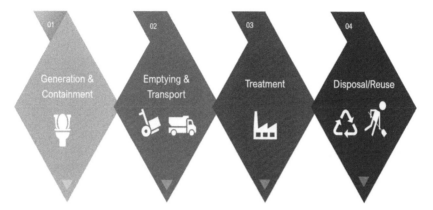

Figure 5.2 Containerized sanitation service chain. (Source: Frost & Sullivan (2021). Under CCA 4.0 license, © 2021 by authors.)

(IV) The service chain taps into the sanitation economy with smart sanitation incorporating technologies such as mobile applications to improve efficiencies in the sanitation service value chain (SSVC). These applications connect service providers and customers, facilitate remote payment, and track health data to prevent spread of diseases (Frost & Sullivan, 2021). SSVC also taps into the residential markets accounting for the majority of installed user interfaces, either connected to sewered sanitation systems or loosely connected to the non-sewered sanitation service chain. There is a vast commercial market for service provision, which includes the office and banking sectors, retail sectors and industrial establishments, warehousing spaces and public spaces such as schools, open spaces, markets, parks and gardens (Frost & Sullivan, 2021; Koottatep *et al.*, 2019c).

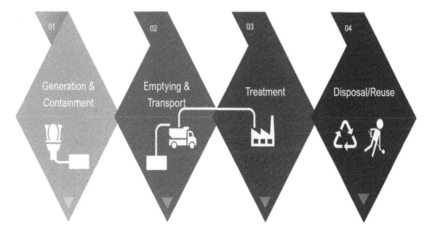

Figure 5.3 On-site and/or non-sewered sanitation service chain. (Source: Frost & Sullivan (2021). Under CCA 4.0 license, © 2021 by authors.)

5.2 SANITATION SERVICE ARRANGEMENT

Sanitation services entail the payment of money for the provision of technology or waste-related service. These services exist throughout the sanitation system, from households through the collection and transport of waste to treatment plants (O'Keefe *et al.*, 2015), see Figure 5.4. It aims to address both health and environmental issues. The stages of generation, capture and storage of excreta and or faecal sludge are primarily associated with improvement of household health levels, in order to collect and remove the excreta either through transportation or through networks of pipes.

Businesses/entrepreneurs need to provide toilets and/or latrines through improved designs that effectively capture and store the excreta and/or faecal sludge until it can be emptied or conveyed by networks of pipes. Faecal sludge and/or sewage can only be treated if it has been collected. The stages of transportation/conveyance, treatment and disposal or end-use have a wider environmental focus (Medland *et al.*, 2016). Sanitation services are delivered from sanitation infrastructure whether sewered (centralized or off-site) and/or non-sewered (decentralized or on-site) at public and private levels and must take into account all the technical aspects of the services, and all economic, social, organizational, institutional and environmental aspects (Koottatep *et al.*, 2019a,b,c; van Welie *et al.*, 2019).

Sanitation services cover series of businesses and entrepreneurs delivering value-added services to customers/users and relates to infrastructure operations and maintenance, collection, emptying, transport, treatment and disposal/reuse. It also includes value-added businesses that provide facilities and services for the safe management of human excreta from the toilet to containment and storage and treatment onsite or conveyance, treatment and eventual safe end use or disposal.

In addition, providers that offer sanitation services (e.g., building latrines, emptying pits) and those that sell sanitation products (e.g., manufacture of plastic toilets, making soap and other hygienic products) are part of this group. The off-site services are delivered with a sequence of several individual sanitation processes where each individual process represents a step of the sanitation service value chain (SSVC) that is mainly operated and/or delivered by public business/enterprise organizations. On the other hand, the

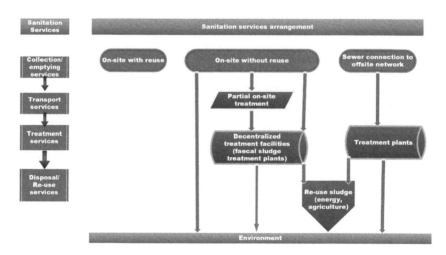

Figure 5.4 Sanitation service chain (Source: authors).

Box 5.1

Household expenditure on sanitation services in Nigeria (FMWR *et al.*, 2020). Access to sanitation indicators across Nigeria shows a slight improvement from 42% in 2018 to 44% in 2019, which is equivalent to a 6.6 million increase in the number of persons accessing basic sanitation services. Of the number of people using at least a basic sanitation facility, 21% use safely managed sanitation services, a two-percentage point increase from the 2018 findings (FMWR *et al.*, 2020). Households spent a total of ₦61 billion ($199 million) on sanitation services in 2018, whereby 38% was spent on bills and levies, 30% on toilet emptying and 21% on construction. In total, households spent the sum of ₦2 trillion ($6.6 bn) in 2018 on hygiene services. The bulk of the hygiene expenditure was on purchasing or replenishing bathing and laundry soaps (44%) and washing materials and equipment (23%) (FMWR *et al.*, 2020).

This is indicative that people are not just willing to, but actually do pay for sanitation services in Nigeria. The critical focus is on providing sufficient coverage and quality of on-site sanitation in order to improve the health and living conditions of the population and also reduce contamination of groundwater and water bodies. Continuous efforts are needed towards understanding and collapsing the barriers and drivers to adoption and usage of improved sanitation facilities and developing effective behaviour change interventions to curb open defecation.

Source: https://www.nigerianstat.gov.ng/pdfuploads/WASH%20NORM%20II%20Final%20 Report%202019.pdf

---

on-site sanitation services are made up of several private businesses/enterprises, and in some cases public utilities (treatment plant operations), which usually synergistically combine to provide a complete sanitation services system to end-users (Koottatep *et al.*, 2019a), see Box 5.1. Services can be divided into two types: operational services and support services. Operational services directly support or perform outsourced business operations of sanitation enterprises. They are made up of the business models of the sanitation service value chain (SSVC). Support services in turn provide services that benefit groups of service value chain operators (Springer-Heinze, 2018b).

If sanitation services are not well developed and are inadequate and unsustainable, service providers will find it difficult to deliver the right services in terms of quantity, quality and price to service end-users. Service market failure means that service costs will be too high and capable of limiting economic and technical efficiency (Springer-Heinze, 2018b) and negatively impact service arrangements. Sanitation services are conceived of as a system of at least three elements (Albert, 2000): service end-users demanding, paying and receiving service; service providers delivering the service products; and service arrangements defining the organization of service delivery often including third parties providing regulatory and funding functions (Springer-Heinze, 2018a, 2018b). It, therefore, becomes imperative that players on the SSVC should not downplay consumer demand and desire for functionality, practicality, aesthetics and affordability.

Figure 5.5 shows service conceptualization dominated by private enterprises providing services to end-users of non-sewered sanitation systems. These service arrangements in many locations, are handled by the informal and private sectors or a mix of public and private operators. In many settings, the service falls outside regulatory frameworks,

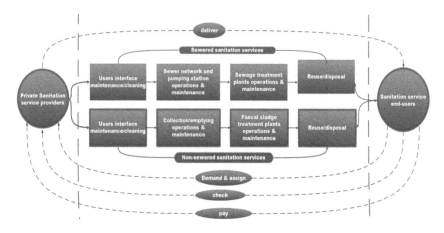

Figure 5.5 Concept of private enterprises' sanitation service arrangement (Source: authors).

policies or utility jurisdictions (Rao *et al.*, 2017). The elements in the service concept are closely interlinked. Essentially, the sanitation service value chain interactions has two parties : the service providers and the end-users (customers). The verbs placed into the arrows (in Figures 5.5 and 5.6) denote the main functions taken by either side. The relationship is a closed cycle in which service providers get clear incentives and clients have control over the service process. The service delivery system works only when both sides (private service providers and service end-users) take their role seriously. In reality, this is not always the case. Government and/or international donor agencies sometimes tend to subsidize particular operational services (Springer-Heinze, 2018b). For example, private emptying enterprises will not expand their offer until and unless potential service end-users express their demand effectively (Springer-Heinze, 2018b). In the same vein, if service end-users are not satisfied with the quality of the service provided or are not

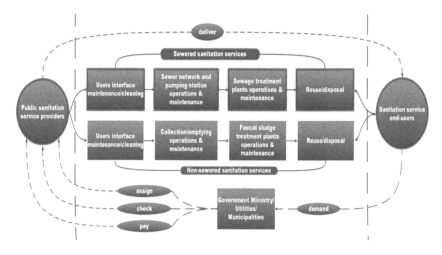

Figure 5.6 Concept of public services' sanitation arrangement (Source: authors).

convinced that they need such services, they will not demand and/or be willing to pay for such services. This service arrangement is common in Africa and South Asia where a larger proportion of faecal waste is handled by non-sewerage systems and highly problematic because businesses *lack the necessary information* to make faecal sludge management a functional component of the sanitation value chain (Rao *et al.*, 2017).

In Figure 5.6 the role is split between three parties which can either be the government, ministries/utilities and/or municipalities providing sanitation services to end-users (Huppert & Urban, 1998) and it is a much more complex form of organizing service delivery. The final recipients of the service have little influence on the service provided and their expectations might differ from those of the public service providers. Hence, there is a gap in the service system because the incentives are impaired (Huppert & Urban, 1998; Springer-Heinze, 2018b). In Figure 5.6, sewerage systems are mostly provided by government agencies that act as public operators (i.e. an organization mainly owned and controlled by the government) who also regulate or operate wastewater treatment plants and establish policies on environmental sanitation (BMGF, 2012; Rao *et al.*, 2017).

5.3 SANITATION SERVICES VALUE CHAIN (SSVC) MAPPING

From environmental and operational perspectives, it is important to assess sanitation services through the full value chain starting with access at household or property level and on to final disposal/reuse. From the generation of excreta to its final disposal or reuse, the journey flows through functional groups including conveyance, treatment, disposal and reuse. This approach helps to see the value chain of sanitation services from functional grouping perspectives of user-interface, collection and storage, conveyance, treatment and disposal/reuse for non-sewered sanitation (BMGF, 2011; Mehta & Mehta, 2013). On the other hand, sewered sanitation services examine the value chain of sanitation services from wastewater collection, treatment of wastewater, disposal and recovery of resources (Chofreh *et al.*, 2019) as shown in Figure 5.7.

The value addition of sanitation services includes sanitation logistics and transportation services, treatment plant maintenance, faecal and sewage sludge collection, emptying and treatment, cleaning and hygiene services, laboratory/analytical services, mobile toilet services, local artisans like plumbers, masons, and so on. Therefore, sanitation service providers range from the masons that build household

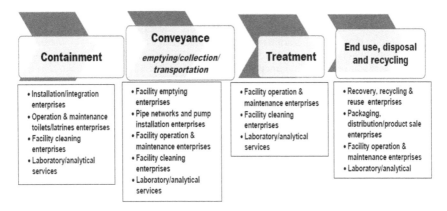

Figure 5.7 Sanitation services value chain. (Source: adapted by authors from Chofreh *et al.* 2019)

latrines to the entrepreneurs that build and run public/communal toilet blocks and from manual pit-emptiers to privately run emptying vacuum trucks to large public utilities providing networks of sewerage infrastructure along with sewage treatment and faecal sludge treatment services.

The key stakeholders that engage in the sanitation service value chain can be categorised across the stages of the chain as presented in Figure 5.7.

(I) *Containment*, including the user-interface, which is the starting point for the delivery of sanitation services. The key stakeholders involved in the provision of these services that come with access to toilet/latrines services are householders, businesses and institutions (public and private) (UNEP/IWMI, 2020). The operations and maintenance services for the containment systems are provided by local masons, plumbers, construction and installation enterprises. Cleaning services are carried out by different private enterprises, while in households these services are carried out by members of the household. Lavatory cleaning enterprises in some cases may be mechanized systems that automatically clean the user interfaces providing more hygienic conditions to users and cleaning service operators.

(II) *Conveyance services*, (emptying, collection and transportation) differ depending on the sanitation service arrangement. For sewered systems the networks of pipes and pumps convey sewage from the user interfaces directly to the treatment plants. In most regions, it is the responsibility of the municipality/ public utilities. The non-sewered systems, however, have municipalities and/or specialized agencies providing emptying and transportation services. In some cases, public institutions and private companies provide sanitation services including collection and disposal of faecal sludge. Also, private companies may sometimes be contracted by municipalities/public utilities to carry out desludging services, depending on the region. In addition, private companies operate independently in areas where public entities are unable to provide reliable timely services (Blackett & Hawkins, 2017; Parkinson *et al.*, 2014; Rao *et al.*, 2017). Also, households may be difficult to access by trucks, so the sludge is disposed of manually by operators. Community Based Organizations (CBOs) or Non-Governmental Organizations (NGOs) also play a key role in providing emptying services and transportation in underserved communities (Blackett & Hawkins, 2017; Opel & Bashar, 2013; Rao *et al.*, 2017; Strande *et al.*, 2014; UN-HABITAT, 1995; WASH PLUS, 2011).

(III) The *treatment and end-use, disposal and recycling services* are provided by either the municipality or public utility. In a very few cases, the municipality/public utility contracts operations to a private company (Rao *et al.*, 2017). Meanwhile, reuse involves stakeholders, depending on the type of resource to be recovered – energy or nutrient (Bolomey, 2003; Rao *et al.*, 2017; Takouleu, 2020).

5.4 ECONOMIC ANALYSIS OF SANITATION SERVICES VALUE CHAIN

Value chain is short for 'value added chain', which points to the fact that the value chain is a system that adds value at every stage (Springer-Heinze, 2018a). The subject of the economic analysis is the creation of value and its distribution along the SSVC and the assessment of parameters of chain competitiveness and efficiency. The economic analysis of a value chain includes: calculation of total value added; composition of value added along the SSVC; and the assessment of parameters of chain competitiveness and efficiency (Springer-Heinze, 2018a). Economic analysis means attaching numbers to the elements of the value chain map – end markets, service providers and business linkages (Springer-Heinze, 2018a). This analysis requires very important and sensitive financial data that is not always easy to collect from value chain operators.

The calculation *of total value added (total value generated)* is based on the sales price and volume sold. The formula to calculate total value of SSVC is set as the total volume of services sold to the end-users' market multiplied by the price paid by the end-user of sanitation services. The total value generated (or value added) is the single most important number in the economic analysis of the SSVCs:

$$(\text{Total value generated}) = (\text{end price of service per unit}) \times (\text{number of units sold}) \quad (5.1)$$

Another SSVC economic indicator is the estimation of the *composition of the value generated* in each stage of the SSVC, which is indicative of the actual value captured by the chain operators at each stage of the SSVC. For instance, a large share of the value generated from service providers is often transferred from retailers to internal and external suppliers. Therefore, the total value generated at each stage of the SSVC includes the value of intermediate products delivered by the service providers at preceding stages of the value chain or by external companies. The value actually captured by chain service providers at each stage of the SSVC can be obtained by deducting total value generated from the cost of bought-in materials used in the service delivery and components of equipment deployed for service delivery (Springer-Heinze, 2018a). The value added is then calculated by equation (5.2):

$$\begin{aligned}(\text{Value added}) = (\text{Value generated}) &- (\text{value of intermediate products}) \\ &- (\text{value of other inputs and services})\end{aligned} \quad (5.2)$$

The third indicator of economic analysis parameter is the *assessment of chain competitiveness and efficiency* to determine the specific strategies used by the SSVC service provider that enable them to gain more clients. Cost, quality, availability and innovation are very important variables used in assessing the competitiveness of the SSVC.

Table 5.1 shows the composition of the value generated in visual form. The value generated is composed of the value of the intermediate services that the service providers at one stage of the value chain obtain from their suppliers, the value of other inputs and

Table 5.1 Components of SSCV (adapted from Springer-Heinze, 2018a).

VALUE GENERATED by SSCV or by each stages	Value added
At each stage of the SSCV the value generated must take into account the value added by the operator of each of the stages, the intermediate products and other input products and services used for operation. Value generated will be the sum of the three parameters. For all SSCV the total value generated can be determined by multiplying the unit price by the quantity of the service sold.	• Wages • Interests and rents • Depreciation • Direct taxes • Profits and so on. **Intermediate products** Transferred to the operator by the previous SSCV stage operator: • Fresh or transported/semi-transformed faecal sludge **Other input products and services** Transferred to the SSCV operator by external suppliers: • Equipment • Transportation • Energy consumption • Water consumption and so on.

services they utilize, and the value added. The value added by service providers includes wages, interests and rents, depreciation, direct taxes, profits, and so on. As a result, high value added does not necessarily imply high profits for the service providers. The value added is used to pay for the production factors labour, land and capital as well as for the owners and management of the enterprise – in the form of profits (Springer-Heinze, 2018a).

5.5 ENVIRONMENTAL AND SOCIAL ANALYSIS

The economic performance of the value chain is the basis for its success, but that success may not last long if it is detrimental to the natural environment. Unsafe sanitation management harms overall human health and child health and also damages the quality of air, soil, surface water and groundwater (Hyun *et al.*, 2019; UN WATER, 2021). While the economic objective is the primary focus of the SSVC, it is not sufficient on its own to report on SSVC's environmental sustainability. In fact, all sanitation value chain activities must meet sustainability criteria. The environmental analysis places the SSVC into an ecosystem context to identify negative environmental impacts of the value chain on the environment and vice versa; and the impact of natural resource scarcity and climate change on business operations (Springer-Heinze, 2018a). The main elements that make up the environmental analyses are material consumption, energy consumption, water consumption, greenhouse gas emissions, land erosion/pollution, air pollution, water pollution, waste and biodiversity. The climatic parameters to be taken into account are in particular the change in temperature, the weather, changes in rainfall patterns and climate variability. Studies in Dakar have estimated the consumption of a vacuum truck at 15 litres of fuel for a distance of 20 km (Gning *et al.*, 2017). Tricycles consume an average of 4.5 litres over a distance of 100 km (Huanghe Motors, 2020). CO_2 emissions are estimated at 2 kg of CO_2/km for vacuum trucks and 0.1 kg of CO_2/km for tricycles (Ecoscore, 2020).

Environmental pollution caused by the disposal of faecal sludge into the natural environment affects the usability of ground and surface water leading to serious disruption of environmental processes and the destruction of ecosystems. In some areas, along rivers and streams, upstream residents normally enjoy better quality water, while downstream users are often obliged to take water with the properties of "diluted sewage" (UN WATER, 2021). In South-East Asia, 13 million tonnes of faeces are emitted into inland waters every year, as well as 122 million cubic metres of urine and 11 billion cubic metres of grey water. Water pollution costs Southeast Asia more than $2 bn per year. In Indonesia and Vietnam, it creates environmental costs of more than $200 m per year; mainly due to the loss of productive land (UN WATER, 2021).

The promotion of SSVC is only justified if it generates social benefits and contributes to reducing poverty. SSVC development should seek to support market-driven economic development that is inclusive of the poor and other vulnerable social groups as well as addressing gender gaps and providing better income opportunities (Springer-Heinze, 2018a). It is this social character that makes it easier for the public sector to promote a certain sector of private activity. The parameters used for social analysis are general working conditions, social security, training and education, workers' health and safety, human rights, living wages, consumer health and safety, product quality, and gender involvement in the SSC. For instance, a study was conducted in nine countries – Bangladesh, Bolivia, Burkina Faso, Haiti, India, Kenya, Senegal, South Africa and Uganda – on sanitation workers' conditions (health, safety and dignity) and it was found that:

(I)   sanitation workers are exposed to multiple occupational and environmental hazards;

(II) sanitation workers have weak legal protection resulting from working informally, a lack of occupational and health standards, and weak agency to demand their rights;

(III) financial insecurity is a great concern because typically, informal and temporary sanitation workers are poorly paid, and income can be unpredictable – some sanitation workers report being only paid in food;

(IV) social stigma and discrimination exist, and in some cases, are experienced as total and intergenerational exclusion (Ren, 2019).

These working conditions are not necessarily applied to all sanitation workers. A minority of these sanitation workers do, however, enjoy good social, economic or both conditions. These include the best organized entrepreneurs in the sector who often manage to hire workers for services.

5.6 SANITATION SERVICE VALUE CHAIN CASES

5.6.1 Non-sewered sanitation service value chain, Ouagadougou (Burkina Faso)

Ouagadougou is the capital city of Burkina Faso with approximately 2.5 million inhabitants, and it accounts for about 14 percent of the nation's population with a growth rate of 3 percent per annum (INSD, 2013; UNEP/IWMI, 2020; WSUP, 2014). Less than 2 percent of the population are connected to a sewer network. The remaining 73 percent use pit latrines and 15 percent use septic tanks for faecal sludge *containment* (ONEA, 2015). The case illustration in Ouagadougou (Burkina Faso) is designed to provide a better understanding of the sanitation service value chain (SSVC) activities and their enterprises (Figure 5.8). There are about 36 cleaning enterprises providing *containment*

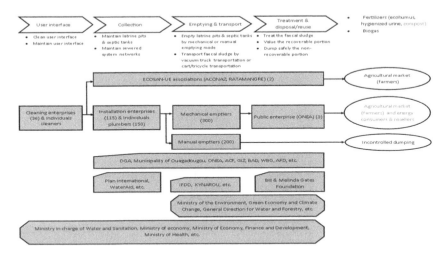

Figure 5.8 SSC mapping in Ouagadougou, adapted from structural analysis – value chain mapping. (Source: authors). *Note*: ONEA – Office National de l'Eau et de l'Assainissement, ACF - Action Contre la Faim, GIZ - Deutsche Gesellschaft für Internationale Zusammenarbeit, BAD – Banque Africaine pouur le Développement, WBG - Word Bank Group, DGA –' Direction Générale de l'Assainissement, ACONAZ - Association Communautaire Namanebg-Zanga, RATAMANGRE - association specialised in ECOSAN, IFDD - Institut de la Francophonie pour le Développement Durable.

services in Ouagadougou. Some of these services include: cleaning & maintenance services, collection (facility maintenance) services, emptying, treatment and disposal/ reuse services.

The major players in the sector are ECONAF enterprises, BZ and Service d'Hygiène du Faso (SHF). Cleaning and maintenance enterprises provide services to establishments, institutions, commercial enterprises and some households and are one of the main sanitation services at this stage of the value chain. In most cases, households' user-interface cleaning and maintenance services are carried out by members of the household. Other value stage players are construction and installation enterprises. Local masons and plumbers are also contracted for maintenance services when the need arises. For on-site sanitation systems, the *conveyance services* which include provisions of services such as emptying, collection, transportation and maintenance services, are carried out by construction enterprises and other private business concerns. The emptying of onsite sanitation systems is either manually or mechanically done. Users choose based on what they can pay for and/or their location or the availability of emptiers. A formal SSVC for Ouagadougou is presented in Figure 5.8. There are about 250 manual emptiers throughout the city of Ouagadougou, and more than 300 vacuum trucks involved in emptying and transportation of faecal sludge. For the sewered sanitation system, maintenance and upkeep (of the only existing sewerage system in Ouagadougou) is managed by the technicians in the Office National de l'Eau et de l'Assainissement (ONEA). To present a strong front for favourable bargain the emptiers (mechanical and manual) have joined forces to form two associations: Association des Vidangeurs du Faso (AVIF) and Association Burkinabé pour l'Assainissement et la Sauvegarde de l'Environnement (ABASE) for mechanical and manual emptiers respectively.

Faecal sludge emptied are transported to one of the three *treatment plants* for treatment/reuse/disposal at Zagtouli, Sourgoubila and Kossodo. The treatment plants are equipped with drying beds to separate liquids and solids and ponds to treat liquids. The three plants are managed by ONEA (see Figures 5.7 and 5.8). The Kossodo faecal sludge treatment plant has a biogas production unit of 400 m^3/d of sludge with a production of electricity expected to reach 2 160 MWh per year. Unfortunately, no valorization is yet applied to this sludge to allow its *reuse* (except during the pilot phase when it produced electricity).

The SSCV enablers in Ouagadougou are generally the ministries. The core ministries are the Ministry of Environment, Green Economy and Climate Change and they are responsible for developing policies and regulations for environmental protection. They ensure that ONEA and emptiers comply with the General Direction of Water and Forestry, which deploys its agents in the field for monitoring missions. The Ministry of Water, Sanitation and Hydro-agricultural Installations, the Ministry of Economy, Finance and Development and the Ministry of Health are also enablers who intervene at all stages of the SSC.

5.6.1.1 Economic analysis of SSCV in Ouagadougou

The economic analysis of SSCV in Ouagadougou was carried out to determine the value addition of each service provider and total value generation in the entire SSCV. This analysis was based on each service provider (enterprises and entrepreneurs) at each stage of the SSVC delivery services such as cleaning, maintenance, emptying, (manual and mechanical emptiers) and ONEA treatment facilities providing treatment and disposal/reuse services. The *cleaning services* are directly related to the user interfaces and so without intermediate services contributing any value addition, rather they create value themselves. Cleaning enterprises are able to deliver services to multiple clients at a time (about eight or more) because of the number of employees engaged which depends on the size of the firm as well as their market coverage. They earned an average of $163 per client per month.

On the other hand, the *maintenance service* providers such as plumbers and installation enterprises are able to deliver services to about 12 and 35 clients in a month at an average price of $6 and $10 per client respectively. *Emptying services* are provided mechanically and/or manually. In the case of mechanical emptiers, the most common vacuum trucks used in Ouagadougou have a capacity of 8 m³. The cost for mechanical emptying is fixed at $37 which is for filling of 4m³ of the truck and they empty an average of 3 pits per day. Hence, the cost can increase if the volume of emptied faecal sludge decreases, but not up to 4m³. In the case of manual emptiers, the estimated cost of emptying takes into account several parameters such as the volume of sludge emptied, the depth of the pits and also the customers' ability to negotiate. In any case, they charge an average of $47 per client and are available every day to provide services to their customers and undertake emptying of an average of 3 pits in a week. *Treatment services* are provided at ONEA's treatment plants and the volume of faecal sludge received at ONEA's treatment plants per day varies from 250 m³ (Sourgoubila's faecal sludge treatment plant) to 800 m³ (Kossodo's faecal sludge treatment plant), which averages about 500 m³ for each plant. For each cubic metre deposited, the emptiers pay an average of $0.55 which is $2.2 for 4 m³ (Table 5.2).

Also, households that benefit from the ECOSAN pit-emptying and urine canister collection services provided by the ECOSAN-UE associations are charged a monthly fee of $0.95. There is no rigorous follow-up and households unfortunately do not pay these charges. Biofertilizer is sold to the farmers in 50 kg bags at a price of $5 and 20 litres of treated urine at a price of $0.5. The total added value created by the whole SSCV in Ouagadougou is currently put at $ 12 97 811. The best value-addition is derived from the mechanical and manual emptying (Figure 5.9).

5.6.1.2 Environmental analyss

In the case of SSCV, direct utilization of natural resources is very low and can be minimised. Nevertheless, faecal sludge transportation to the treatment plant using vacuum trucks and tricycles consume fuel. Studies in Dakar have estimated that vacuum trucks use 15 litres of fuel for a distance of 20 km (Gning *et al.*, 2017). Tricycles consume an average of 4.5 litres over a distance of 100 km (Huanghe Motors, 2020). CO_2 emissions are estimated at 2 kg of CO_2/km for vacuum trucks and 0.1 kg of CO_2/km for tricycles (Ecoscore, 2020). Operationalization of the SSCV in Ouagadougou releases annual CO_2 emissions from vacuum estimated at 120 kg per day for an average distance of 60 km. These emissions are not to be ignored because of the negative impacts they can have on the climate. In addition, the faecal sludge treatment plants' open system makes them another major source of air pollution in the city because of the odour nuisance. The environment also influences the activities of the SSCV as the demand for emptying services by users becomes more frequent during the rainy season. Also, high rainfall leads to flooding which affects the operation of sanitation systems such as the treatment plants.

5.6.2 Sewered sanitation service value chain, Khuzestan (Iran)

To reduce wastewater management problems and to provide better value for end-users, the sewage industry requires the transformation of conventional system into sustainable sewage management systems (Wei *et al.*, 2017). In this process operators of sewage companies need to re-analyse their value chain to create sustainable value for customers (Chofreh *et al.*, 2019). Sanitation service value chain analysis enables sewage companies to evaluate business processes so that they can provide the greatest opportunities to

Table 5.2 SSCV total value generated and value-added determination in Ouagadougou.

Operators	Quantity/ Month/ Operator	Unit	Price Per Unit ($)	Operator's Number	Value Generated/ Month ($)	Intermediate Product ($)	Other input Products/ Services ($)	Added Value/ Month ($)
Cleaning enterprises	8	contracts (clients)	163	36	46 933	–	–	46 933
Individual plumbers	12	clients	6	150	10 000	–	–	10 000
Installation enterprises	35	clients	10	115	37 269	–	–	37 269
Manual emptiers	12	clients	47	250	1 94 444	–	215	1 94 230
Mechanical emptiers	89	clients	37	300	9 90 476	–	460	9 90 016
ONEA's treatment plant	12 276	m³ of sludge	0.55	3	20 461	–	1 086	19 375
ECOSAN-UE associations (RATAMANGRE)	2	50 kg of Birg-koenga	5	2	19	–	47	– 12
	17	20 litres of Birg-koom	0.5		16			
Total value generated/month					12 99 617	Total added value/month		12 97 811

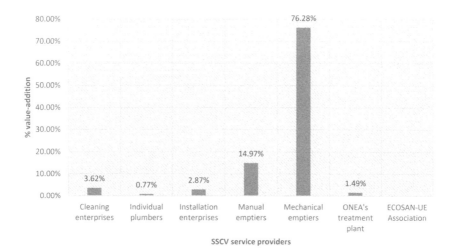

Figure 5.9 Contribution of the value-addition of each SSCV service provider to the total value generated in Ouagadougou. (Source: authors).

reduce operational costs, optimise efforts, eliminate waste, improve health and safety, and increase profitability (Reese *et al.*, 2016).

The Khuzestan Urban Water and Sewage utility companies is one of the largest water and wastewater companies in Iran and provides water and sewage services in the Khuzestan Province. The company was established in Ahvaz in 1992 and has 57 branches spread throughout the cities in Khuzestan (Khuzestan Water and Sewage, 2018). The company manages water and wastewater systems in Khuzestan's urban area with a population of approximately 2 180 301 people in 57 cities of the Khuzestan Province. The main activity of Khuzestan Urban Water and Sewage is to distribute clean water to households and industries, monitor and control water treatment and purification, and manage wastewater systems in households and industries. Operation and maintenance of water supply and distribution facilities include catchment basins, underground utilities, refineries, pumping stations, transmission lines, water supply networks, control systems, and distribution networks. All municipalities in Khuzestan except Ahvaz have similar water supply and wastewater systems (Chofreh *et al.*, 2019), see Figure 5.10.

Wastewater from households and industries are transferred using sewage pipelines and collected into the central wastewater tanks. The wastewater is then directed to the treatment plants using an underground drainage system for the screening stage. The process consists of several stages including a screening process to remove large objects that can damage equipment, primary, secondary treatment, and final treatment (Chofreh *et al.*, 2019). The main stakeholders of sanitation services value chain in Khuzestan Urban Water and Sewage include consultants and contractors who design wastewater treatment systems and provide equipment to the project operators and the end-users.

The water and wastewater agency are also involved in the enforcement of local government policies, rules and regulations on sewage and urban water management. The systems and equipment design process is conducted by developers and contractors who have agreements with the water and wastewater company (Figure 5.6). According to Chofreh *et al.* (2019), the value chain mapping results of Khuzestan Urban Water and Sewage utility company, indicate a need to effectively embed sustainability initiatives

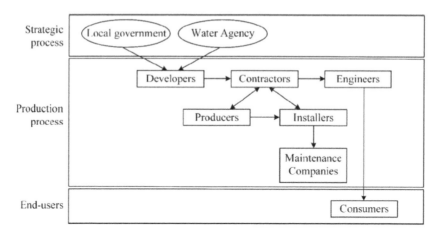

Figure 5.10 Main actors involved in the Khuzestan sewered sanitation service value chain. (Source: Chofreh, *et al.* (2019) with permission from Elsevier Ltd.)

into the business processes of the company, since water production and wastewater systems are not sustainably managed by the operators.

5.6.3 Container-based sanitation service value chain

Container-based sanitation (CBS) consists of an end-to-end service in which toilets collect excreta in sealable, removable containers (also called cartridges). The containers are regularly collected and transported to treatment facilities when full (Russel *et al.*, 2019), see Figure 5.11.

CBS services are typically provided by social enterprises or NGOs (e.g., Sanergy, Nairobi, Kenya; Clean Team Ghana Ltd, Kumasi, Ghana; Loowatt Ltd, London, United

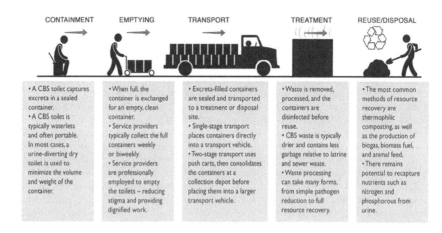

Figure 5.11 CBS sanitation value chain. (Source: Russel *et al.* (2019), under CCA 4.0 license, © 2019 by the authors.)

Box 5.2: Some CBS service providers

Sanergy, Nairobi, Kenya
Sanergy builds affordable CBS products designed specifically for urban slums, and franchises them to community members to serve all residents. The toilets are high-quality, low-cost sanitation units. With a urine-diverting squat plate, they source-separate solid and liquid waste, making collection and conversion safe and easy. For the comfort and convenience of owners and users, amenities include a handwashing station with soap, water and a bin for feminine hygiene products. They also professionally collect sanitation waste from the community by hand-carts and trucks. Handcarts ensure that toilets are installed deep in slums. The company also converts the waste at a centralized facility into valuable end-products such as organic fertilizer and insect-based animal feed. So far, the company serves 123 768 residents per day in Kenya (http://www.sanergy.com/)

Clean Team Ghana Ltd, Kumasi, Ghana
Clean Team Ghana Ltd provides CBS that are safe, affordable in-home toilets for low-income families. End-users pay a small weekly fee for the service and are provided with toilets and wastes are collected from end-user homes weekly in sealed containers and taken away for safe disposal (https://www.cleanteamtoilets.com/)

Loowatt Ltd, London, United Kingdom
Loowatt's patented toilet integrates revolutionary waterless flush technology with a 360-degree waste processing system to deliver hygienic and safe toilets for both domestic and commercial installations around the world. From the manual flush Home toilet, designed for domestic use, to the electric flush Pro toilet for commercial applications, there is a Loowatt solution designed to fit. The company has already served over a quarter of a million people with waterless flush toilets from the U.K to Madagascar, delighting customers and sustainably generating energy (https://www.loowatt.com/)

X-Runner, Lima, Peru
X-runner offers a sustainable sanitation solution that provides urban households with a portable dry toilet and a responsible service that removes and converts a human waste into compost, thus improving the daily lives, health and environment of thousands of individuals. The non-conventional sanitation solution consists of a Container-Based Sanitation (CBS) system which provides homes with a dry toilet of high technology and also includes the collection and responsible treatment of the generated waste and a continuous customer service support. Once a week we pick up the faeces from every household or respective collection point and bring them to the treatment plant, where they are transformed into compost, an ecological fertilizer. After successful implementation of a pilot project with 40 families in 2012, X-Runner extended their operations to serve currently over 900 families in three districts in Lima (https://www.x-runner.org/)

Kingdom; X-Runner, Lima, Peru), and several city utilities (such as Cape Town and Manila) are adopting CBS as part of their approach to citywide inclusive sanitation (CWIS), see Box 5.2. Revenue comes from customer service fees and the sale of waste treatment by-products, such as compost, protein (for animal feed), and energy (Foote *et al.*, 2017; Preneta *et al.*, 2013; Russel *et al.*, 2019).

The main target market for CBS services are the urban poor, who typically live in rented accommodation in densely packed settlements or with no formal land title. The portable nature of CBS as a sanitation approach makes it appealing in these contexts, as it requires little space and limited or no in-house construction (World Bank, 2019). CBS has typically been used where it is infeasible or inappropriate to install sewerage systems, such as in densely populated urban neighbourhoods, informal settlements, displaced person camps, or areas with high water tables or risk of frequent flooding (Russel *et al.*, 2019). The CBS service providers operate across the whole sanitation service chain, treating the faeces for reuse, whereas others focus on collection and emptying. Also, CBS approaches can be deployed with different types of toilets (seated or squat, shared or private) to respond to customer preferences (World Bank, 2019). Most CBS toilets are seated portable units that are placed inside the household. Sanergy's service involves shared squat toilets, and Loowatt has solutions for use in homes and in external superstructures. CBS service providers have to adapt their businesses to the limited and fragile cash flows of the urban poor, so they have developed a variety of payment plans to smooth out sanitation payments over time (World Bank, 2019).

5.7 WAY FORWARD FOR SANITATION SERVICES

Improving sanitation services requires huge capital investment, and government, private sector and development agencies finances are not sufficient. Therefore, promotion of the sanitation service value chain (SSVC) is the necessary step to take towards large-scale comprehensive private sector involvement in delivering quality sanitation services. This would require proper coordination of the actors in the SSVC as well as the introduction of innovative tools for effective and efficient service delivery such as the introduction of digital tools, including GPS tracking conveyance systems and technologies, mapping of operators and operation of call centre services (UNEP/IWMI, 2020). Special attention must be paid to performance improvement and sustainability of the treatment stage of the SSVC through the adoption of the circular economy approach and low-cost biological methods for faecal sludge management that will contribute to the end-value satisfaction. Treatment plants can generate revenue and value-addition/creation to sustain its operations by implementing appropriate business models in collaboration with the private sector, which can add value to the by-products generated (UNEP/IWMI, 2020).

From a citywide sanitation perspective, improving the SSVC will depend on extending the partnerships between the public and private sector as well as the penetration to more aspects of the SSVC. While the state could have a direct involvement in service delivery through public entities, it could also engage indirect provisions by delegating the service provision to non-state (or private) providers. Thus, a dimension involving the degree of 'state penetration' is deemed necessary for SSVC improvement and promotion (ADB Institute, 2020). Thus, significant improvement in the SSVC will require more investments in all stages of the value chain and mechanisms should be created to direct more investments to the poor and vulnerable segments of the population. In addition, the state will need to address the various bottlenecks and inefficiencies inherent in the SSVC.

5.8 Take Action

(I)   Visit your municipality office and find out registered sanitation services enterprises in your area and how they operate.
(II)   Assess the competitive advantage between public and private sanitation services delivery actors.

5.9 Journal Entry

(I) Identify and review existing sanitation services and delivery formal guidelines and standards as well as regulations at your local, regional and national levels.

(II) What are the economic indicators of the sanitation services value chain and how useful are such information in a value chain analysis?

5.10 Reflection

What is the level of quality and standards of sanitation service delivery in your area and how does it affect the sanitation landscape of your locality?

5.11 Guiding Questions

(I)   What are sanitation services and what are the difference between sanitation products and sanitation services?

(II) What are the three ways sanitation services value chain can be conceptualized?

(III) Define a sanitation service arrangement?

(IV) Differentiate between private enterprises and public service sanitation arrangements?

(V)   Describe the sanitation service value chain with a sample value chain overview map.

REFERENCES

ADB Institute (2020). Revisiting the Public–Private Partnership for Rapid Progress on the Sanitation-Related Sustainable Development Goals. Policy Brief No. 2020-2 (May). https://www.adb.org/sites/default/files/publication/603931/adbi-pb2020-2.pdf (accessed 07/05/2021)

Albert H. (2000). Agricultural Service Systems: A Framework for Orientation. GTZ, Eschborn.

Bhadwal R. (2015). Porter's Value Chain – Does it serve the service industry? Published 4 November 2015. https://www.linkedin.com/pulse/porters-value-chain-does-serve-service-industry-rajesh-bhadwal/ (accessed 20/04/2021)

Blackett I. and Hawkins P. (2017). FSM Innovation: Case Studies on the Business, Policy and Technology of Faecal Sludge Management, 2nd ed, Bill & Melinda Gates Foundation, Seattle, USA, p. 159. ISBN 978-1-5136-2513-3

BMGF (Bill & Melinda Gates Foundation) (2011). Water, Sanitation and Hygiene: Strategy Overview. Global Development Program, Seattle, WA.

BMGF (Bill & Melinda Gates Foundation) (2012). Business Analysis of Fecal Sludge Management: Emptying and Transportation Services in Africa and Asia. Bill & Melinda Gates Foundation, Seattle, USA.

Bolomey S. (2003). Amélioration de la gestion des boues de vidange par le renforcement du secteur privé local, 55. https://www.fsmtoolbox.com/assets/pdf/pdf-04April2019-original-learnpdfsource/Am%82lioration_de_la_gestion_des_boues_de_vidange_cas_de_Bamako,_Mali.pdf

Chofreh A. G., Goni F. A., Zeinalnezhad M., Navidar S., Shayestehzadeh H. and Klemeš J. J. (2019). Value chain mapping of the water and sewage treatment to contribute to sustainability. *Journal of Environmental Management*, **239**, 38–47, https://doi.org/10.1016/j.jenvman.2019.03.023

ECOSCORE (2020). Ecoscore. November 17, 2020. https://ecoscore.be/fr/info/ecoscore/co2?path=info%2Fecoscore%2Fco2

Federal Ministry of Water Resources (FMWR), Government of Nigeria, National Bureau of Statistics (NBS) and UNICEF (2020). Water, Sanitation and Hygiene: National Outcome Routine Mapping (WASH NORM) 2019: A Report of Findings. FCT Abuja, Nigeria.

Foote A. M., Woods E., Fredes F. and Leon J. S. (2017). Rendering fecal waste safe for reuse via a cost-effective solar concentrator. *Journal of Water, Sanitation and Hygiene for Development*, **7**, 252–259. https://doi.org/10.2166/washdev.2017.112

Frost and Sullivan (2021). Sanitation economy value chain opportunity assessment [version 1; not peer reviewed]. *Gates Open Research*, **5**, 68. https://doi.org/10.21955/gatesopenres.1116767.1 (accessed 16/05/2021)

Gning J. B., Diop C. and Dongo K. (2017). Facteurs déterminants le tarif de la vidange mécanique des matières de boues d'assainissement à Dakar," 20

Hamilton J. (2004). The Virtual Service Value-Chain: Disruptive Technology Delivering Competitive Advantage for the Services Industry. The Fourth International Conference on Electronic Business (ICEB2004)/Beijing. https://www.researchgate.net/publication/221365796_The_Virtual_Service_Value-Chain_Disruptive_Technology_Delivering_Competitive_Advantage_for_the_Services_Industry (accessed 20/04/2021)

Huanghe Motors (2020). Repassez La Consommation de Carburant Motorisée de L Du Trois-Roues 4,5 Du Tricycle 250cc de Cargaison/100 km. http://french.cargo-motorcycle.com/sale-11132774-iron-motorized-cargo-tricycle-250cc-three-wheeler-4-5-l-100km-fuel-consumption.html

Huppert W. and Urban K. (1998). Analysing Service Provision – Instruments for Development Cooperation by Examples From Irrigation. GTZ, Eschborn.

Hyun C., Burt Z., Crider Y., Nelson K. L., Prasad C. S. S., Rayasam S. D. G., Tarpeh W. and Ray I. (2019). Sanitation for low-income regions: a cross-disciplinary review. *Annual Review of Environment and Resources*, **44**(1), 287–318, https://doi.org/10.1146/annurev-environ-101718-033327

INSD (Institut National de la Statistique et de la Démographie) (2013). Recensement généraux de la population et l'habitation de 1985, 1996, 2006 [General Census of Housing and Population 1985, 1996 and 2006]. http://www.insd.bf/n/nada/index.php/catalog/RGPH

Khuzestan Water and Sewage (2018). Introduction of a Company. https://www.abfakhz.ir (accessed 29 November 2018)

Koottatep T., Cookey P. E. and Chongrak P. (2019a). Technological system. In: Regenerative Sanitation: A New Paradigm For Sanitation 4.0. IWA Publishing, London, UK, pp. 141–207. https://doi.org/10.2166/9781780409689_0141

Koottatep T., Cookey P. E. and Chongrak P. (2019b). Resource system. In: Regenerative Sanitation: A New Paradigm For Sanitation 4.0. IWA Publishing, London, UK, pp. 210–282, https://doi.org/10.2166/9781780409689_0209

Koottatep T., Cookey P. E. and Chongrak P. (2019c). Sanitation 4.0. In: Regenerative Sanitation: A New Paradigm for Sanitation 4.0. IWA Publishing, London, UK, pp. 283–322, https://doi.org/10.2166/9781780409689_0283

Medland L. S., Scott R. E. and Cotton A. P. (2016). Achieving sustainable sanitation chains through better informed and more systematic improvements: lessons from multi-city research in Sub-Saharan Africa. *Environmental Science: Water Research Technology*, **2**(3), 492–501, https://doi.org/10.1039/C5EW00255A

Mehta M. and Mehta D. (2013). City sanitation ladder: moving from household to citywide sanitation assessment. *Journal of Water, Sanitation and Hygiene for Development*, **3**(4), 481–488, https://doi.org/10.2166/washdev.2013.134

O'Keefe M., Lüthi C., Tumwebaze I. K. and Tobias R. (2015). Opportunities and limits to market-driven sanitation services: evidence from urban informal settlements in east Africa. *Environment & Urbanization*, **27**(2), 421–440, https://doi.org/10.1177/0956247815581758

ONEA (Office National de l'Eau et de l'Assainissement – National Water and Sanitation Office) (2015). Résultats d'analyses [Analysis Results]. Burkina Faso. ONEA unpublished monitoring sheets.

Opel A. and Bashar M. K. (2013). Inefficient technology or misperceived demand: the failure of vacutug-based pit-emptying services in Bangladesh. *Waterlines*, **32**(3), 213–220. https://doi. org/10.3362/1756-3488.2013.022

Parkinson J., Lüthi C. and Walther D. (2014). Sanitation21 – A Planning Framework for Improving City-Wide Sanitation Services. IWA, Eawag-Sandec, GIZ. https://iwa-network.org/ wp-content/uploads/2016/03/IWA-Sanitation-21_22_09_14-LR.pdf (accessed 23/04/2021)

Preneta N., Kramer S., Magloire B. and Noel J. M. (2013). Thermophilic co-composting of human wastes in Haiti. *Journal of Water, Sanitation and Hygiene for Development*, **3**, 649–654, https://doi.org/10.2166/washdev.2013.145

Rao K. C., Otoo M., Drechsel P. and Hanjra M. A. (2017). Resource recovery and reuse as an incentive for a more viable sanitation service chain. *Water Alternatives*, **10**(2), 493–512. https://www. water-alternatives.org/index.php/alldoc/articles/vol10/v10issue2/367-a10-2-17/file (accessed 24/04/2021)

Reese J., Gerwin K., Waage M. and Koch S. (2016). Value Chain Analysis: Conceptual Framework and Simulation Experiments. Nomos Verlagsgesellschaft, Baden-Baden, Germany.

Ren G. (2019). Toxic Conditions Expose Milions of Sanitation Workers to Infectious Disease & Death. *Health Policy Watch*. November 2019. https://healthpolicy-watch.news/ toxic-conditions-expose-millions-of-sanitation-workers-to-infectious-disease-death/

Russel K. C., Hughes K., Roach M., Auerbach D., Foote A., Kramer S. and Briceño R. (2019). Taking container-based sanitation to scale: opportunities and challenges. *Frontiers in Environmental Science*, **7**, 190, https://doi.org/10.3389/fenvs.2019.00190

Springer-Heinze A. (2018a). ValueLinks 2.0: Manual on Sustainable Value Chain Development. Vol. 1: Value Chain Analysis, Strategy and Implementation. Eschborn, Germany. https:// beamexchange.org/uploads/filer_public/f3/31/f331d6ec-74da-4857-bea1-ca1e4e5a43e5/ valuelinks-manual-20-vol-1-january-2018_compressed.pdf (accessed 26/04/2021)

Springer-Heinze A. (2018b). ValueLinks 2.0: ValueLinks 2.0: Manual on Sustainable Value Chain Development. Vol. 2: Value Chain Solutions. Eschborn, Germany. https://beamexchange.org/ uploads/filer_public/d3/a4/d3a4882e-eb14-4c30-8f7e-6ba4f51f6ec9/valuelinks-manual-20- vol-2-january-2018_compressed.pdf (accessed 26/04/2021)

Strande L., Ronteltap M. and Brdjanovic D. (2014). Faecal Sludge Management: Systems Approach for Implementation and Operation. IWA Publishing, London, UK. www.iwapublishing.com

Takouleu J. M. (2020). KENYA: In Naivasha, Stantec recycles sewage sludge into biomass briquettes. *Afrik21*. September 2020. https://www.afrik21.africa/en/ kenya-in-naivasha-stantec-recycles-sewage-sludge-into-biomass-briquettes/

TNUSSP (2018). Knowledge Management and Exchange Strategy for Urban Sanitation. Tamil Nadu Urban Sanitation Support Programme by Indian Institute for Human Settlements, Bengaluru, India, https://doi.org/10.24943/tnusspkme.20180901. https://www.susana.org/_resources/ documents/default/3-3766-226-1615554411.pdf (accessed 16/05/2021)

UNEP (United Nations Environment Programme) and IWMI (International Water Management Institute) (2020). Faecal Sludge Management in Africa: Socioeconomic Aspects and Human and Environmental Health Implications. UNEP-IWMI, Nairobi. ISBN: 978-92-807-3811-7. https://wedocs.unep.org/handle/20.500.11822/34350 (accessed 01/05/2021)

UN-HABITAT (1995). "The Vacutug." *Enginering for Change*. https://www.engineeringforchange. org/solutions/product/the-vacutug/

UN WATER (2021). L'assainissement favorise un environnement propre. UN WATER (Accessed 30 January 2021). https://www.un.org/fr/events/toiletday/pdf/factsheet-5.pdf

van Welie M. J., Truffer B. and Yap X.-S. (2019). Towards sustainable urban basic services in low-income countries: a technological innovation system analysis of sanitation value chains in Nairobi. *Environmental Innovation and Societal Transitions*, **33**, 196–214, https://doi. org/10.1016/j.eist.2019.06.002

WASH PLUS (2011). Systèmes à moindre coût pour la gestion des boues de blocs sanitaires, Cas d'Ambositra et de Mahanoro (Madagascar), Techniques courantes et Options améliorées. Rapport d'activité, Madagascar. https://www.pseau.org/outils/ouvrages/usaid_systemes_a_ moindre_cout_pour_la_gestion_des_boues_de_blocs_sanitaires_2011.pdf

Wei J., Wei Y. and Western A. (2017). Evolution of the societal value of water resources for economic development versus environmental sustainability in Australia from 1843 to 2011. *Global Environmental Change*, **42**, 82–92, https://doi.org/10.1016/j.gloenvcha.2016.12.005

Wikipedia contributors (2021). Service (economics). In Wikipedia, the Free Encyclopedia. Retrieved 07:09, 21 April 2021, from https://en.wikipedia.org/w/index.php?title=Service_(economics)&oldid=1016334998

World Bank (2019). Evaluating the Potential of Container-Based Sanitation. World Bank, Washington, DC.

WSUP (Water & Sanitation for the Urban Poor) (2014). World Urbanization Prospects – the 2014 Revision Highlights. United Nations, New York, Department of Economic and Social Affairs.

doi: 10.2166/9781789061840 _0125

Chapter 6

Sanitation biomass recovery and conversion

Peter Emmanuel Cookey, Olufunke Cofie, Thammarat Koottatep and Chongrak Polprasert

---

Chapter objectives

The aim of this chapter is to present the sanitation biomass recovery and conversion value chain (SBRCVC) activities that show how enterprises offer resources that enhance the emerging low-carbon circular bioeconomy and in turn reduce reliance on virgin raw materials. Furthermore, it intends to explore better understanding of enterprises and businesses that valorise secondary organic resource-materials from excreta, wastewater, sewage and faecal sludge with other blended organic-waste-derived biomass and then those ventures that could convert these bioresources into different valuable products, ranging from high-value amino acids and proteins, short-chain fatty acids, enzymes, biopesticides, bioplastics, bioflocculants, biofertilizer and biosurfactants as well as those that use them to produce other kinds of commodities.

6.1 INTRODUCTION

A value chain that addresses sanitation biomass recovery and conversion (SBRC) could offer resources that enhance the emerging low-carbon circular bioeconomy in developing and developed countries and in turn reduce the reliance on virgin raw materials as a result of being biomass drawn from secondary materials (Panoutsou *et al.*, 2020). This could also mitigate climate change and contribute to local economic growth such as creating skilled employment opportunities (BFG, 2012; Panoutsou *et al.*, 2020). It could focus on recovery and reuse of resources from excreta and wastewater fractions that do not interfere with natural ecosystems or human food chains, but rather on recovered resources such as soil conditioners, compost and effluent for irrigations which are well established end-products. Wastewater treatment for resource recovery is a rational solution to avoid problems derived from droughts and water shortages, especially for countries with water restrictions (Jodar-Abellan *et al.*, 2019; Zarei, 2020a), while wastewater management including safe reuse of water and the recovery of vital resources, could open remarkable opportunities for commercial markets. Recently, nanomaterials gained significant attention for widespread applications in biosensing,

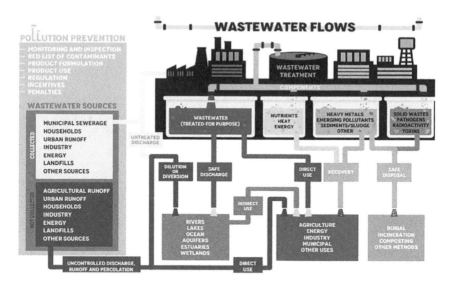

Figure 6.1 Wastewater flows and sources. (Source: WWAP (UN-Water), (2017). CC-BY-SA 3.0 IGO, license © 2017 by the authors)

water splitting, energy recovery, environmental remediation, and wastewater treatment (Kadam *et al.*, 2020; Wang *et al.*, 2020; Zarei, 2020a, 2020b, 2020c; Zarei & Aalaie, 2019). In particular, treated wastewater can be reused for multiple purposes in the industrial sector and for agricultural purposes, irrigation, groundwater recharge for effluent quality improvement; it can also be used for domestic purposes – fire protection, car wash, and toilet flushing (Zarei, 2020a; see Figure 6.1). Other possibilities that are starting to be implemented include sewage sludge and faecal sludge biomass composting used to produce animal feed from black soldier fly larvae or fodder crops, incorporating building materials such as bricks, tiles, cements, concretes, mortar, and so on, and also energy in the form of fuel, electricity and heat (Andriessen *et al.*, 2019; Zarei, 2020a). In addition to these recovered end-products, all of these could also support solutions for coverage and access problems of safely managed sanitation while providing appropriate incentives for faecal sludge management (Diener *et al.*, 2014; Wielemaker *et al.*, 2018). Revenues from resource recovery and reuse after conversion could partially offset operation costs, incentivize proper operation and maintenance, and stimulate regular emptying and delivery of faecal sludge, particularly in developing countries (Andriessen *et al.*, 2019). Studies have confirmed the emergence of viable business models for value chains around sanitation biomass resource recovery and reuse that in turn help ensure sustainable provision of safely managed sanitation (Diener *et al.*, 2014; Murray & Ray, 2010).

This could be viewed as the sanitation biomass recovery and conversion value chain (SBRCVC) and it would deal with enterprises and businesses that valorise secondary organic resource-materials from excreta, wastewater, sewage sludge, and faecal sludge with other blended organic-waste-derived biomass as well as those that could convert these bioresources into different valuable products, ranging from high-value amino acids and proteins, short-chain fatty acids, enzymes, biopesticides, bioplastics, bioflocculants, biofertilizer and biosurfactants (Zhang *et al.*, 2018). Value chains encompass the full range of activities required to bring a product or service from conception, through different phases of production that involve a combination of physical transformation and input from various producers and services to delivery, to the final consumer and final disposal after use (Maaß & Grundmann, 2016). Unlike conventional value chains the SBRCVC is not necessarily made up of sequential and linear activities; rather it is viewed as manifold connections in which value is co-created by a combination of players and enterprises (Maaß & Grundmann, 2016; Peppard & Rylander, 2006) comprising environment, social, economic and governance actors interacting through institutions, technology and other relevant stakeholders to:

- Co-produce product and service offering;
- Exchange product and service offering, and
- Co-create value along the biomass recovery transformation chain (Lusch *et al.*, 2010; Maaß & Grundmann, 2016).

The economic value created from value chains is commonly measured by the added value, that is a success indicator that describes the performance of a firm, business or the increase in value resulting from production, processing, marketing and other economic activities (Haller, 1997; Maaß & Grundmann, 2016). It can also be understood as the difference between the value of goods and/or services delivered from one business to another, and the value of all inputs received by this business from other businesses and/ or enterprises for producing that particular good or service. It is a set of interlinked activities that deliver products/services by adding value to bulk materials (feedstock through the process of conversion to high-value products). In such a bio-based sanitation-waste value chain, the feedstocks tend to be biomass drawn from by-products of existing primary production or secondary origins like sanitation-derived biomass (Lokesh *et al.*, 2018; see Figure 6.2). Bioresources value chains that valorise secondary resources are designed to turn available organic materials into different valuable products that range from high-value chemicals, fertilizers, and biochar to secondary-used by-products and renewable energy; and are capable of transforming waste/secondary feedstock

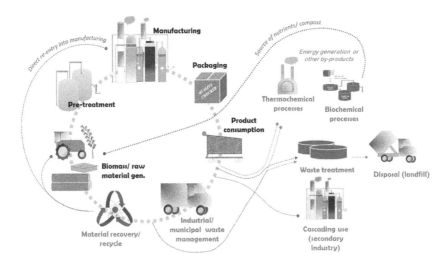

Figure 6.2 A generalised map of a bio-based value chain. (Source: Lokesh *et al.* (2018). Under CCA 4.0 license, © 2018 the authors)

into arrays of high-value products called integrated biorefineries (Lokesh *et al.*, 2018; Pan *et al.*, 2015). Integrated biorefineries contain a pre-treatment plant that prepares the feedstock for biomass conversion/transformation within the value chain before packaging and distribution (Greene, 2014; Lokesh *et al.*, 2018) to the final consumer or end-user.

In addition, biomass recovery value chains contribute to materials recycling, climate mitigation, and greenhouse-gas (GHG) emission reduction as well as the development and implementation of several outstanding technologies like combustion, gasification and anaerobic digestion hinged upon sustainable strategies that could overcome some obvious challenges. Pan *et al.* (2015), for instance, proposed such related strategies with options that cover technology, finance, institutions, public concerns and regulations (Zarei, 2020a). However, in Malaysia Lam *et al.* (2013) developed a two-stage biomass model waste-to-energy (WTE) process with the first stage being a micro-stage waste-biomass optimization and allocation integrated waste-biomass processing hub; and the second stage being a macro-stage designed to handle the synthesis and optimization for the WTE (Zarei, 2020a). These strategies provided both the analysis of economic value and sustainable solutions for the utilization of waste-biomass for resource recovery.

Normally, the value chain theory recognizes the stages and activities of the value chain's competitive advantages or disadvantages, and where cost advantage strategy optimization should focus mostly on activities that contribute the most to cost reduction (Darmawan *et al.*, 2014). Furthermore, types and forms of sanitation biomass resource recovery should always meet local conditions and user acceptance, and, whenever possible, should be decided early in the planning process, so that appropriate treatment objectives can be set to ensure public health protection of the end-users (Reymond, 2014). Also, a market-driven assessment can help to inform which end-product is most marketable in a specific location (Andriessen *et al.*, 2017, 2019). For instance, research indicates that there is a high demand for solid fuels in urban areas of Sub-Saharan Africa, particularly from manufacturing industries like the brick and cement industries (Diener *et al.*, 2014). Also, wastewater sludge is used as fuel in co-combustion with coal or other solid fuels in industrial setups, both in carbonized and dried forms (Fytili & Zabaniotou, 2008; Werther & Ogada, 1999).

In fact, there are nine ways for the recovery of energy from sewage/faecal sludge (Rulkens, 2008):

- anaerobic digestion of sewage/faecal sludge;
- production of biofuels from sewage/faecal sludge;
- direct production of electricity from sewage/faecal sludge in microbial fuel cells;
- incineration of sewage/faecal sludge with energy recovery;
- co-incineration of sewage/faecal sludge in coal-fired power plants;
- gasification and pyrolysis of sewage/faecal sludge;
- use of sludge as energy and raw material source in the production of portland cement and building materials;
- supercritical wet oxidation of sewage/faecal sludge; and
- hydrothermal treatment of sewage/faecal sludge.

The SBRCVC value chain comprises all stages and activities that input a resource flow from sanitation systems (sewered and non-sewered sanitation) such as urine, faeces, excreta, anal cleansing materials (dry and water), flushwater, brown water, black water, greywater, wastewater, sewage, faecal sludge, and so on, (McConville *et al.*, 2020) to recover and reuse products. It can be analysed in such a way that all important connections are balanced in a circular manner to achieve resource efficiency and sustainability from the very beginning (Koottatep *et al.*, 2019; Panoutsou *et al.*, 2020). The SBRCVC is the sum of the remuneration received from all value-added activities of all stakeholders participating in the primary treatment of excreta and wastewater, pre-treatment of recovered biomass from sewered and non-sewered sanitation systems, biomass conversion and transportation, and biomass products packaging, as well as the biomass end-use market (Haller, 1997; Maaß and Grundmann, 2016). In other words, SBRCVC takes a look at the remunerations of participating businesses and/or enterprises (public and/or private) involved in the treatment, recovery and reuse of excreta, wastewater, sewage sludge and faecal sludge, crop production, bioenergy generation, and so on. The added value also reveals some social distributional implications of the value chain. The parameters differ from the conventional profit calculation because the remunerations paid to employees, creditors, and the State are considered as part of the added-value and not as value-reducing components (Möller, 2006).

6.2 THE SANITATION BIOMASS RECOVERY AND CONVERSION VALUE CHAIN AND CIRCULAR BIOECONOMY

Escalating environmental and economic pressure to use resources responsibly and add value to the used material/products in the commercial sphere has helped the development of technology routes and material circularity in the sanitation and waste biomass sector (Lokesh *et al.*, 2018). The aim of such systems thinking is to 'close the loop' by becoming resource efficient through developing and establishing a sanitation symbiosis to reduce the pressure on virgin biomass (Lokesh *et al.*, 2018). The SBRCVC aligns with the implementation of a circular bioeconomy and water–food–energy nexus approaches, that is, a coordinated integration approach that cuts across natural-resources-related sectors and sanitation, which is expedient for solving water, energy and food supply security. Conventional sanitation systems often dispose large loads of nutrients into water bodies, and this causes eutrophication (Mallory *et al.*, 2020; Wang *et al.*, 2017); global wastewater has enough nutrients to replace 50 million tonnes of fertilizer (CGIAR, 2013; Mallory *et al.*, 2020), which represents a significant proportion of the estimated 262 million tonnes supplied per year (FAO, 2019; Mallory *et al.*, 2020). The core argument of the nexus approach and circular bioeconomy for sanitation is that the multiplicity of feedbacks and interdependencies resulting from linkages among subsystems, such as sanitation, water, food and energy, jointly affect the sustainability of the broader social-ecological systems (Ganter, 2011; Hellegers *et al.*, 2008; Hussey & Pittock, 2012; Villamayor-Tomas *et al.*, 2015; Waughray, 2011). The integration of the circular bioeconomy, the nexus and sanitation value chain expands the base of sanitation natural resources which is capable of enhancing water, food and energy security on a local and global scale (Maaß & Grundmann, 2016). This extends the water–food–energy nexus approach to take into account not only the linkages between single resources, but also the connections between whole biomass recovery value chains that use these resources. The benefit of the economic impacts of reducing virgin natural-resource utilization and turning sanitation input–inflow materials to generate desirable out-products complies with the core principles of a circular bioeconomy. The added advantage is that these complex linkages and integration resulting from the adoption of the circular economy for sanitation can further enhance the recovery of resources like faecal sludge, wastewater and sewage sludge through products like animal-feed, energy, biogas, compost and recycled water (Ddiba *et al.*, 2020; Diener *et al.*, 2014; Mallory *et al.*, 2020).

The combination of sanitation biomass recovery and a circular bioeconomy has the potential to directly contribute to 12 out of the 17 UN Sustainable Development

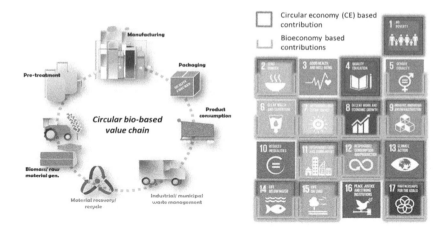

Figure 6.3 Potential for circular biobased value chain to contribute to achieving UN's SDGs and the potential of value chain mapping and analysis in quantifying these goals. (Source: Lokesh *et al.* (2018). Under CCA 4.0 license, © 2018 the authors)

Goals (SDGs) (Figure 6.3). These make a direct contribution to access to water and sanitation for all (SDG 6), sustainable consumption and production (SDG 12), and reducing pressure on the environment, air, water, and land (SDGs 13, 14 and 15) (Blair *et al.*, 2021; Lokesh *et al.*, 2018). There is also a contribution to SDGs related to food security and sustainable agriculture (SDG 2), decent work and economic growth (SDG 8), resilient infrastructure and sustainable industry (SDG 9), climate action (SDG 13), terrestrial ecosystems (SDG 15) and SDG 7 (affordable and clean energy) (Blair *et al.*, 2021). In addition, further contributions can be seen in utilizing the rural knowledge pool and alleviating poverty (SDG 1), good health and well-being (SDG 3), reducing inequalities (SDG 10), guarding the local ecosystem services that encourage sustainable cities and communities (SDG 11), creating jobs and socio-economic opportunities (SDG 8), forging skills among communities through quality education (SDG 4), and working in partnership with rural communities and local biobased biomass recovery infrastructure (SDG 17) (Blair *et al.*, 2021; Lokesh *et al.*, 2018). The use of biomass recovery requires devising smart strategies and value-chain pathways to lock the chains of GHG emissions, either via carbon capture or soil incorporation of high-quality biochar (Blair *et al.*, 2021; Lokesh *et al.*, 2018).

The circular bioeconomy is, therefore, the intersection of the bioeconomy and the circular eco-economy which is the regenerative system for resource input, waste emission and energy leakage formed by closing material and energy loops (Geissdoerfer *et al.*, 2017; Koottatep *et al.*, 2019; Morone & Imbert, 2020). Thus, the sustainable bioeconomy represents the renewable segment of the circular economy (European Commission, 2018) while the circular bioeconomy focuses on the sustainable, resource-efficient valorization of biomass in integrated multi-output production chains while also making use of residues and wastes and optimizing the value of biomass over time via cascading (Feleke *et al.*, 2021). The key elements of the circular bioeconomy (Feleke *et al.*, 2021; Stegmann *et al.*, 2020) include:

- sustainable biomass sourcing;
- circular and durable product design;

- use of residues and waste;
- integrated, multi-output production chains;
- bioenergy and biofuels;
- biobased products, food, and feed;
- prolong and shared use;
- energy recovery and composting; and
- recycling and cascading

The circular economy (CE) is an economic system that is based on business models that replace the 'end-of-life' concept with reducing, reusing, recycling and recovering materials in production/distribution and consumption processes to accomplish sustainable development (Ddiba *et al.*, 2020). The circular economy aims to promote the maximum use of resources and reduce waste by closing economic and ecological loops of resource flows (Haas *et al.*, 2015) and eliminates waste by design, keeping the added value of a product for as long as possible (Sariatli, 2017). Waste is viewed as a resource in a production process, which suggest less extraction of fresh materials and energy consumption (Feleke *et al.*, 2021). On the other hand, the bioeconomy involves production of renewable biological resources and converting these resources and waste streams into value-added products, such as food, feed, biobased products, and bioenergy (European Commission, 2012a). An important feature of the bioeconomy is extending biomass production and processing beyond food, feed, and fibre to include a range of value-added products with potential applications in many sectors, for example, the food, health and energy sectors (East African Science & Technology Commission, 2019). Therefore, implementing circularity within the sanitation system (sewered and non-sewered) forms a biological materials cycle involving recovering water, nutrients, energy and other materials which are typically managed within different resource management sectors (Ellen MacArthur Foundation, 2017).

An analysis of 56 of the world's largest cities found that closing the nutrient loops in large urban cities is most feasible in Africa, Asia and Europe due to cropland density local to these cities (Moya *et al.*, 2019; Trimmer & Guest, 2018). And so a circular bioeconomy within the context of a sanitation biomass recovery value chain could: create an opportunity for incentivizing and stimulating sustainable sanitation by providing additional income streams and reducing the sanitation service cost to the user (Moya *et al.*, 2019); contribute to keeping the added value in products for as long as possible (Maaß & Grundmann, 2016; Smol *et al.*, 2015); and to ensuring higher regional and domestic competitiveness by increasing the effectiveness of resource allocation, resource utilization and productivity (Maaß & Grundmann, 2016; Su *et al.*, 2013). Other potential benefits of circular approaches to a sanitation biomass recovery marketplace include mitigating greenhouse gas emission, securing water, food and energy resources, and providing employment opportunities in growing cities (Andersson *et al.*, 2016).

However, the main determinants of sanitation biomass recovery and conversion products and services in a circular bioeconomy (CBE) are volume of waste collected, integration of faecal sludge (FS) and sewage sludge (SS) with other waste streams, enabling policies and subsidies, and marketing. Also, a number of technical, social and political transformations would need to take place to make a CBE conducive for sanitation businesses that could drive the sanitation biomass recovery value chain (Mallory *et al.*, 2020). Some studies have revealed that, technically, businesses often struggle to collect sufficient waste to make their model of reuse viable, and large increases in financial viability can be achieved by increased collection (Ddiba, 2016). Furthermore, literature looking at the circular bioeconomy for sanitation biomass recovery mostly focuses solely on sewage or faecal sludge, but business models are often driven by the integration of organic solid waste and other biomass (Moya *et al.*, 2019; Otoo & Drechsel, 2018; Remington *et al.*, 2018; World Bank, 2019). On this basis the Toilet Board Coalition (TBC) argues that FS/SS should be seen as part of a biological waste stream encompassing all biodegradable or organic waste to really enable a CBE for sanitation biomass recovery (TBC, 2017). Thus, when considering the circular bioeconomy for a sanitation biomass recovery value chain, it is essential to assess the contribution of other sources of biomass to the development of intended products and the market for the potential products or they will not provide the additional income stream that is desired (Dumontet *et al.*, 1999). This is because in terms of social transformation, marketing and awareness of products also have a large influence in the ability of enterprises and businesses in the value chain to recover economic benefits from the CBE (Mallory *et al.*, 2020).

6.3 SANITATION BIOMASS RECOVERY AND CONVERSION VALUE CHAIN MAPPING

Value chain mapping describes stages of value creation by enterprises and other organizations as part of the process of designing and delivering goods and services for their end-users (European Commission, 2012b; Lokesh *et al.*, 2018). Value chain maps are a valuable, flexible and convenient tool to develop and analyse the scope and performance potential of a biobased business model by breaking down the various process dynamics into logistics, sectors of application and embedded stakeholders. The strengths, weakness, costs and competition from other value chains in the production of specific commodities can be visualized via a value chain map (Lokesh *et al.*, 2018). In essence, value chain mapping provides a generalized yet visual schematic of the dynamics including the resource flow and actors integrated within the SBRVC that are actively playing crucial roles in the delivery of relevant sanitation-derived products (SDPs) to the end-user markets. They involve a network of technologies and infrastructures to convert low-value biomass raw materials to high-value products; activities that safely recycle excreta and organic waste while minimizing the use of non-renewable resources such as water and chemicals. Safely recycling means that waste flows are managed to ensure that physical, microbial and chemical risks are minimized so that recycled products do not pose any significant health threat or environmental impact when correctly used (McConville *et al.*, 2020; Tapia *et al.*, 2019).

The SBRVC main activities and enterprises are broken down into biomass feedstock, biomass pre-treatment, biomass conversion, ancillary services (transportation, storage, product packaging services) and end-user market; together they make up the entire value chain. As such, the value chain involves physical attributes and needs to be designed with a focus on minimizing physical challenges throughout raw material production and conversion (Panoutsou *et al.*, 2020), which is described as a physically efficient value chain. The market assets refer to the delivery of biobased products to the end-users and this adds an innovative nature to the value chain (Panoutsou *et al.*, 2020). The system design of the SBRCVC integrates other value chain activities and enterprises within the IFSVC such as sanitation service (chapter 5), product design & development (chapter 2), and product and equipment manufacturing (chapter 3) as major contributors to the operationalization of the SBRCVC. There are five competitive priorities that have to be considered to ensure that the value chain delivers the required/expected value-added specific targets. These competitive priorities are: (i) flexibility, (ii) quality, (iii) cost, (iv) innovation, and (v) transparency (Panoutsou *et al.*, 2020) (see Figure 6.4). Meanwhile, the activities associated with these technologies and infrastructure include sourcing raw

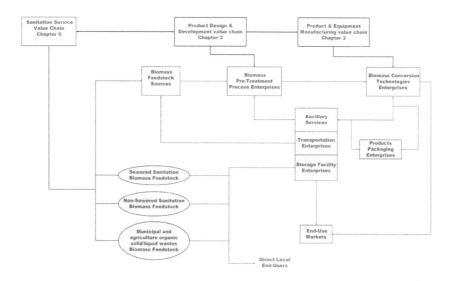

Figure 6.4 Sanitation biomass recovery and conversion value chain enterprises' activities (Source: Authors).

materials, processing, logistics, inventory management and waste management (Jarvis & Samsatli, 2018).

6.3.1 Biomass feedstocks

Waste biomass forms the feedstocks for the sanitation-derived resource recovery value chain. They are heterogeneous and chemically combined renewable-source waste products and/or by-products of either plants and/or animal origin (Siwal *et al.*, 2021). In other words they are any organic materials derived from plants and animals that are classified as biomass feedstock. Biomass can also broadly be classified according to origin and source: biomass generated in rural (agriculture, forestry, and livestock), urban (sewage and municipal solid wastes), and industrial (cellulose and agri-food industries) areas (Ahmed *et al.*, 2019; IREA, 2014; Saxena *et al.*, 2009). Due to the usual abundance, sustainability and low price of biomass, these forms have proved to be possible options for the replacement of non-renewable energy and other useful products (Anukam & Berghel, 2021; Anukam *et al.*, 2016). Sanitation biomass recovery (SBR) feedstocks belong to the non-lignocellulosic biomass (NLB) class of biomass – waste derived from sewage sludge, faecal sludge and organic solid wastes (McConville *et al.*, 2020; Rulkens, 2008; Siwal *et al.*, 2021), see Figure 6.5. The blending of other classes of NLB is to enhance the quality of the raw materials for the production of high-value bioproducts such as biomass derived from municipal organic solid wastes, animal and human wastes, and agricultural waste (Begum *et al.*, 2013).

However, biomass varies owing to a number of factors such as the heterogeneity of biomass, its application and origin (Ahmed *et al.*, 2019). Any organic materials directly or indirectly derived from the process of photosynthesis is considered biomass (Anukam & Berghel, 2021). The chemical composition of biomass depends strongly on it sources (Ahmed *et al.*, 2019; Bajpai, 2016; Popa, 2018).

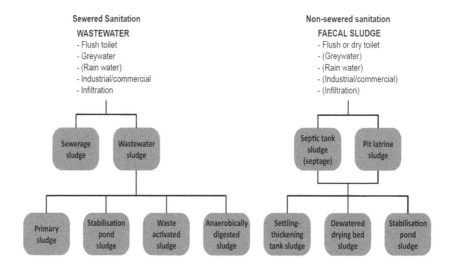

Figure 6.5 Examples of terminology used for different types of sludge relating from sanitation systems. (Adapted from Englund and Strande (2019) by McConville *et al.* (2020). ©Swedish University of Agricultural Sciences (SLU), Department of Energy and Technology. Uppsala, Sweden.)

6.3.1.1 Plant biomass feedstocks

In plants, biomass is formed through conversion of carbon dioxide in the atmosphere into carbohydrates in the presence of the sun's energy.

$$6CO_{2(g)} + 6H_2O_g \rightarrow C_6H_{12}O_{6(s)} + 6O_{2(g)} \tag{6.1}$$

Biological species will then grow by consuming these botanical or other biological species, adding to the biomass chain (Basu, 2018; Mamvura & Danha, 2020). The plant origin biomasses are commonly referred to as lignocellulosic biomass (LB) which is composed of an aromatic polymer (lignin) and carbohydrate polymers (cellulose, hemicellulose) (Anukam & Berghel, 2021; Li & Jiang, 2017).

The internal structure of LB reveals a crystalline fibrous structure of cellulose, which forms the core of the complex structure of biomass. The position between the micro- and microfibrils of the cellulose matrix is occupied by hemicellulose, while lignin plays a structural role that encapsulates both cellulose and hemicellulose. However, their complex structure greatly hinders their utilization due to the high level of crystallinity of cellulose as well as the cross-linking of carbohydrates and lignin (Ahmed *et al.*, 2019; Chang & Holtzapple, 2000) which results in their stable and recalcitrant structure that make them resistant to enzymatic attack (Ahmed *et al.*, 2019; Taherzadeh & Karimi, 2008; Tursi, 2019), see Figure 6.6. To overcome this challenge, pre-treatments of the feedstock become crucial.

(I) ***Cellulose (40–50%)*** is a linear polymer and a complex carbohydrate (or polysaccharide) with a high molecular weight and a maximum of 10 000 monomeric units of D-glucose, linked by β-1,4-glycosidic bonds. The molecular formula of cellulose is $(C_6H_{12}O_6)_n$ (*n* indicates the degree of polymerization) and its structural base is cellobiose (i.e. 4-o-β-D-glucopyranosyl-D-glucopyranose, see Figure 6.7). Cellulose is the most abundant organic compound found in nature and plays a structural function in plant cell walls. The reactivity and

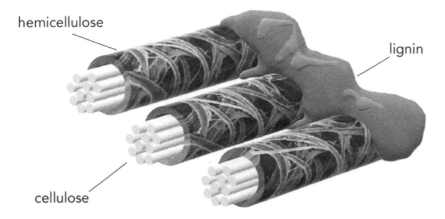

Figure 6.6 Structure of lignocellulose biomass (Source: Tursi, 2019).

morphology of cellulose chains are structurally influenced by the intermolecular hydrogen bond between the hydroxyl group on C-3 carbon and the oxygen of the nearby glycosidic ring. The formation of these bonds makes the molecules more stable and rigid (Ahmed *et al.*, 2019; Bernal *et al.*, 2017; Smith *et al.*, 2010; Tursi, 2019).

(II) ***Hemicellulose (25–35%)*** is one of the major constituents of plant cell walls and consists of heterogeneous branched polysaccharides. It is strongly linked to the surface of cellulose microfibrils. The content and structure of hemicellulose are different depending on the type of plant (Bala *et al.*, 2016). The various sugar units are arranged with different substituents and in different proportions. Hemicellulose decomposes thermally between 180 and 350°C, thereby producing non-condensable gas, coal and a variety of ketones, aldehydes, acids and furans (Ahmed *et al.*, 2019; Bernal *et al.*, 2017; Carpenter *et al.*, 2014; Smith *et al.*, 2010; Tursi, 2019). In nature, hemicellulose is amorphous and has adhesive properties, with a high tendency to toughen when it is dehydrated. Hemicellulose almost entirely consists of sugars with five carbon atoms (xylose and arabinose) and six carbon atoms (glucose, galactose, mannose and rhamnose) with an average molecular weight of <30 000 (Bonechi *et al.*, 2017; Jindal & Jha, 2016; McKendry, 2002; Tursi, 2019). The different groups of molecules making hemicellulose include xylans, mannans and arabinogalactan (Tursi, 2019), see Figure 6.8).

(III) ***Lignin (15–20%)*** is also contained in plant cell walls, with the function of binding, cementing, and putting the fibres together in order to enhance the compactness and resistance of the plant structure. Lignin is also recognized for

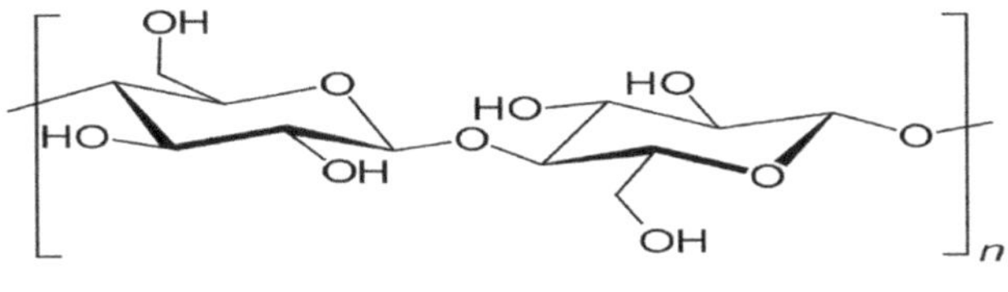

Figure 6.7 The structural formula of cellulose (Source: Tursi, 2019).

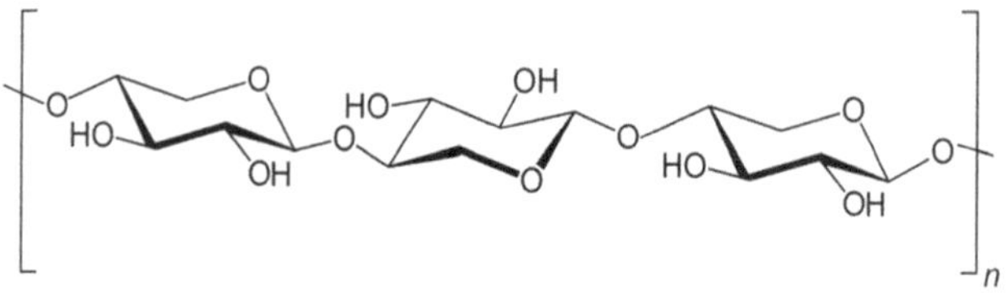

Figure 6.8 The structural formula of hemicellulose (Source: Tursi, 2019).

its encrusting effect as it protects fibres and prevents degradation (Tursi, 2019). Its elemental composition is approximately 61–65% carbon, 5–6% hydrogen and the remainder is oxygen (Fromm *et al.*, 2003). Structurally, it is a complex amorphous aromatic polymer with a three-dimensional network composed of phenylpropane units linked together. The monomeric units are held together in different ways: through oxygen bridges between two propyl and phenyl groups, between a phenyl and a propyl group, or through carbon–carbon bonds between the same groups. In particular, this macromolecule is formed through the radical oxidative polymerization of three hydroxycinnamyl alcohols representing the basic structural monomers: p-phenyl monomer (type H), guaiacyl monomer (type G) and siringyl monomer (type S), deriving from coumarinic, coniferyl and synapyl alcohol respectively (Ahmed *et al.*, 2019; Bernal *et al.*, 2017; Smith *et al.*, 2010; Tursi, 2019), see Figure 6.9. These compounds differ from each other due to the different degrees of methoxylation. Overall, given the considerably high global availability of lignin, that is 300 billion tons, with an annual increase of about 20 billion tons (Hodásová *et al.*, 2015), development of innovative technologies for lignin valorization is essential.

The two main components of biomass are lignocellulosic biomass (LB) and non-lignocellulosic biomass (NLB) which can be in the form of cellulose, hemicellulose and lignin (Ahmed *et al.*, 2019; Bernal *et al.*, 2017; Smith *et al.*, 2010; Tursi, 2019). Other minor components of biomass are extractives, proteins, water and inorganic

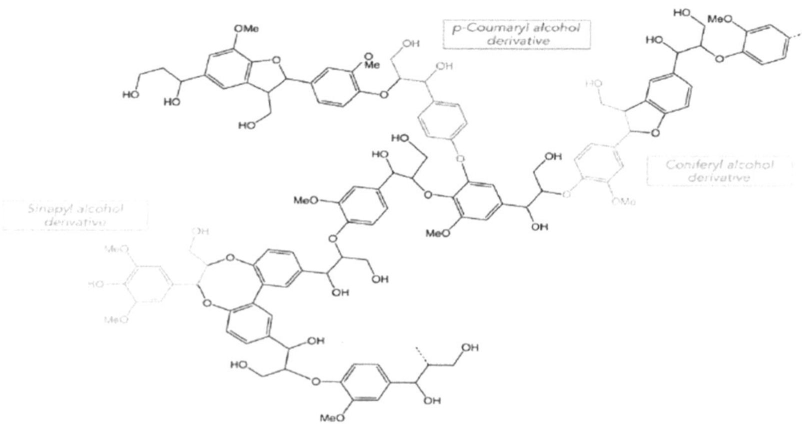

Figure 6.9 The structural formula of lignin and its precursors (Source: Tursi, 2019).

components such as silicon (Si), sodium (Na), potassium (K), calcium (Ca), magnesium (Mg) and aluminium (Al). These minor constituents do not significantly contribute to the formation of the total structure of the biomass (Anukam & Berghel, 2021; Raven *et al.*, 1992; Tursi, 2019).

6.3.1.2 Animal biomass feedstocks

Besides the plants and their derivatives, biomass also contain animals, microorganisms, and a portion of plants and materials derived from them, which are defined as non-lignocellulosic biomass (NLB) (Li & Jiang, 2017); the components mainly include lipids, proteins, saccharides, inorganics, minerals as well as a fraction of lignin and cellulose (Anukam & Berghel, 2021; Li & Jiang, 2017). NLB also includes resources such as sewage sludge, faecal sludge, plants and animal organic wastes, manure, algae, animal hair and bone and so on (Anukam *et al.*, 2016; Li & Jiang, 2017). It is chemically composed of C, H, O and N, comparable to non-renewable resources (Rana *et al.*, 2021; Siwal *et al.*, 2021; Thakur *et al.*, 2012). NLB is an excellent architectural material, arranging various atoms in an orderly manner to build units. Compared with LB, the NLB usually contains more miscellaneous elements such as N, P, S and metals, which are embedded in the skeleton of its structural unit. During heat treatment, the heteroatom of its structural unit can act as an activator or catalyst for biomass pyrogenic decomposition (Li & Jiang, 2017). Also, different compositions of NLB can lead to different thermochemical conversion behaviour in comparison with LB; understanding NLB behaviour during heat treatment and its physicochemical properties is essential to optimizing the conversion process for efficient waste disposal, resource recovery, and preparation of functional NLB materials (Li & Jiang, 2017; Liu *et al.*, 2015a; Yoshida & Antal, 2009).

(I) **Sewage sludge (SS),** a product of sewered sanitation can be described as any solid, semi-solid or liquid waste generated from a wastewater treatment facility. This wastewater can be sourced from municipal, commercial or industrial processes. The physical properties (low ratio of solid to liquid matter) of sewage mean it requires thickening and mechanical dewatering to help increase the solid particles to about 10–25 wt% from the original predominantly liquid (<3 wt% solid) state (Cieslik *et al.*, 2015; Li & Jiang, 2017; Magdziarz *et al.*, 2016; Oladejo *et al.*, 2019; Seiple *et al.*, 2017; Syed-Hassan *et al.*, 2017). The solid phase in sludge is made up of an inhomogeneous mix of proteins, carbohydrates, oils, inorganic matter and micro-organisms. This mixture of organic, inorganic and living organisms results in an unstable, volatile and putrid matter with toxic elements (Cieslik *et al.*, 2015; Li & Jiang, 2017; McConville *et al.*, 2020; Oladejo *et al.*, 2019; Rulkens, 2008; Siwal *et al.*, 2021; Wang *et al.*, 2016). Sewage treatment or stabilization involves biological (composting or digestion), physical (e.g., pressure, heat, vibration, microwaves) or chemical (oxidation, alkalinity adjustment) methods to stabilize the organic matter (including destruction of pathogens, odour elimination and reduction of volatile contents) contained in the primary sludge in order to improve the quality of effluent, maximize nutrient recovery and/or for safer disposal. The product of this stabilization process can be referred to as secondary sludge if it undergoes further biological processes (Chan & Wang, 2016; Mulchandani & Westerhoff, 2016; Oladejo *et al.*, 2019; Seiple *et al.*, 2017; Vaxelaire & Cézac, 2004). Anaerobic digestion is one example of such stabilization techniques: its secondary sludge can be used as fertilizer while the biogas produced from the digester can form part of the energy recovery capabilities of the process (Oladejo *et al.*, 2019; Winkler *et al.*, 2013). The elemental composition of sewage sludge differs greatly from case to case despite the common elements like C, O, H and

N. The C content varies between 25% and 70%, caused by the high ash content varying from 15% to 50%. The high ash content in sewage sludge is usually linked with the significant levels of other elements such as P, Ca, K, Mg, Fe, Si, Na, and so on (Li & Jiang, 2017). On the other hand, sewage sludge contains many easily available plant nutrients such as N, P, K and organic matter, which raises wide interest in its use as a fertilizer in agriculture or as a regenerator for soil (Khan *et al.*, 2013; Li & Jiang, 2017). Furthermore, it also can be used to produce renewable biofuel owing to its high decomposable organic content (Li & Jiang, 2017; Xie *et al.*, 2014). Sustainability measures have increased focus on the recovery and reuse of sludge after treatment to reduce landfill requirements and environmental footprints, and to lessen impacts on the land, groundwater and food supply (Li & Jiang, 2017; Oladejo *et al.*, 2019).

(II) **Faecal sludge (FS),** a product of non-sewered sanitation, is the raw or partially digested semisolid material that is produced primarily from human excreta and blackwater, but also includes anything else that goes into onsite sanitation systems such as flush-water, cleansing materials, menstrual hygiene products, greywater (i.e. bathing or kitchen water, including fats, oils, and grease), and solid wastes, and which needs to be removed periodically and transported to a faecal sludge treatment plant, followed by safe disposal or end-use (Barani *et al.*, 2020; Strande, 2021). Faecal sludge is grouped by consistency according to Strande (2021) and Velkushanova (2021) as:

- *liquid (TS <5%):* which is relatively diluted with the consistency of water or domestic wastewater, is readily pumpable and usually collected from wet containments such as leach pits, septic tanks or wet pit latrines;
- *slurry (TS 5–15%):* normally thicker than liquid, but still watery with a wet mud consistency, pumpable in lower ranges and thus difficult to shovel; it is common in pit latrines (improved or unimproved) with a frequent input of greywater or subject to infiltration;
- *semi-solid (TS, 15–25%):* soft paste-like materials, not pumpable, but can be spadable at the higher end of the range; it is collected from onsite containments such as pit latrines, composting toilets and leach pits, or from dewatering treatment technologies; and
- *solid (TS >25%):* the majority of free water has been removed; it can come from dry toilet systems or dewatering treatment technologies.

FS recovery may support the development of viable business models for sustainable sanitation (Barani *et al.*, 2020; Diener *et al.*, 2014). The most common form of resource recovery from faecal sludge solids has been that of soil conditioning. However, more promising options have recently emerged including the use of faecal sludge as a component of building materials, as source of protein for animal feed and as industrial fuel (Barani *et al.*, 2020; Diener *et al.*, 2014). Other approaches for energy recovery from non-sewered sanitation systems are combustion (Sellgren *et al.*, 2017), gasification (Onabanjo *et al.*, 2016), smouldering (Yermán *et al.*, 2015), hydrothermal oxidation (Miller *et al.*, 2015) and hydrothermal carbonization (Afolabi *et al.*, 2017).

Other notable contributions of NLB feedstock for the SBRCVC as biomass include:

(III) **Livestock manure (LM)**, a predictable side-product of animal husbandry that adds to greenhouse gases through the release of CH_4 to the environment if not regularly captured (Siwal *et al.*, 2021);

(IV) **Food waste (FW)**, valorisation through AD, fermentation and composting processes can create high-value products such as biofuels, biomass, and biofertilizers (Siwal *et al.*, 2021);

(V)   ***Agricultural waste (AW)***, which is a standard classification of carbon-rich biomass overflowing with cellulose, hemicellulose and lignin as lignocellulose (Siwal *et al.*, 2021; Thakur *et al.*, 2012; Zielinska *et al.*, 2021);

(VI)   ***Forestry residue (FR)*** is an essential lignocellulose raw material for bioenergy generation; pyrolysis of FR has been used to generate bio-oil and biochar (Demirbas & Balat, 2006; Singh *et al.*, 2018; Siwal *et al.*, 2021);

(VII)   ***Marine processing waste*** (**MPW**), includes fish production trash such as scales, skin, visceral mass, air bladders, gonads, head, tails and fins, crab shells and shellfish waste, head, and body carapace, and much more;

(VIII) ***Manure*** is an important nutrients source containing abundant organic matters, N, P, K and other trace elements; manure from humans and animals is widely used as plant fertilizer. The proportions of C, O, H and N in manure are usually 40–50%, 30–35%, 5–7%, and 2–5% respectively; and

(IX)   ***Fermentation processing waste (FPW)***, which includes lipids, proteins, and carbohydrates that can be converted into products such as fatty acids (acetic, propionic and butyric acid) and alcohols (ethanol and butanol) by the fermentation process (Chohan *et al.*, 2020); and food processing waste (Siwal *et al.*, 2021).

Other sources of feedstock are organic wastes derived from municipal activities such as restaurant and kitchen wastes, food processing industry waste, and agricultural and crop processing (crop and garden waste, sawdust, fruit, chicken and other animal manure and abattoir waste) (Polprasert & Koottatep, 2017).These classes of waste can either be reduced, recycled or transformed through the application of new and innovative approaches and technologies into energy, organic fertilizers, and animal feed as well as other useful products (Polprasert & Koottatep, 2017).

As noted earlier, although feedstock sourcing seems a simple process, technically, businesses still find it difficult to access enough of the right quality waste biomass to achieve a viable reuse business model; stronger financial viability improved feedstock collection (Ddiba, 2016; Koottatep *et al.*, 2019; Polprasert & Koottatep, 2017). It is crucial that the SBRCVC is flexible enough to use variety of feedstock to produce high-value goods (Hennig *et al.*, 2016; Lokesh *et al.*, 2018). Feedstock end-of-life characteristics play a prominent role at any given stage of a value chain because of the capability of utilizing waste biomass for raw feedstock (also called 'cascading use'), which makes it a sustainable business model as there will be a regular influx of low-cost feedstock that promises a continuous product supply to the market (Budzinski *et al.*, 2017; Lokesh *et al.*, 2018). This strategic management and utilization of sanitation-derived feedstocks and organic waste could deliver three-fold benefits: environmentally through reduction of waste treatment and disposal; economically by enabling resource efficiency and through transformation of waste (as low-cost raw materials for a secondary industry); and socially through creation of jobs, new value chains and social equity (Lokesh *et al.*, 2018; Pagotto & Halog, 2016).

6.3.2 Biomass Pre-treatment processes

There are various options for enterprises in the value chain to be involved in pre-treatment of biomass; and the most appropriate one or the most appropriate combination mainly depends on the subsequent conversion and utilization of that biomass, that is for thermochemical or biochemical conversion technologies (Papadokonstantakis & Johnsson, 2020). Collected biomass is subjected to pre-treatment and/or pre-processing to increase its resource value as well as enhance its conversion to high-quality and high-value bioproducts. Some common pre-processing/pre-treatment steps are mainly related to removing moisture by drying and decreasing the size of biomass particle, typically by grinding, milling, balling and pelletizing. These steps may also influence the efficiency of the subsequent biomass utilization processes (Papadokonstantakis & Johnsson, 2020). Such processes make subsequent biomass conversion more economical and environmentally friendly for transportation and storage (Tapia *et al.*, 2019).

Pre-treatment is a necessary process step for both biochemical and thermochemical conversion of biomass and involves structural alteration aimed at overcoming the recalcitrant nature of biomass. It is required to improve biomass characteristics in order to enhance their efficient utilization for production of high-value bioproducts (Anukam & Berghel, 2021; Anukam *et al.*, 2016; Chiang *et al.*, 2012). The main goal of biomass pre-treatment is to facilitate microbial digestion by removing barriers and making the organic content of the substrate easily accessible and usable for producing high-value bioproducts (Kasinath *et al.*, 2021). Thus, complex organic matter (e.g., cellulose, hemicellulose, and lignin, proteins, polysaccharides and lipids) need to be solubilized and hydrolysed into simple components such as long-chain fatty acids, sugars and alcohols (Kasinath *et al.*, 2021; Zhen *et al.*, 2017). Pre-treatment processes (also known as conditioning) are used to speed up and enhance digestion as well as improve dewatering and the quality of the digestate (Kasinath *et al.*, 2021). For example, in pre-treatment processes requiring the use of heat, the degradation ability of lignocellulosic biomass (LB) is controlled by its polymeric and aromatic constituents (cellulose, hemicellulose and lignin), while the heteroatoms and inorganic elemental components of non-lignocellulosic biomass (NLB) could act as catalysts to facilitate decomposition. This then forms a product that has a carbon framework with a change in the original structure that increases the performance of the pre-treated material during bioconversion processes (Anukam & Berghel, 2021; Anukam *et al.*, 2017; Liu *et al.*, 2015a; Yoshida & Antal, 2009).

Pre-treatment technologies can be classified into physical, chemical, physicochemical and biological pre-treatment methods (E4tech (UK) Ltd *et al.*, 2015; Papadokonstantakis & Johnsson, 2020):

(I) ***Physical pre-treatment*** aims to increase the accessible surface area and pole volume and decrease the degree of polymerisation of cellulose and its crystallinity.

(II) ***Chemical pre-treatment*** mostly uses alkalis, acids, ozonation, Fenton or Fe (II)-activated persulfate oxidation to delignify the biomass and decrease the polymerisation and crystallinity of cellulose (Kasinath *et al.*, 2021; Patinvoh *et al.*, 2017). The most commonly used acid is H_2SO_4 (Morales *et al.*, 2017; Papadokonstantakis & Johnsson, 2020) and the most common alkali is NaOH. These are applied to solubilise the hemicellulose fraction of biomass and make the cellulose accessible to enzymes. Organic acids can also be used to enhance cellulose hydrolysis and reduce production of inhibitors (Papadokonstantakis & Johnsson, 2020).

(III) ***Physicochemical pre-treatment*** affects both physical and chemical properties of the biomass; among such techniques are steam explosion, ammonia fiber explosion (AFEX), wet explosion, CO_2 explosion, and so on.

(IV) ***Biological pre-treatment*** is carried out using microorganisms (temperature-phased anaerobic digestion and microbial electrolysis cells) such as white rot fungi (Kasinath *et al.*, 2021; Papadokonstantakis & Johnsson, 2020; Patinvoh *et al.*, 2017; Sarkar *et al.*, 2012). This alters the structure of lignin and cellulose, separating them from the lignocellulosic matrix.

Although biological pre-treatment is typically carried out under mild conditions, the rates of hydrolysis are low and current efforts focus on combining this technology with other pre-treatment methods and developing new microorganisms for rapid hydrolysis (Kasinath *et al.*, 2021; Papadokonstantakis & Johnsson, 2020; Patinvoh *et al.*, 2017). Other pre-treatment methods can make use of mechanical techniques such as ultrasonic, microwave, electrokinetic and high-pressure homogenization (Kasinath *et al.*, 2021; Patinvoh *et al.*, 2017).

Pre-treating sewage sludge and faecal sludge (being the main feedstock for the SBR) is usually characterized by high concentrations of solid and organic matter and a significant presence of pathogens, nutrients, and organic and inorganic pollutants, and involves single or combined physical, chemical and biological means to disrupt the floc structure of sludge and hydrolyse organic matter, as well as provide significant enhancements in terms of solid reduction to produce the required high-value bioproducts (Neumann *et al.*, 2016). Pre-treatment can be applied to primary, secondary and/or mixed sludges and has been known to significantly improve pathogen deactivation and sludge quality. Therefore, its application to mixed and primary sludge can be attractive depending on the main objective (Wilson & Novak, 2009). Pre-treatment of sludge is expected to rupture the floc structure as well as some bacterial cell walls, resulting in the release of intercellular matter in the aqueous phase (Kasinath *et al.*, 2021), and so helps to reduce its high resistance to both dewatering and biodegradation. The increase in nutrient accessible to microbes enhance the digestion rates and reduces the retention time of conversion of biomass to high-value bioproducts (Kasinath *et al.*, 2021; Khanal *et al.*, 2007; Pilli *et al.*, 2011). The first commercially used thermal pre-treatments for SS were Porteous and Zimpro which were implemented in the 1960s and the early 1970s; a modified lower temperature was subsequently used to enhance the dewaterability of SS (Camacho *et al.*, 2008). During the 1980s, however, various combinations of thermal and pH-based (acid and alkaline) technologies were tested (e.g. Synox and Protox), but none were successfully commercialized owing to insufficient cost-effectiveness. In 1996 the CambiTHP™ process, a combination of thermal hydrolysis and high pressure, was implemented to increase biogas production and digester loading (Neyens & Baeyens, 2003). Then in 2006 Veolia, following their batch process Biothelys®, introduced a continuous-flow process called Exelys – a pre-treatment thermal hydrolysis process for municipal and industrial sludge, as well as for sludges containing fats, oils and grease (Kasinath *et al.*, 2021).

A successful SBRCVC depends on a business model often driven by blending agricultural waste, food waste and organic (biodegradable) fractions of municipal solid waste biomass and sanitation-derived biomass (Moya *et al.*, 2019; Otoo & Drechsel, 2018; Remington *et al.*, 2018; World Bank, 2019). It should also be noted that any type of agricultural, food or organic fraction of municipal-waste biomass that consists of lignocellulose fibres will require pre-treatment (Kasinath *et al.*, 2021). This pre-treatment is most frequently a combination of elevated temperature and chemical treatment, while thermal and other mechanical pretreatment methods are also considered (Fernandes *et al.*, 2009; Kasinath *et al.*, 2021). The pretreatment efficiency with respect to lignocellulose biomass depends mainly on the lignin content of the treated materials (Fernandes *et al.*, 2009; Kasinath *et al.*, 2021).

The detrimental effects of pretreatment for these classes of biomass include the formation of refractory compounds, mainly from high-thermal pretreatment. Thermoacid pretreatment may also generate biomass conversion inhibitors such as furans and phenolic compounds, which may hinder microbial activity (Taherzadeh & Karimi, 2008; Vavouraki *et al.*, 2013).

6.3.3 Biomass conversion technologies
The enterprises and actors in the conversion processes generate the needed revenue for the SBRVC by transforming biomass resources such as collected and/or pre-processed

biomass into valuable products (Papadokonstantakis & Johnsson, 2020; Tapia *et al.*, 2019). The conversion pathways that transfer sanitation biomass to high-value biobased products include biochemical (photobiological hydrogen production, anaerobic digestion, and fermentation); thermochemical (combustion, pyrolysis, gasification, and liquefaction); mechanical extraction; and physical or chemical (Panoutsou *et al.*, 2020; Papadokonstantakis & Johnsson, 2020). All of these allow low-value biomass resources to gain economic value when transformed into high-value products such as biofuels (biogas, biohydrogen, biodiesel), power, heat, oleochemicals that serve as substitutes for petroleum-based products known as petrochemicals (Papadokonstantakis & Johnsson, 2020; Wikipedia contributors, 2022), single-cell proteins, animal proteins, building materials, soil conditioners, biofertilizers, short-chain fatty acids, enzymes, biopesticides, bioplastics, bioflocculants and biosurfactants (Diener *et al.*, 2014; Eze, 2004; Koottatep *et al.*, 2019; Mafakheri & Nasiri, 2014; Otoo & Drechsel, 2018; Papadokonstantakis & Johnsson, 2020; Polprasert & Koottatep, 2017; Puyol *et al.*, 2017; Zhang *et al.*, 2018).

The two most important physical properties of biomass, regardless of conversion process, are particle size and moisture content. Practically all conversion methods require some degree of size reduction (Williams *et al.*, 2017). For instance, biochemical conversion processes can accept a greater range of particle sizes, and the final size needed tends to be dependent on the processing system utilized (Dibble *et al.*, 2011; Van-Walsum *et al.*, 1996; Williams *et al.*, 2017). Hydrothermal liquefaction is much more insensitive to particle size owing to high heating rates in the liquid media (Akhtar & Amin, 2011; Williams *et al.*, 2017), but a significant amount of size reduction is needed to pump biomass sludges in a continuous system (Jazrawi *et al.*, 2013; Williams *et al.*, 2017). On the other hand, moisture increases heating rates during steam pre-treatment for biological conversion (Brownell *et al.*, 1986; Williams *et al.*, 2017) and also reduces bio-oil quality and thermochemical conversion (Bridgwater *et al.*, 1999; Williams *et al.*, 2017) and causes low thermal efficiency in combustion processes (Jenkins *et al.*, 1998; Williams *et al.*, 2017). Aside from particle size and moisture content, other physical properties of interest include bulk density, elastic properties, and microstructure. Bulk density has a strong effect on transportation and handling costs because lower densities greatly increase transportation cost. Biomass chemical properties also have a large influence on the best conversion process and the quality of the final product. The three primary compounds of interest in biomass conversion are ash content, volatiles and lignin. High ash content generally has a negative effect on biomass conversion across the board by reducing the effectiveness of dilute acid pre-treatment for biological processes (Weiss *et al.*, 2010; Williams *et al.*, 2017) and increasing char yields and fouling in thermochemical processes such as hydrothermal liquefaction (HTL) (Toor *et al.*, 2011; Williams *et al.*, 2017), pyrolysis (Tumuluru *et al.*, 2012; Williams *et al.*, 2017), and combustion (Jenkins *et al.*, 1996; Williams *et al.*, 2017). Conversion technologies covered in this chapter with reference to the SBRCVC are presented below:

6.3.3.1 Thermochemical conversion technologies

The shortage of conventional energy resources, as well as environmental issues related to landfilling the considerable amount of excess sewage sludge and faecal sludge, raised interest in developing methods for the utilization of sludge for energy purposes (Smoliński *et al.*, 2018). Thermochemical technology involves a high-temperature chemical reformation process that requires bond breaking and reforming of organic matter into biochar (solid), synthesis gas and highly oxygenated bio-oil (liquid). Within thermochemical conversion, there are three main process alternatives available:

gasification, pyrolysis and liquefaction (Lee *et al.*, 2019). This conversion involves the complete oxidative ignition of sanitation-derived (i.e. faecal sludge) and other organic-waste biomass with the primary aim being to produce high-temperature energy. (Siwal *et al.*, 2021). Also, attention is given to thermochemical co-processing of SS and/or FS with fossil fuels and biomass (Garrido-Baserba *et al.*, 2015; Kokalj *et al.*, 2017) and pyrolysis of SS and/or FS then blended with organic solid waste as well as with biomass from other sources (Deng *et al.*, 2017; Ma *et al.*, 2017; Zhang *et al.*, 2015). The selection of conversion type can be influenced by the nature and quantity of biomass feedstock, and the preferred type of energy, for example end-use conditions, environmental principles, financial circumstances and the precise nature of the project (Siwal *et al.*, 2021). Thermal conversion technologies have gained extra attention due to the availability of industrial infrastructure to supply thermochemical transformation equipment that is highly developed, short processing times, reduced water usage and the added advantage of producing energy from other forms of waste that cannot be digested by microbial activity (Uzoejinwa *et al.*, 2018). The main business activities are the construction and operation of conversion installations, ensuring conversion processes' efficiencies and optimization of conversion technologies (Panoutsou *et al.*, 2020; Tapia *et al.*, 2019). The challenges with regards to construction include site selection and access to technology, and for operations, low emission performance, handling mixed volumes of feedstocks and improving synergies for valorisation of residues and co-products (Panoutsou *et al.*, 2020).

6.3.3.1.1 Combustion technology

The combustion of all solid fuels is similar to that of sewage sludge and faecal sludge biomass. In the combustion process, biomass and oxygen are combined in a high-temperature environment to form carbon dioxide, water vapour, heat and trace gases (Oladejo *et al.*, 2019), see equations (2) and (3). This process is known to produce approximately 90% of the total renewable energy from biomass. The use of combustion technology for waste materials such as sewage sludge and faecal sludge can be used primarily to reduce the volume of sanitation-waste materials, and later heat generation as well as electric generation was added as a resource recovery strategy.

$$\text{Biomass} + \text{Oxygen} \rightarrow \text{Carbon Dioxide} + \text{Water} + \text{Heat} \tag{6.2}$$

The approximate chemical equation for biomass combustion is:

$$CH_{1.44}O_{0.66} + 1.0\ 3O_2 \rightarrow CO_2 + 0.72\ H_2O + \text{Heat} \tag{6.3}$$

The amount of generated heat depends on many factors, but mainly on the types and quality of biomass used in the process, although the average thermal energy produced is 20 MJ/kg of biomass (Nussbaumer, 2003). As shown by equations (2) and (3), the combustion process is an exothermic reaction, that is, the biomass is burnt in the presence of air with subsequent release of chemical energy that could be converted into mechanical and electrical energy (Kaushika *et al.*, 2016; Lebaka, 2013).

The principle of solid fuel combustion involves drying, pyrolysis, volatiles combustion, char combustion, ash melting and agglomeration. These stages occur sequentially or simultaneously depending on the configuration, reactor conditions and fuel properties. For example, some sludge and biomass could start pyrolysis at low temperatures (~150°C) typical for fuel drying (Ogada & Werther, 1996; Oladejo *et al.*, 2019; Urciuolo *et al.*, 2012), see Figure 6.10.

The release and burning of volatiles from this stage generate heat, CO, H_2O, CO_2, NO_x and SO_x, which further interact with the solid char particles in the fuel and increase

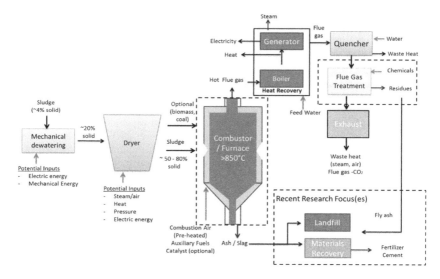

Figure 6.10 Schematic representation of the combustion of sludge. (Source: Oladejo *et al.* (2019) under CCA 4.0 license, ©, 2018 by the authors)

surface temperature (Oladejo *et al.*, 2019). The furnace operates at temperature >850°C for the complete oxidation of sludge which may be done separately or blended with other solid fuels (coal or biomass) (Chen *et al.*, 2018; Rong *et al.*, 2017). This process would require excess air for completion while auxiliary fuel and catalyst might be needed for initiation and maintaining reactor stability for operational efficiency. Ash and flue gas are the main output from this reactor. The flue gas is made up primarily of oxides of carbon, nitrogen, sulphur and particulate matter which act as the thermal store that allows heat transfer from itself to feed water. This aids heat generation for direct use (industrial or residential heating) or electricity generation via steam turbines and generators. After the heat recovery process the flue gas has to undergo treatment for elimination of pollutants before releasing exhaust gas (mostly CO_2 and water vapour) into the atmosphere. The ash generated from this process can be reused in agricultural or construction applications. However, this depends primarily on its chemical contents, particularly the heavy metal content of the ash (Oladejo *et al.*, 2019).

Combustion reactors use various technologies such as multiple hearth, rotary kiln and cyclone and fluidized bed furnace with different fuel needs and operating mode. The major challenge with combustion of sewage/faecal sludge is mostly moisture and ash content that influences the thermal characteristic of the fuel and the design requirements of the combustor. High moisture content is not only a deterrent for increasing the bulk density of the fuel, oxidant and energy for drying the sludge and has the potential of forming erosive sulphuric compounds (Han *et al.*, 2012). The use of ash and slags for other applications contributes to high phosphorus contents and negligible toxic compounds such as heavy metals or polycyclic aromatic hydrocarbons (PAHs) make it suitable for agricultural purposes or raw materials for the construction industry. Co-use of sludge with other fuel such as coal, biomass, other solid waste, fuel oil or gas has been investigated as a means of avoiding the high cost associated with dedicated reactors and also an avenue for reducing net carbon emissions (Oladejo *et al.*, 2019).

6.3.3.1.2 Gasification technology

The thermochemical conversion of sewage sludge/faecal sludge's organic content into high value gases such as H_2 and CO known as synthesis gas as well as CO_2, CH_4, H_2O and other hydrocarbon is the main basis for gasification (Oladejo *et al.*, 2019). The gasification technique comprises chemical reactions in an environment that is oxygen deficient. This process involves biomass heating at extreme temperatures (500–1400°C), from atmospheric pressure up to 33 bar and with low/absent oxygen content to yield combustible gas mixtures. Often described as an incomplete anodic process of organic materials at a high temperature (500–1800°C) to generate synthetic gas (Siwal *et al.*, 2019, 2021), biomass gasification happens to be where the char acts including CO_2 and water stream to create CO and H_2. Also, the volumes of CO, steam, CO_2 and H_2 are compared very quickly on temperatures inside a reactor. Produced gas may be applied as fuel towards the adequate generation of power and/or heat (Colmenares *et al.*, 2016; Siwal *et al.*, 2021). The gasification process transforms carbonaceous constituents into syngas comprising hydrogen, carbon monoxides, carbon dioxide, methane, higher hydrocarbons and nitrogen with the presence of gasification agent and catalyst. By utilizing this syngas, various types of energy or energy carriers are supplied, for example, biofuel, hydrogen gas, biomethane gas, heat, power and chemicals (Lee *et al.*, 2019).

This process is very similar to combustion with the exception of the lower moisture tolerance in the reactor (<15 wt%) and the deficit in stoichiometric oxidants required for complete combustion. The main outputs from the reactor are gases and ash. Depending on the chemical and mechanical properties, as well as heavy metal contents, the ash generated from the process can be reused in agricultural or in construction applications. The product gases require further processing and clean-up for either use in heat and electricity generation or upgrading of synthesis gas for liquid fuels and chemical synthesis (Oladejo *et al.*, 2019), see Figure 6.11. Gasification reactions can be

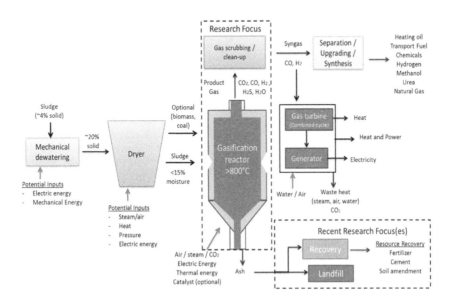

Figure 6.11 Schematic representation of the gasification of sludge (Source: Oladejo *et al.* (2019) under CCA 4.0 license, ©, 2018 by the authors).

divided into sub-stages which are drying of sample (70–200°C), devolatilization (350–600°C), oxidation of volatiles and char gasification. Hence, it can also be termed as an incomplete combustion or extended pyrolysis reaction in which gas–solid, gas–gas and liquid cracking reactions are required to maximise the gaseous product yield.

There are essentially two types of gasification technologies: autothermal (direct) and allothermal (indirect) gasification. In direct gasification, the heat required by the process is only internally generated by the partial combustion of the feedstock, whereas in indirect gasification, energy is also delivered to the process via the gasification agent (steam). Furthermore, in direct gasification all reactions occur in the same device while in indirect gasification, combustion reactions occur in a separate chamber that communicates with the gasification chamber both with mass streams (bed material, char, ashes and feedstock to be combusted) and energy streams (heat carried by the thermal inertia of the bed material itself) (Papadokonstantakis & Johnsson, 2020; Sette et al., 2015). Several types of equipment are usually used for gasification: fixed bed, fluidized bed, including entrained flow gasifier (Papadokonstantakis & Johnsson, 2020; Siwal et al., 2021).

The raw material NLB substance must be well granulated for applications in reactors. Therefore, a trial is required, particularly for sewage/faecal sludges, municipal solid waste (MSW), and so on. (Siwal et al., 2021). Depending on the technology and biomass used, impurities may include dust, ash, bed material, sulphur and chloride compounds. Various types of filters (e.g. textile bag filters such as GoBiGas, Gothenburg) (Papadokonstantakis & Johnsson, 2020; Thunman et al., 2018) can be used to remove the particles from the product gas; the maximum allowable temperature of the filter is an important parameter for avoiding fouling in the heat exchangers cooling the raw product gas. Also, gas composition produced from the gasification process varies according to the type of gasifier, gasification agent, catalyst type and size of particle (Lee et al., 2019) and the technique is considered to be independent autothermic route based on energy balance. It is revealed that biomass gasification is able to recover more energy and higher heat capacity compared to combustion and pyrolysis, probably due to optimal exploitation of existing biomass feedstock for heat and power production (Lee et al., 2019).

6.3.3.1.3 Pyrolysis technology

Pyrolysis is one of the thermochemical technologies for converting biomass in the absence of oxygen into energy and chemical products consisting of liquid bio-oil (also referred to as pyrolysis oil, pyrolysis tar, biocrude, wood liquid, wood oil or wood distillate), solid biochar (also referred to as charcoal), and pyrolytic gas (Papadokonstantakis & Johnsson, 2020). It involves the conversion of sewage sludge/faecal sludge without air at moderate operating temperature (350–600°C), although some pyrolysis reactors that operate at higher temperature up to 900°C exist (Oladejo et al., 2019; Ruiz et al., 2013; Zhang et al., 2010). The output product of this process depends on the process temperature where char yield decreases with an increase in temperature (Oladejo et al., 2019). There are three types of pyrolysis process that differ according to their operational conditions, namely slow, fast and flash pyrolysis (Lee et al., 2019), see Figure 6.12.

It should be noted that high residence time of the fuel in the reactor at low temperature with low heating rates promotes char production, while low or high residence time at high temperature promotes liquid and gas production respectively (Oladejo et al., 2019). The application of this technology is mostly used to maximise liquid fuel yield and energy recovery from sludge. The drying requirements here are greater than for combustion with <10% moisture tolerance in the input sludge fed into the reactor. The pyrolysis of sludge takes place in an inert environment at high temperature, hence an external heat source

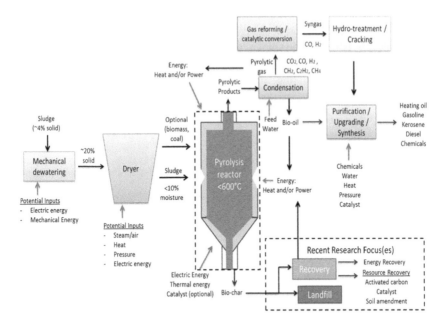

Figure 6.12 Schematic representation of the pyrolysis of sludge (Source: Oladejo *et al.* (2019) under CCA 4.0 license, ©, 2018 by the authors).

(electric or thermal) would be required to supply heat for initiation of the reaction. The utilization of heat sourced from the partial combustion of biogas, or bio-oil derived from the process itself has been critically explored for ensuring self-sustainability of pyrolysis, particularly in waste-to-energy applications (Oladejo *et al.*, 2019).

Pyrolysis technology can be classified based on heating rate and residence time, whether fast or slow pyrolysis. Fast pyrolysis generally uses a high heating rate above 300°C/s and a short vapour residence time below 10 s, while slow pyrolysis adopts a relatively low heating rate (Liu *et al.*, 2015b) and a long vapour residence time and is a promising technology to efficiently treat and sanitize faecal sludge from dry toilets (Mašek *et al.*, 2016). Compared with slow pyrolysis, fast pyrolysis with medium temperatures in the range 400–600°C usually has a higher bio-oil yield (Li & Jiang, 2017). Inside the pyrolysis zone, biomass is exposed to an ideal heat of 700°C during a deficiency of O_2 appearing with the production of bio-oil, char, and syngas. Synthetic gas is a hybrid mainly of CO, CO_2, H_2, CH_4. These may be applied as a subsequent fuel to produce power. Bio-oil yields can be as high as 50–70% wt% of the dry biomass (Lee *et al.*, 2019). Even higher heating rates of 1000–10 000°C/s can achieve bio-oil yields of up to 80 wt% (Amutio *et al.*, 2012). Gas and biochar yields amount to 13–25% and 12–15% of dry biomass feed, respectively (Papadokonstantakis & Johnsson, 2020). In a standard method, the biomass is converted fuel-efficiently without producing slag or transmitting massive amounts of flue gas. The necessary methods and steps of biomass pyrolysis are presented below (Siwal *et al.*, 2021):

(I) crushing to improve the exterior area to enhance heat transmission effect;
(II) dehydrating to improve the effectiveness of gas–solid resources inside the reactor;

(III) anoxic thermal degeneration of organics to produce pyrolysis outcomes (syngas, bio-oil, and char); and

(IV) final subsequent processing of syngas and char.

Biomass pyrolysis reactors can be fixed bed, fluidised bed, heated kiln, rotating cone, screw feeder/auger and vacuum pyrolysis (Bridgwater, 2012). From these reactor types, bubbling and circulating fluidised beds, heated kiln and rotating cone have been commercialized, while others remain at the demonstration or pilot stages. Typical capacities for commercial scale are in the range of 0.2–20 tonnes/hours, at feed moisture less than 10 wt%, feed size of 0.2–50 mm and bio-oil yields of 70–75% wt% (Papadokonstantakis & Johnsson, 2020). Pyrolysis processes decompose organic matter into a solid, liquid and gas mixture. Pokorna *et al.* (2009) classified the condensable pyrolysis products of sewage and faecal sludge into five groups:

(I)   mainly containing oxygenated compounds (fatty acids, alcohols, phenols, etc,);

(II)   nitrogenated compounds;

(III) sulphur compounds;

(IV) hydrocarbons; and

(V)   steroids.

The gas products principally consist of carbon dioxide, carbon monoxide, methane, hydrogen and some volatile liquids like small fractions of phenols, 1H-indols and fatty carboxylic acids (Tsai *et al.*, 2009). The difference between gasification and pyrolysis is that gasification produces fuel gas that can be combusted for heat generation, whereas pyrolysis produces liquid fuel known as pyrolysis oil (py-oil) or bio-oil that can be an alternative to fuel oil in static heating or in the generation of electricity (Lee *et al.*, 2019). Py-oil is dark brown, with high viscosity and low calorific value and is comprised of several chemical components that include acids, alcohols, aldehydes, phenols, and oligomers that originate from lignin (Lee *et al.*, 2019).

Converting sewage and faecal sludge to biochar addresses the stigma of fertilizer obtained from human excreta, since pyrolysis guarantees 100% elimination of pathogens with enriched nutrients in faecal-sludge biochar (Nuagah *et al.*, 2020). Biochar is a rich material obtained by a thermal process (pyrolysis of biomass) in an environment low in oxygen, mostly for the purpose of a soil enhancer. The addition of biochar to soils enhances its properties and filters and retains nutrients from permeating soil water (Crombie *et al.*, 2013; Nuagah *et al.*, 2020). The biochar from sewage and faecal sludge decreases plant accessibility to heavy metals and the danger associated with the probable filtering of heavy metals into the soil that is linked with raw sewage and faecal sludge (Marshall & Eng, 2013; Nuagah *et al.*, 2020).

Pyrolysis has been proven to be an effective technology for treating heavy-metal-polluted biomass, keeping most of the metals inherently. Studies focusing on metal behaviour during pyrolysis of sewage sludge and/or faecal sludge demonstrated that most of the common heavy metals (e.g. Cr, Ni, Cu, Zn and Pb) are retained in the biochar with pyrolysis temperature below 800°C (Jin *et al.*, 2017; Van Wesenbeeck *et al.*, 2014). If the faecal sludge is not dry, the initial energy input will go toward volatilizing the water in the sludge before pyrolysis proceeds (Andriessen *et al.*, 2019). Also, the NLB's biochar mainly contains C, O, H, N, P, and minerals, with percentage ratios highly affected by the mineral contents (Li & Jiang, 2017; Marshall & Eng, 2013; Nuagah *et al.*, 2020). Compared with the thermochemical conversion of LB, an important difference in the pyrolysis process of NLB is attributed to the massive existence of heteroatoms and metals (Li & Jiang, 2017).

In recent years, improvements to py-oil properties have become a major concern. The enhancement of py-oil is desired so that it could be utilized as a substitute for crude oil. There are several routes for upgrading the py-oil that include physical, chemical and catalytical approaches (Lee *et al.*, 2019). Hot vapour filtration is the most frequent method for physical upgrading of py-oil to get better bio-oil. It enables a reduction in the initial molecular weight of the oil and slows down the rate of bio-oil aging. Hot gas filtration eliminates char and inorganic materials from the oil, which is initiated due to the removal of highly unstable compounds of ring-conjugated olefinic substituents and the conversion of guaiacol-type compounds to catechol- and phenol-type compounds (Case *et al.*, 2014).

Hydrodeoxygenation upgradation (HDO), also known as hydrotreatment, is another strategy that offers enhanced oil yield, high oil quality and higher carbon recovery. This process involves the removal of oxygen from oxygenated hydrocarbons via catalytic reactions at high pressure (up to 200 bar (20 MPa)), hydrogen supply and moderate temperature (up to 400°C) (Lee *et al.*, 2019; Zhang *et al.*, 2013). It is stated that the HDO process is able to improve the py-oil quality by refining oil stability and increases energy density (Furimsky, 2000; Huber *et al.*, 2006; Lee *et al.*, 2019; Li *et al.*, 2010). According to Lee *et al.* (2019), there are four main reactions that affect the HDO of py-oil:

- hydrogenation of C-O, C=O and C=C bonds;
- dehydration of C-OH group;
- condensation and decarbonylation of C-C bond cleavage using retro-aldol
- hydrogenolysis of C–O–C bonds.

The main challenge in HDO of py-oil is deactivation of the catalyst which is necessary for effective synthesis for the HDO process (Lee *et al.*, 2019). An alternative method in upgrading py-oil is the use of catalysts and involves the use of methods for enhancing pyrolysis oil quality: (i) the use of downstream process by means of metallic or bi-functional (hydrogenating and acidic) catalysts; and (ii) *in situ* upgrading by integrated catalytic pyrolysis (Dhyani & Bhaskar, 2018). In a catalytic process, the vapour that is produced by pyrolysis will go through extra cracking within the catalyst pores for formation of desirable low-molecular weight compounds (Lee *et al.*, 2019).

6.3.3.1.4 Hydrothermal liquefaction technology

Hydrothermal liquefaction (HTL), also known as hydrothermal carbonization (HTC), involves chemical and physical transformations of carbohydrates into a carbonaceous residue under conditions of wet, high temperature (180–350°C) and autogenous pressure (Li & Jiang, 2017). In the hydrothermal system, water that exists in a subcritical or supercritical state simultaneously acts as medium, reactant and catalyst at a medium temperature range of 250–374°C for 1–12 hours and operating pressure of 40 to 220 bar (4–22 MPa) to convert biomass into bio-oil and biochar (Lee *et al.*, 2019). The HTL process comprises decomposition and repolymerization reactions for bio-oil conversion, aqueous dissolved chemicals, solid deposits and gas. The high pressure in the HTL process helps to maintain water in a liquid state, whilst the blending of elevated pressure and temperature leads to a decrease in the electric constant and density, which influence the hydrocarbons to be water soluble (Pambudi *et al.*, 2017; Tursi, 2019), see Figure 6.13.

This process has shown more advantages and potential than dry carbonization processes (e.g. pyrolysis) for feedstocks containing high moisture. It could be a viable way to dispose of waste streams and realize the value-added utilization (Berge *et al.*, 2011). For example, the process of dehydrating sewage sludge/faecal sludge is time-consuming and costly, owing to the high moisture content. In order to solve this problem, the hydrothermal treatment method was employed to change the physical and chemical properties of SS/FS to yield bio-oil and biochar (Andriessen *et al.*, 2019; Vardon *et al.*, 2011, 2012). A variety of feedstock can be converted to biochar with carbon content

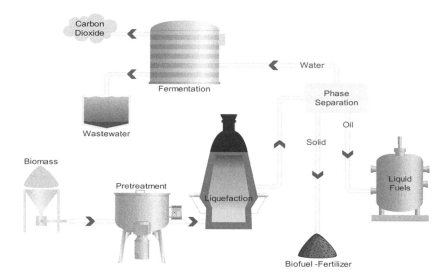

Figure 6.13 Biomass liquefaction scheme. (Source: Tursi, 2019)

similar to lignite with a mass yield of 35–60% via HTL/HTC processes (Kruse *et al.*, 2013; Vardon *et al.*, 2012). A biochar yield of 50–80% was observed with faecal sludge, and higher-value products were obtained even at a lower temperature (Afolabi *et al.*, 2017). The HTC process was found to improve the calorific value of faecal sludge fuel from 16 to 19 MJ/kg as well as to eliminate long drying times on drying beds (Fakkaew *et al.*, 2015a, 2015b; Koottatep *et al.*, 2016). More HTC reactors exist at a pilot scale, but few full-scale examples exist at present (Román *et al.*, 2018). Sewage and faecal sludge are promising feedstocks for HTL/HTC processes as they are readily available in large volumes. In addition, compared to dry sludge, exploiting wet sludge is able to decrease the consumption of energy by 30% (Li *et al.*, 2009).

6.3.3.1.5 Torrefaction technology

Torrefaction can be described as the thermal treatment of biomass to create an output that can be densified by palletization to produce a more energy-dense output called torrefied pellets (TOPs) or pieces, sharing related features to coal (Batidzirai *et al.*, 2013; Siwal *et al.*, 2021). Torrefaction is usually a first stage that is followed by pyrolysis and finally gasification during biomass heat treatment or biomass decomposition (Lange, 2007). It is a low-temperature biomass thermal decomposition process that produces carbon-rich biochar (Mimmo *et al.*, 2014). Biomass partly decomposes during this process generating both condensable and non-condensable gases; the resulting product is a solid substance rich in carbon that is referred to as biochar, torrefied biomass or biocarbon (Lehmann *et al.*, 2011). The torrefaction process is also referred to as roasting, slow and mild pyrolysis, wood-cooking and high-temperature drying (Bergman & Kiel, 2005). As reported in several studies (Agar & Wihersaari, 2012; Bridgeman *et al.*, 2010; Chew & Doshi, 2011; Mamvura & Danha, 2020; Nunes, 2020; Prins *et al.*, 2006), and as shown in Figure 6.14, torrefaction leads to:

(I)   Improved energy density;
(II) Better ignition;

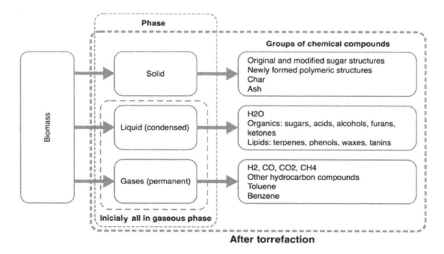

Figure 6.14 Main constituent compounds of each of the fractions formed during the torrefaction process. Nunes, L.J.R. (2020) A case study about biomass torrefaction on an industrial scale: solutions to problems related to self-heating, difficulties in pelletizing, and excessive wear of production equipment. *Applied Sciences* by MDPI under CCA 4.0 license, ©, 2020 by the authors.

(III)   Less moisture;
(IV)   Higher C/O and C/H ratio;
(V)   Improved grind-ability thereby reducing energy required for grinding;
(VI)   Biomass that is hydrophobic that is that has less affinity for water;
(VII)   More homogenized biomass that is torrefaction devolatilizes, depolymerizes and carbonizes the biomass; and
(VIII)   Reduces microbial activity.

This technology enhances combustion performance, particularly in boilers for energy production and for pyrolysis and gasification applications (Basu, 2018), and also leads to better storability of the treated biomass (Mamvura & Danha, 2020). Temperature and retention time are two main parameters that influence torrefaction process efficiency (Wannapeera *et al.*, 2011). Torrefaction is usually conducted at temperatures between 200 and 300°C (Eseltine *et al.*, 2013), and the process temperature is maintained for 15–60 minutes (Verhoeff *et al.*, 2011). Choosing the specific value of those two key parameters for different types of biomass is essential to develop cost-effective biomass treatment (Pulka *et al.*, 2019). Sewage sludge (SS) and faecal sludge (FS) can be valorised via a torrefaction also known as low-temperature pyrolysis. SS/FS are suitable substrates for the torrefaction process in the production of low-quality fuel and/or a source of nutrients essential for plant growth (Nunes, 2020; Pulka *et al.*, 2020). Torrefaction of SS/FS increases the C density and produced biochar that contains a smaller amount of O and H in its structure (Nunes, 2020; Poudel *et al.*, 2015; Pulka *et al.*, 2020). It could also be used as pre-treatment for SS/FS by easing its grindability and improving some of its fuel properties (Atienza-Martínez *et al.*, 2015; Nunes, 2020). The method involves cutting down the biomass to achieve sufficient drying and over 20% humidity, and then a tiny portion of the raw biomass is applied as fuel to the humid content during aeration and torrefaction. The torrefied biomass can then be used as a replacement for charcoal since it is hydrophobic and resistant to degeneration. (Agar & Wihersaari, 2012; Nunes, 2020; Siwal *et al.*, 2021).

6.3.3.1.6 Plasma gasification technology

Plasma gasification of waste biomass is a technologically advanced non-incineration thermal process that uses extremely high temperatures in an oxygen-starved environment to decompose input waste materials completely into very simple molecules (Mountouris *et al.*, 2008). Plasma which consists of free electrons, ions, and neutral particles is defined as the fourth state of matter. Also, the presence of electrons and charged particles is what allows plasma to be considered as neutral. Plasma is thermally and electrically conductive due to the charged particles and can be described as an ionized gas (Roth, 1994). Plasma can be partially ionized as well as fully ionized (Bogaerts *et al.*, 2002). It can occur at different temperatures and densities and there should be sufficient energy in the medium to form plasma from the gas. Also, energy in the medium should be continuous to sustain the plasma, as without sufficient energy to form plasma, the particles will turn to neutral gases. The energy used here can be electrical, thermal, or ultraviolet light, and so on (Sanlisoy & Carpinlioglu, 2017). The unconventional method found in plasma gasification system can be used to convert sanitation-biomass such as SS/FS into synthesis gas and an inert vitreous by-product material known as slag, an efficient energy form (Imris *et al.*, 2005; Sanlisoy & Carpinlioglu, 2017).

This technology utilizes the conversion of a variety of fuels such as sewage sludge, faecal sludge, industrial, medical or municipal wastes and low-grade coals into syngas that mainly include CO, H_2, and CO_2. The produced syngas can be used as fuel in combustion systems, for the generation of electricity and for the production of hydrogen as well as slag and ash (Sanlisoy & Carpinlioglu, 2017). A standard plasma gasification technology reactor is operated within the range of 400–850°C and does not use any external heat source, relying on the process itself to sustain the reaction (Littlewood, 1977; Mountouris *et al.*, 2008). Normal gasifiers are really partial combustors, and a substantial portion of carbon is combusted just to support the reaction (Mountouris *et al.*, 2008). Plasma at high temperature breaks down nearly all the materials to their elemental form excluding the radioactive materials (Lemmens *et al.*, 2007; Mountouris *et al.*, 2008), see Figure 6.15. As a result of the high temperature, toxic compounds decompose to harmless chemical elements. In fact, this is the advantage it offers in comparison with conventional methods of gasification.

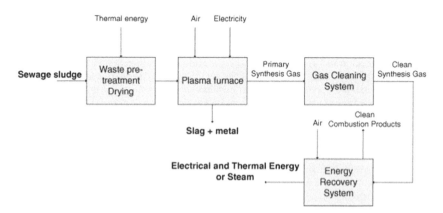

Figure 6.15 Block diagram of plasma gasification process. Mountouris, A., Voutsas, E., and Tassios, D. (2008) Plasma gasification of sewage sludge: process development and energy optimization. Energy Conversion and Management ©, 2008 Elsevier Ltd.

The plasma furnace is the central component of the system where the gasification/vitrification process takes place. Two graphite electrodes, as a part of two transferred arc torches, extend into the plasma furnace. An electric current is passed through the electrodes and the conducting receiver, that is the slag in the furnace bottom. The gas introduced between the electrode and the slag that becomes plasma can be oxygen, helium or other, but the use of air is very common due to its low cost (Mountouris *et al.*, 2008). As the temperature is maintained within the plasma furnace, the organic molecules contained in the sewage sludge begin to break down and react with air to form carbon monoxide, hydrogen and carbon dioxide. Water contained in the sludge feed also dissociates and reacts with other organic molecules. As a result of these reactions, all organic constituents and water are transformed into a synthesis gas containing mostly hydrogen, carbon monoxide and nitrogen (Mountouris *et al.*, 2008). The basic types of plasma reactors are:

- plasma fixed bed reactor;
- plasma moving bed reactor;
- plasma entrained bed reactor or plasma spout bed reactor (Sanlisoy & Carpinlioglu, 2017; Tang *et al.*, 2013).

6.3.3.2 *Biochemical conversion processes*

Biochemical conversion processes allow the decomposition of biomass to available carbohydrates, which could be converted into liquid fuels and biogas, as well as different types of bioproducts, using biological agents such as bacteria, enzymes, and so on (Mahalaxmi & Williford, 2014; Tursi, 2019). Biochemical transformation is mainly the process of enzyme secretion released by microorganisms to control energy production and conversion into solid fuel (Siwal *et al.*, 2021). They can also be referred to as biological pre-treatments aimed to turn biomass into a number of products and intermediates through selection of different microorganisms or enzymes. The process provides a platform to obtain fuels and chemicals such as biogas, hydrogen, ethanol, butanol, acetone and a wide range of organic acids (Chen & Qiu, 2010; Garba, 2020). This process is used when the intention is to make products that could replace petroleum-based products and those obtained from grain. Biomass biochemical conversion technologies are clean, pure and efficient when compared with other conversion technologies (Chen & Wang, 2016; Garba, 2020); classical options are composting and other sanitation-derived nutrients for agriculture, and anaerobic digestion.

6.3.3.2.1 Composting and other sanitation-derived nutrients

Compost is a soil-like substance resulting from controlled aerobic degradation of the organic material in sewage sludge, faecal sludge and/or co-combined with some other biomass conversion composting facility to support agricultural productivity (McConville *et al.*, 2020; Nikiema *et al.*, 2020; Otoo & Drechsel, 2018; Otoo *et al.*, 2018). It is a fertilizing process that can be described as the natural breakdown of biomass through the process of biodegradation with the aid of a microbial population in an aerobic environment to CO_2, H_2O, heat and a further stable output named fertilizer (Siwal *et al.*, 2021). The fertilizer is trouble-less, simple to manage and may be harmlessly employed in farming to improve the soil (Irvine *et al.*, 2010; Kalyani & Pandey, 2014). Compost is a soil conditioner that contains nutrients and organic matter and it contributes to the formation of humus in the soil, thus improving soil structure and water retention capacity. By adding carbon to soil, compost also contributes to soil carbon storage capacity, which supports climate change mitigation (McConville *et al.*, 2020; Nikiema *et al.*, 2020). The composting prices provides significant amounts of the three main

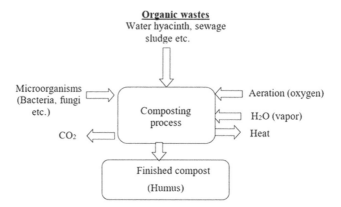

Figure 6.16 Outline of composting process. Singh, J., Kalamdhad, A.S., and Lee, B.K., 2016: published in effects of natural zeolites on bioavailability and leachability of heavy metals in the composting process of biodegradable wastes. Useful Minerals by IntechOpen under CCA 3.0 license, ©, 2016 by the authors.

components of agricultural fertilizer: nitrogen (N), potassium (K), and phosphorus (P; in the form of phosphate), and when sanitation materials are processed along with other organic waste even more N, P, and K can be recovered. Sanitation materials also contain micronutrients such as iron, chlorine, boron, copper and zinc, which are vital for plant and human or animal nutrition, but not generally found in synthetic fertilizer (Andersson *et al.*, 2020; Singh *et al.*, 2016- Figure 6.16).

There has been enhanced consideration provided to heat healing through aerobic composting operations as a process to develop their commercial feasibility (Siwal *et al.*, 2021; Smith & Aber, 2017). The composting progression is driven by the C/N proportion of the biomass, pile wetness, oxygen stages and heat which are strictly observed (Fan *et al.*, 1981; Siwal *et al.*, 2021). Three classes of microorganisms called bacteria, actinomycetes and fungi are extravagant during the fertilizing method (Polprasert & Koottatep, 2017; Siwal *et al.*, 2021). Other composting conversion technologies that provide nutrients for agriculture are:

(I)   *Vermicomposting and vermifiltration* are two low-cost options for human and organic biomass treatment in which earthworms are used as biofilters under aerobic conditions. The end product is worm cast or compost that is a nutrient-rich organic fertilizer and soil conditioner. Also the worms can be harvested from the system, depending on the processes and earthworms can reduce the volume of the faecal sludge by 60 to 90%. The two important parameters are moisture content and the carbon to nitrogen (C:N) ratio. The most commonly used method of vermicomposting is the in-vessel method in which the compost is held in an open vessel. Vermifiltration happens in a watertight container that can receive more liquid inputs such as blackwater or water sludge (McConville *et al.*, 2020);

(II) *Black soldier fly composting and/or black soldier fly larvae (BSFL)* treatment technology is a biological process that relies on the natural growing cycle of the black soldier fly (*Hermetia illucens* (L.), Diptera Stratiomyidae. The BSFL feed only during the larvae stage, then migrate for pupation and do not feed any more, even during the adult stage. The treatment residue, comprised of the larval droppings and undegraded material appears as a compost-like material that can be used as soil conditioner. The larvae can be harvested as a source of protein for animal feed (McConville *et al.*, 2020; Polprasert & Koottatep, 2017);

(III) *Composting toilet conversion technology* is also known as composting based-sanitation systems, dry toilets, biological toilets, biotoilets or waterless toilets (Anand & Apul, 2014; Del Porto & Steinfeld, 1998; Polprasert & Koottatep, 2017). A composting toilet has two primary components, the toilet and the composting tank. The other parts of a composting system often include a fan and vent pipe to remove any odour. The toilet in composting is a waste collector whereby the waste is collected into the composting tank and digested aerobically. Some systems may use earthworms (vermicomposting) as an alternative to aerobic composting (Hill & Baldwin, 2012; Polprasert & Koottatep, 2017; Yadav *et al.*, 2010). Bulking agent or amendments (e.g. sawdust, leaves, and food waste) are often added to help co-manage different types of waste, adjust carbon to nitrogen ratio, and increase porosity of the compost. These toilets are often equipped with mechanical mixers that homogenizes the compost matrix to maintain conditions favourable to aerobic digestion where organic matter is oxidized into ammonia, carbon dioxide, and humus. The end product from these toilets contain stable, high molecular weight dissolved organic matter (Narita *et al.*, 2005; Polprasert & Koottatep, 2017) that can be recycled as soil fertilizers (Anand & Apul, 2014; Polprasert & Koottatep, 2017).

The practice of composting has the ultimate objective of a closed-loop approach that promotes the circular bioeconomy paradigm through the collection, transportation, treatment and recovery of bioresources from sanitation materials using technologies such as urine deviated vacuum toilets, anaerobic digesters, struvite ($Mg(NH_3)PO_4$) precipitation to recover high-value products like water, nutrients, organic matter, energy, and so on; and offers sustainable solutions to sanitation management (Kujawa-Roeleveld & Zeeman, 2006; Lens *et al.*, 2001; Maurer *et al.*, 2012; Polprasert & Koottatep, 2017; Wielemaker *et al.*, 2018; Zeeman, 2012). Also, the organic matter in wastewater and excreta mainly consist of proteins, carbohydrates and fats, that is captured and processed through composting or fermentation process, it could be used as a potent soil conditioner and source of energy when supplemented with food waste and agricultural residues (Lal, 2008; Polprasert & Koottatep, 2017). Increasing soil organic matter (SOM) supports soil functions such as retaining nitrogen and other nutrients, retaining water, protecting roots from diseases and parasites, and making retained nutrients available to the plants (Bot & Benites, 2005; Polprasert & Koottatep, 2017).

Other sanitation-derived biomass nutrients bioproducts include:

(I) *Stored urine from urine-diverting sanitation systems* – primarily made of nitrogen and phosphorus in their mineralized forms and are directly accessible to plants. It can be applied as a liquid fertilizer in agriculture or as an additive to enrich compost (McConville *et al.*, 2020; Polprasert & Koottatep, 2017);

(II) *Concentrated urine* – a nutrient solution obtained by removing water from urine. Water removal is achieved through evaporation, distillation or reverse/forward osmosis of urine. The finished product is between 3–7% of the initial volume. In order to ensure that nitrogen is not lost in the process, nitrification or acidification of the urine is done prior to volume reduction. Depending on the pretreatment process, the majority of the nutrients are retained (McConville *et al.*, 2020; Polprasert & Koottatep, 2017);

(III) *Dry urine* – a nutrient-rich solid fertilizer produced by dehydrating and concentrating human urine in an alkaline substrate (pH $>$, 10). Dry urine's treatment technology (i.e. alkaline urine dehydration) can be implemented using different alkaline substrates, which will determine the composition and physicochemical properties of the dried product. The dried urine captures nearly all of the fertilizing nutrients in urine (McConville *et al.*, 2020);

(IV) *Sanitised blackwater* – refers to blackwater that has been treated in order to reduce microbial risks. Since black water is toilet waste collected with flush water, the water content is rather high since excreta have a low volume of total solid (TS) ($\sim$4%) even without flushwater. Lime treatment can be done by the addition of quick lime (CaO) or slaked lime [Ca(OH)$_2$]. Ammonia sanitization is done by adding urea or aqueous ammonia (NH$_3$) solution to increase the NH$_3$ concentration so that it inactivates pathogens. The addition of urea or ammonia also increases the nitrogen concentration of the blackwater (McConville *et al.*, 2020; Polprasert & Koottatep, 2017);

(V) *Digestate* – material remaining after the anaerobic digestion of any feedstock. The feedstock can consist of foodwaste, agricultural or industrial organic waste, sludge or wastewater fractions. The digestate in this context is the liquid, non-dewatered digestate from wet fermentation of sludge, possibly mixed with other feedstocks. Digestate in this form is a mixture of liquid and particles/solids and can also be called 'slurry'. It is often applied as fertilizer or soil conditioner in agriculture. To be a soil conditioner, it should contain organic material to increase the organic carbon (McConville *et al.*, 2020);

(VI) Struvite – often referred to as magnesium ammonium phosphate hexa-hydrate (MAP), is a phosphate mineral that occurs naturally in sanitation systems. It is a common precipitate in pipes and heat exchangers and can also be purposefully extracted from waste streams through the addition of magnesium to urine. Struvite precipitation can be applied to reduce phosphorus concentrations in effluents while at the same time generating a product that can be applied as a fertilizer or industrial raw material (McConville *et al.*, 2020); and so on.

Consequently, composting could be an attractive solution for treating faecal/sewage sludge and other organic waste when blended together. It provides an opportunity to sanitize the sludge, recover nutrients from sanitation biomass and then return them back to soil especially in areas where soil organic matter is depleted due to poor agricultural practices or a lack of fertilizer use (Cofie *et al.*, 2009; Moya *et al.*, 2019). Also, several container-based sanitation companies successfully produce sanitation-derived fertilizer and sell their full production in the local market (Moya *et al.*, 2019). As a result of the high nutrient value of the compost and/or co-compost as well as other sanitation-derived nutrients, many farmers in Africa, Asia and Latin America are very eager to use it in crop production because it also offers a cheaper alternative source to nutrients and is much more readily available (Cofie & Adamtey, 2009; Nikiema *et al.*, 2013). The World Health Organization (WHO) has developed guidelines to promote the safe use of human excreta in agriculture, realizing its resource value and nutrient content for crop production. This has resulted in recent developments of technology and pre-agricultural use of sanitation materials such as composting of dried, faecal sludge, sewage sludge, co-composting with other organic matter and enriched with inorganic fertilizer (Nikiema *et al.*, 2013, 2014).

6.3.3.2.2 Anaerobic digestion (AD)

Anaerobic digestion (AD) is one of the most sustainable and cost-effective technology for sanitation-derived biomass and other organic waste biomass as well as other form of waste treatment for energy in the form of biofuels. This process does not only minimize the amount of waste, but also transforms such waste into bioenergy. Also, the digestates produced during the process are rich in nutrients and can serve as fertilizers for agricultural

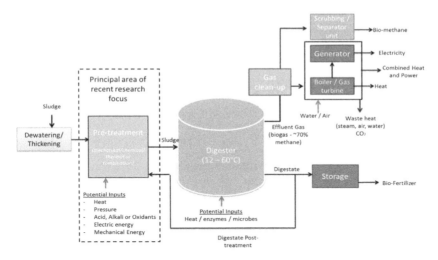

Figure 6.17 Schematic representation of the anaerobic digestion of sludge. Oladejo J, Shi K, Luo X, Yang G, Wu T., 2019: published in A review of sludge-to-energy recovery methods. *Energies* by MDPI under CCA 4.0 license, ©, 2018 by the authors.

purposes (Garba, 2020; Li *et al.*, 2019; Polprasert & Koottatep, 2017). AD is a common profitable process owing to its vast energy improvement into the formation of CH₄ and its inadequate ecological influences; and is additionally capable of deactivating pathogens and stabilizing solid fuel production (Polprasert & Koottatep, 2017; Sawatdeenarunat *et al.*, 2015; Siwal *et al.*, 2021; Zhen *et al.*, 2017). This is a biological process that occurs in an inert environment that converts organic compounds into biogas by using microorganisms. The use of naturally occurring bacteria for biodegradation involves a series of biochemical stages, for example, hydrolysis, acidogenesis (fermentation), acetogenesis and methanogenesis (Lee *et al.*, 2019; Oladejo *et al.*, 2019; Polprasert & Koottatep, 2017; Rulkens, 2008; Siwal *et al.*, 2021 – Figure 6.17).

The metabolic stages is used for mass and volume reduction of the sludge while the organic contents are converted to biogas by the pathogens. The hydrolysis stage involves the conversion of the non-toxic organics into simple sugars, fatty acid and amino acids. Afterwards, the acidogenesis and acetogenesis stages aid the fermentation of the hydrolysis products into acetate, carbon dioxide and hydrogen gas, which are further converted to methane through methanogenesis (Lee *et al.*, 2011; Polprasert & Koottatep, 2017). Each stage of the process affects the performance of the digester. The dewatered sludge can be used directly for energy recovery and aids the conversion of volatile organic solids in the digester. Parameters that affect the yield and energy content of the biogas include nutrient profile of biomass, operating temperature, operating pH, biomass loading rate, as well as hydraulic and solid retention time. The hydraulic and solid retention time must be optimized so that the hydrolysis process (rate-determining step) is not limited by slow loading rate and the methanogenesis process is not bounded by rapid loading rate (Lee *et al.*, 2019; Sialve *et al.*, 2009).

The digester is an air-tight tank where micro-organisms are aided by physical, biological or chemical catalysts (heat, enzymes and/or solvents) for the decomposition of organic matter (Oladejo *et al.*, 2019; Polprasert & Koottatep, 2017). Chemical pre-treatment mainly involves the use of strong reagents such as acid and alkali and oxidants for adjusting the pH of the sludge such that the yield of biogas is maximised by increasing the soluble organic fraction (Devlin *et al.*, 2011; Oladejo *et al.*, 2019; Polprasert & Koottatep, 2017; Valo *et al.*, 2004). Mechanical pre-treatments involve the use of mechanical vibrations such as ultra-sonication to disrupt of the organic solid in the sewage sludge (Devlin *et al.*, 2011; Oladejo *et al.*, 2019; Polprasert & Koottatep, 2017; Valo *et al.*, 2004). Physicochemical pre-treatment such as microwave radiation quickens biological, chemical and physical processes due to heat and/or pressure treatment for improving sludge digestibility and is currently commercially available (Nielsen *et al.*, 2011; Oladejo *et al.*, 2019; Polprasert & Koottatep, 2017).

The effluent gas is biogas which is made up of 60–70% methane, 30–40% carbon dioxide and trace elements of other gases (H_2S) with total calorific value of up to, 28.03–38.92 MJ/Nm^3 (Aryal & Kvist, 2018; Oladejo *et al.*, 2019; Polprasert & Koottatep, 2017; Sivagurunathan *et al.*, 2017; Syed-Hassan *et al.*, 2017). The biogas with its high methane content can be recovered for heat and electricity production using boilers, turbines and generators or alternatively upgraded for use as biomethane. There is also the potential of upgrading biogas to 97.55% methane through the use of water scrubbers. These increases the calorific value of the biogas from, 28.03 to 51.31 MJ/Nm^3 (Aryal & Kvist, 2018; Polprasert & Koottatep, 2017). The remnant, after the digestion process, has high nutritional contents (phosphorus, potassium and nitrogen) that could be used as compost and/or fertilizers for agricultural and soil reclamation purposes (Oladejo *et al.*, 2019; Polprasert & Koottatep, 2017). Biogas energy can offset about 50% of the operational energy used in wastewater treatment facilities. The energy can be used at other sources or sold to the grid. The utilization of this biogas contributes to the reduction of greenhouse gases emissions (Mills *et al.*, 2014; Oladejo *et al.*, 2019; Xu *et al.*, 2014).

6.3.4 Ancillary services

Ancillary services are support activities provided by the enterprises in the SBRVC to ensure operational reliability and maintenance of the value chain. They also create the conditions within which the main activities of the operators are carried out. These services are storage, transportation, and product packaging services.

(I) *Storage facilities*: Storage ensures that the pre-processed biomass is either transported to conversion processes or stored for future demand (Tapia *et al.*, 2019). The SBRCVC enterprises require blended feedstock with other organic waste materials and this becomes a challenge as more types of feedstock are introduced into the systems. Practically, storage facility stocking is required to align with the biomass conversion plan. Therefore, storage facilities are essential to the smooth operations of the SBRVC (Tapia *et al.*, 2019). They include simple stacks in the biomass generation plants or sites and in centralised storage sites. These activities also require energy for preservation of feedstock (Tapia *et al.*, 2019).

(II) *Transportation services:* Transportation infrastructure enable demand satisfaction of one or many resources through its movement from one geographic region to another. In the SBRCVC pre-processed biomass is transported to storage sites and to conversion plants as well as to end-users' market (Tapia *et al.*, 2019). This is done through any means of adequate transportation infrastructure and services available such as road, rail, waterways or any combination of them, but must be based on the type of biomass, path shape and distance of distribution as well as the demand of customers (Tapia *et al.*, 2019).

(III) *Product packaging services:* Product packaging is the act of containing, protecting and presenting the contents through the long chain of production, handling and transportation to their destination as good as they were, at the time of production (Adebisi & Akinruwa, 2019). It is the overall feature that underlines the uniqueness and originality of the product and becomes an ultimate selling proposition, which stimulates the impulse buying behaviour (Adebisi & Akinruwa, 2019; Silayoi & Speece, 2005) Packaging provides physical protection, information transmission, convenience, barrier protection, security and marketing to biomass products after conversion (Pongrácz, 2007). In addition to the above, packaging provides protection and preservation to products while at the same time supporting distribution and sales of the products (Pongrácz, 2007). Indicators of safety and usage instructions that describe how end-users should use the product are provided on packages along with information about the contents, the products, as more or less a message from the manufacturer to the customer (Pongrácz, 2007; Selke, 1990). Being biobased products requires packaging that assure preservation and helps in loading, collection, and product stabilization during transportation and storage. This keeps products from shifting and falling as well as reduces damages, breakage and keeping waste as well as related cost to a minimum (Alexander, 1997; Pongrácz, 2007). Distributing bulk and liquid biobased products is virtually impossible without packaging; and packaging should help make a favourable impression, aid identification, and stimulate purchase as well as provide visual pleasing that attracts attention, which is important in an increasingly competitive environment (Pongrácz, 2007; Young, 2002). Also, a wide range of materials are used for packaging applications, including metal, glass, wood, paper or pulp-based material, plastics, ceramics, or a combination of more than one materials as composites (Pongrácz, 2007).

6.3.5 End-Use markets/direct local End-users

The end-use of biomass-based products include activities related to distribution and final consumers' use. Products should be compatible with existing infrastrcture, standards and distribution channels (Panoutsou *et al.*, 2020). Customer acceptance and successful market uptake will be subjected to their fitness as substitute for existing products and commodities in sectors (e.g., chemical, food, energy etc.) (Panoutsou *et al.*, 2020). Thus, end-use market depends on social feasibility because technology and product for sanitation-derived biomass products should meet social acceptance to ensure that such products find a place in the market (Tyagi & Lo, 2013). The biomass market includes farmers who make use of the biofertilizers and other soil amendment organic matters. Others are the industrial markets of refined biomass finished products such as biofuels, which may include the chemical industry, pharmaceutical industry, fertilizer manufacturers and food producers (Ruamsook & Thomchick, 2014; Tyagi & Lo, 2013), as well as the biobased industrial products which are the end-markets' customers of biobased products, such as building materials, animal protein, biogas, and so on (Ruamsook & Thomchick, 2014).

6.4 SBRVC COMPETITIVE PERFORMANCE PRIORITIES

The five competitive performance priorities of (i) flexibility, (ii) quality, (iii) cost, (iv) innovation, and (v) transparency are factors that the SBRCVC requires to operate well to achieve performance-based competitive advantages in a sustainable and resource efficient manner (Panoutsou *et al.*, 2020):

(I) *Flexibility* – refers to how the SBRCVC operations responds to external factors, and adjust capacity and product design to meet end-users expectations (Henshall, 2018; Panoutsou *et al.*, 2020). Flexibility is essentially to reduce the cost of the impacts of external factors that may negatively affect the value chain. It also ensures that there is all year-round supply of feedstock to meet the requirement of the conversion pathway for quality production and timely delivery of high-value products (Panoutsou *et al.*, 2020).

(II) *Quality* – deals with maintenance and commitment to best standards of systems' and products' performance that ensures the delivery of high-value bioproducts to the consumer. It also focuses on continuous improvement of processes and products performance as well as adherence to quality standards (Díaz-Garrido *et al.*, 2011; Panoutsou *et al.*, 2020). Therefore, quality of feedstocks, practices and end-products are important for successful establishment and uninterrupted operations throughout the value chain (Fritsche & Iriarte, 2014; Panoutsou *et al.*, 2020).

(III) *Cost* – addresses the reduction of production costs of goods sold as well as generating added-value (Panoutsou *et al.*, 2020; Saarijarvi *et al.*, 2012). The competitiveness of the SBRCVC relies on the cost of each stage and biomass conversion accounting for almost half of the total (Fritsche & Iriarte, 2014; Panoutsou *et al.*, 2020). Creating value with innovation and reducing cost along the chain is important for commercial viability of the enterprises and actors within the value chain (Lee, 2002; Panoutsou *et al.*, 2020).

(IV) *Innovation* – addresses new and improved processes and products as well as equipment in each stage of the chain and among enterprises and actors within the value chain (Panoutsou *et al.*, 2020; Torjai *et al.*, 2015). With sanitation and organic-waste biomass being major resource for the sustainability of the value chain, innovation becomes the key in defining which value chain configurations perform best and is resource efficient as well as effective (Fritsche & Iriarte, 2014; Panoutsou *et al.*, 2020); and

(V) **Transparency** – provide current information about the status of the system to avoidance of displacing other activities or product sectors as this is of great importance for the development of the sanitation and organic-waste biomass sector (Panoutsou *et al.*, 2020; Torjai *et al.*, 2015). There is, therefore, the need to provide clarity and awareness of the benefits from the implementation of the value chain as well as create trust among the society's members (Panoutsou *et al.*, 2020).

6.5 CASE STUDIES

6.5.1 Reusing wastewater and sludge in crop production in Braunschweig, Germany

The city of Braunschweig, located in the Federal State of Lower Saxony, Germany has a wastewater reuse scheme managed by the Wastewater Association of Braunschweig since 1954. The members of this association are drawn from the city of Braunschweig, the water association of the neighbouring city of Gifhorn, and 430 owners of land cultivated and/or leased to farmers. The physical and natural conditions in Braunschweig are rather favourable to the reuse of wastewater for agricultural production, since agricultural soils in the region are sandy and poor in nutrients limited water and nutrient retention capacity (Maaß & Grundmann, 2016; Ternes *et al.*, 2007); this means that a continuous

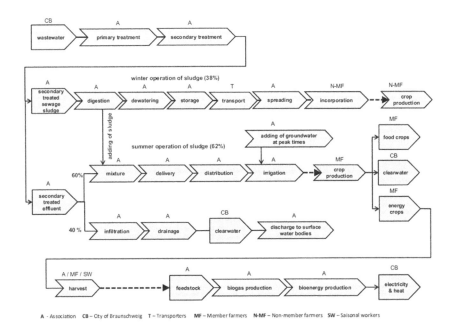

Figure 6.18 Linkages between the value chains of the wastewater reuse scheme in Braunschweig. Reprinted from Maaß, O. and Grundmann, P. (2016) Added-value from linking the value chains of wastewater treatment, crop production and bioenergy production: a case study on reusing wastewater and sludge in crop production in Braunschweig (Germany). *Resources, Conservation and Recycling*, **107**, 195–211, with permission from Elsevier.

additional supply of water and nutrients is essential for crop production. The value chains of wastewater treatment in the city linked crop production and bioenergy production (which are organized by the Braunschweig Wastewater Association), see Figure 6.18. The outputs resulting from the primary and secondary treatment of wastewater, including secondary treated effluent and sewage sludge, are further processed in the value chains of wastewater treatment and reused water as inputs for crop production in the value chains of food and energy. The energy crops are inputs for the anaerobic digestion step in the bioenergy value chain. In this way, the material flows of value chains (including wastewater treatment, crop production and bioenergy production) are linked, based on the agricultural reuse of treated wastewater and sludge. The wastewater of Braunschweig and the surrounding communities is delivered for primary purification to a wastewater treatment plant with a capacity of 60 000 m³d⁻¹ and a population equivalent of 350 000.

The current treatment process includes mechanical treatment, biological phosphate removal, in combination with nitrification and denitrification, and anaerobic stabilization of sludge (Maaß & Grundmann, 2016; Ternes *et al.*, 2007). In addition, a downstream system of irrigation and infiltration fields is used for the final treatment of the secondary effluent. The largest part of the effluent (60%) is used directly for irrigation on croplands of the member farmers (about 2700 ha). The remaining part (40%) is discharged to infiltration fields (about 220 ha) near the treatment plant. These infiltration areas serve as a natural treatment step by using a meandering system and soil passage before the drained water is discharged to the surface water bodies.

The sewage sludge produced is stabilized via anaerobic digestion and utilized in two different value chains. In the winter period, the sewage sludge is dewatered and stored on-site before it is transported in the summer time to croplands (700 ha) of farmers who are not members of the association in the greater Braunschweig area. Subsequently, the sludge is spread by the association's staff and the farmers incorporate the sludge into the croplands. During the vegetation period, the sewage sludge is added to the effluent prior to irrigation. The mix of effluent and sewage sludge is discharged to a gravity sewer system that brings the mixture to the irrigation fields. The mixture is then spread by the association's staff on the croplands of the member farmers. However, due to precautionary hygienic restrictions, farmers are not allowed to produce fruit or vegetables in the association territory for direct consumption (Bezirksregierung Braunschweig, 2001; Maaß & Grundmann, 2016). Therefore, the main crops cultivated in the irrigation area are maize, grain and sugarbeet. The wastewater reuse scheme was enhanced in 2007 by the installation of a biogas plant operated by the association's members.

6.5.2 Commercialization of human excreta derived fertilizer in Haiti and Kenya
6.5.2.1 Sustainable organic integrated livelihoods (SOIL) – Haiti
SOIL started as a not-for-profit organization in Northern Haiti in 2006 with the approach that access to safe sanitation was a human right; their aim was to provide dignified and safe sanitation to deprived communities that were not served by municipal sanitation in two cities of Haiti, Cap Haitian and Port au Prince. SOIL provides household dry toilets on a lease basis with a service fee directly collected from customers. They provide their 6000 customers with urine-diverting toilets at a cost of $3.20 per month, and six collectors collect the faeces weekly (about 350 tonnes per year) and transform it into compost. Faeces are contained in sealed buckets and then collected in carts and transferred to the waste treatment site by truck. Toilet customers add a cover material after each toilet use – sugar cane bagasse or peanut husks, included in the service fee charged by SOIL – to obtain the optimal carbon to nitrogen ratio for composting. The buckets are emptied into large composting bins with walls made up of pallets filled with carbon-rich material such as straw to allow for air to flow through and provide sufficient aeration in the bin. The bin is sealed when full and left untouched for 2–3 months depending on the temperature and pathogen concentration evolution in the compost bins. The compost bin is then emptied, and the material arranged into windrows where further degradation of the material occurs. The piles are turned once a month for about six more months until the compost properties fulfil the quality criteria set internally. Temperature, moisture, pH and *E.coli* concentration are monitored throughout the process to ensure compliance with WHO standards for thermophilic composting and the safety of the final product. SOIL has chosen to sell its fertilizer to NGOs because they can buy it in large quantities and have greater purchasing power than farmers.

6.5.2.2 Sanergy – Kenya
Sanergy is a social enterprise that has provided safe sanitation in urban slums of Nairobi through shared dry toilets since 2011. They use urine-diverting dry toilets as part of a franchise system (called Fresh Life Initiative) which local entrepreneurs join. They invest in a toilet and operate it as a pay-per-use public toilet, at a cost of $0.05 per use. Another model exists where toilets are installed in accommodation compounds and leased to landlords as an extra service provided to tenants. The toilet entrepreneur or tenants (depending on the model) are responsible for the maintenance and cleaning of the toilet, and for sourcing cover material (usually sawdust) and adding it to the faeces. A third model exists for toilets installed in schools, where toilets are sold to head teachers at a subsidized price to ensure adequate sanitation coverage. About 30 000 people are being

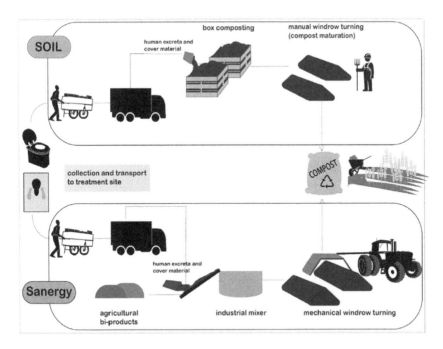

Figure 6.19 Visual summary of SOIL and Sanergy's compost production processes. Moya, B., Sakrabani, R., and Parker, A. (2019) Realizing the circular economy for sanitation: assessing enabling conditions and barriers to the commercialization of human excreta derived fertilizer in Haiti and Kenya. *Sustainability* by MDPI under CCA 4.0 license, ©, 2019 by the authors.

served at the time of reporting, and 60 people are employed in composting and collection. The sanitation and waste management arm of Sanergy are separate: the toilet business, Fresh Life Initiative, being not-for-profit and the waste management arm, Sanergy, is a social enterprise, which collects and treats toilet waste. Similarly to the SOIL system, the waste is collected in sealed containers and transported by truck to the waste treatment facility, about 400 tonnes per month. There the containers are emptied into a mixing tank where additional organic wastes are added, such as agricultural residues. After the mixing phase, the material is laid out in windrows, which are mechanically turned and watered. Process performance is periodically monitored by measuring process parameters (temperature, moisture, pH, CO_2, pathogen concentration, germination tests). The resulting compost is sieved, bagged and sold for agricultural use once the piles meet the WHO guideline standards, which permits their sale to vegetable growers, who receive a good return on investment from the use of fertilizer. The fertilizer production processes are different between the two ventures, as illustrated in Figure 6.19.

6.6 CONCLUSION

Viable business models could emerge from designing SS/FS management systems around resource recovery as this could in turn help ensure sustainable provision of adequate sanitation (Brands, 2014; Murray & Ray, 2010; Puyol *et al.*, 2017; Tyagi & Lo, 2013; Zhang *et al.*, 2018), as sustainable sanitation management involves the recovery and reuse of valuable products and the minimisation of the possible adverse impact of SS/FS on both environmental health and human health (Zhang *et al.*, 2018). Thus, there are two components in SS/FS that are technically and economically feasible to recycle: nutrients (primarily nitrogen and phosphorus) and energy (carbon) (Campbell, 2000). There are several options available for energy recovery from sanitation-waste biomass. The outstanding routes are anaerobic digestion of sludge with biogas recovery; co-digestion, incineration and co-incineration with energy recovery; pyrolysis; gasification; supercritical (wet) oxidation; use in the production of construction materials; production of biofuels (hydrogen, syngas, bio-oil); electricity generation by using specific microbes; and beneficial recovery of heavy metals, nutrients (nitrogen and phosphorus), protein and enzymes (Brands, 2014; Koottatep *et al.*, 2019; Polprasert & Koottatep, 2017; Puyol *et al.*, 2017; Tyagi & Lo, 2013; Zhang *et al.*, 2018). There are global examples of beneficial reuse of resources recovered from SS/FS. The major factors behind this concept are sustainability and environmental concerns, especially those due to resource depletion, soil pollution and global warming. Also, hikes in energy prices, stringent directives for sludge disposal, and increasing protest from environmental authorities and from the public domain (Kalogo & Monteith, 2008; Tyagi & Lo, 2013) contribute effectively. However, the technical feasibility, risks, costs and benefits of the SBRCVC activities and products all need to be assessed to determine viability of each of the value chain pathways and products. The quality of sanitation-biomass-derived products and their market values are important factors with respect to the future feasibility of these processes (Zhang *et al.*, 2018).

6.7 Take action

(I)   Identify sanitation biomass recovery and conversion business enterprises and other actors in your local area and country

(II) Conduct an informal survey to determine the operational and financial viability of such ventures

6.8 Journal entry

(I)   Find out the level of sanitation biomass recovery and conversion activities between non-governmental organizations (NGOs) and private business enterprises in your area and indicate their differences and similarities.

(II) What SDGs does the sanitation biomass recovery and conversion value chain (SBRCVC) have the potential to enhance?

6.9 Reflection

What is your perspective on the sanitation biomass recovery and conversion value chain at global, national and local levels and what can be done to improve and strengthen the value chain?

6.10 Guiding questions

(I)   What is the sanitation biomass recovery and conversion value chain (SBRCVC)?

(II)   What are the main activities on the SBRCVC?

(III) What is the relationship between the sanitation biomass recovery and conversion value chain (SBRCVC) and the circular bioeconomy (CBE)?

(IV) How does the sanitation biomass recovery and conversion value chain (SBRCVC) mitigate climate change hazards?

(V)   What are the major nine ways of recovering energy from sewage/faecal sludge?

(VI) Mention the key elements of the circular bioeconomy.

(VII) What are the conversion pathways that transfer sanitation-biomass to high-value biobased products?

(VIII) What are the major ancillary services in the sanitation biomass recovery and conversion value chain (SBRCVC)?

REFERENCES

Adebisi S. O. and Akinruwa T. E. (2019). Effectiveness of product packaging on customer patronage of Bournvita in Ekiti State. *British Journal of Marketing Studies (BJMS)*, **7**(2), 1–14. https://www.eajournals.org/wp-content/uploads/Effectiveness-of-Product-Packaging-on-Customer-Patronage-of-Bournvita-in-Ekiti-State.pdf (accessed 01 September 2021).

Afolabi O. O. D., Sohail M. and Thomas C. L. P. (2017). Characterization of solid fuel chars recovered from microwave hydrothermal carbonization of human biowaste. *Energy*, **134**, 74–89, https://doi.org/10.1016/j.energy.2017.06.010

Agar D. and Wihersaari M. (2012). Bio-coal, torrefied lignocellulosic resources – key properties for its use in co-firing with fossil coal – their status. *Biomass Bioenergy*, **44**, 107–111, https://doi.org/10.1016/j.biombioe.2012.05.004

Ahmed B., Aboudi K., Tyagi V., Gallego C. J., Güelfo L. A. and García L. I. (2019). Improvement of anaerobic digestion of lignocellulosic biomass by hydrothermal pretreatment. *Applied Sciences*, **9**, 3853, https://doi.org/10.3390/app9183853

Akhtar J. and Amin N. A. S. (2011). A review on process conditions for optimum bio-oil yield in hydrothermal liquefaction of biomass. *Renewable and Sustainable Energy Reviews*, **15**(3), 1615–1624. https://doi.org/10.1016/j.rser.2010.11.054

Alexander J. H. (1997). Solid waste in perspective. In: First Annual Packaging and Government Seminar, The Packaging Institute, Washington DC, pp. 1–235. https://nepis.epa.gov/Exe/ZyPDF.cgi/2000Q54D.PDF?Dockey=2000Q54D.PDF (accessed 01 September 2021).

Amutio M., Lopez G., Aguado R., Bilbao J. and Olazar M. (2012). Biomass oxidative flash pyrolysis: autothermal operation, yields and product properties. *Energy Fuels*, **26**, 1353–1362, https://doi.org/10.1021/ef201662x

Anand C. K. and Apul D. S. (2014). Composting toilets as a sustainable alternative to urban sanitation – a review. *Waste Management*, **34**, 329–343. https://doi.org/10.1016/j.wasman.2013.10.006

Andersson K., Rosemarin A., Lamizana B., Kvarnström E., McConville J., Seidu R., Dickin S. and Trimmer C. (2016). Sanitation, Wastewater Management and Sustainability: From Waste Disposal to Resource Recovery. United Nations Environment Programme and Stockholm Environment Institute, Nairobi and Stockholm.

Andersson K., Rosemarin A., Lamizana B., Kvarnström E., McConville J., Seidu R., Dickin S. and Trimmer C. (2020). Sanitation, Wastewater Management and Sustainability: From Waste Disposal to Resource Recovery. 2nd edn. United Nations Environment Programme and Stockholm Environment Institute, Nairobi, Stockholm. https://www.sei.org/publications/sanitation-wastewater-management-and-sustainability/ (accessed 31 August 2021).

Andriessen N., Schoebitz L., Bassan M., Bollier S. and Strande L. (2017). Market driven approach for Faecal sludge treatment products. In: Paper Presented at the 40th WEDC International Conference, Loughborough, UK.

Andriessen N., Ward B. J. and Strande L. (2019). To char or not to char? Review of technologies to produce solid fuels for resource recovery from Faecal sludge. *Journal of Water, Sanitation and Hygiene for Development*, **9**, 2. https://doi.org/10.2166/washdev.2019.184; https://iwaponline. com/washdev/article/9/2/210/66755/To-char-or-not-to-char-Review-of-technologies-to (accessed 23 August 2021).

Anukam A.I. and Berghel J. (2021). Biomass Pretreatment and Characterization: A Review. *Biotechnological Applications of Biomass*, 1–17. http://dx.doi.org/10.5772/intechopen.93607. https://www.researchgate.net/publication/344452463_Biomass_Pretreatment_and_ Characterization_A_Review (accessed 13 August 2021)

Anukam A., Mamphweli S., Reddy P., Meyer E. and Okoh O. (2016). Pre-processing of sugarcane bagasse for gasification in a downdraft biomass Gasifier system: a comprehensive review. *Renewable and Sustainable Energy Reviews*, **66**, 775–801, https://doi.org/10.1016/j. rser.2016.08.046

Anukam A., Mamphweli S., Okoh O. and Reddy P. (2017). Influence of torrefaction on the conversion efficiency of the gasification process of sugarcane bagasse. *Bioengineering*, **4**, 22–23, https:// doi.org/10.3390/bioengineering4010022

Aryal N. and Kvist T. (2018). Alternative of biogas injection into the Danish gas grid system—a study from demand perspective. *ChemEngineering*, **2**, 43.

Atienza-Martínez M., Mastral J. F., Ábrego J., Ceamanos J. S. and Gea G. (2015). Sewage sludge torrefaction in an auger reactor. *Energy Fuels*, **29**, 160–170, https://doi.org/10.1021/ef501425h

Bajpai P. (2016). Pretreatment of Lignocellulosic Biomass for Biofuel Production. Springer, Berlin/ Heidelberg, Germany, p. 8. ISBN 978-981-10-0686-9.

Bala J. D., Lalung J., Al-Gheethi A. A. S. and Norli I. (2016). A review on biofuel and bioresources for environmental applications. In: Renewable Energy and Sustainable Technologies for Building and Environmental Applications, M. Ahmad, M. Ismail and S. Riffat (eds.), Springer, Cham, pp. 205–225.

Barani V., Hegarty-Craver M., Rosario P., Madhavan P., Perumal P., Sasidaran S., Basil M., Raj A., Berg A. B., Stowell A., Heaton C. and Grego S. (2020).Characterization of fecal sludge as biomass feedstock for thermal treatment in the southern Indian state of Tamil Nadu [version, 2; peer review: 2 approved]. *Gates Open Research*, **2**, 52. https://doi.org/10.12688/ gatesopenres.12870.2

Basu P. (ed.) (2018). Biomass characteristics. In: Biomass Gasification, Pyrolysis and Torrefaction – Practical Design and Theory, 3rd edn. Academic Press, San Diego, USA, pp. 49–91.

Batidzirai B., Mignot A. P. R., Schakel W. B., Junginger H. M. and Faaij A. P. C. (2013). Biomass torrefaction technology: techno-economic status and future prospects. *Energy*, **62**, 196–214, https://doi.org/10.1016/j.energy.2013.09.035

Begum S., Rasul M. G., Akbar D. and Ramzan N. (2013). Performance analysis of an integrated fixed bed gasifier model for different biomass feedstocks. *Energies*, **6**, 6508–6524, https://doi. org/10.3390/en6126508

Berge N. D., Ro K. S., Mao J., Flora J. R. V., Chappell M. A. and Bae S. (2011). Hydrothermal carbonization of municipal waste streams. *Environmental Science & Technology*, **45**(13), 5696–5703, https://doi.org/10.1021/es2004528

Bergman P. C. A. and Kiel J. H. A. (2005). Torrefaction for biomass upgrading. In: Proceedings of the, 14th European Biomass Conference & Exhibition, 17–21 October, 2005, Paris, France.

Bernal M. P., Sommer S. G., Chadwick D., Qing C., Guoxue L. and Michel F. C. (2017). Current approaches and future trends in compost quality criteria for agronomic, environmental, and human health benefits. *Advances in Agronomy*, **144**, 143–233, https://doi.org/10.1016/ bs.agron.2017.03.002

Bezirksregierung Braunschweig (2001). Neufassung der WasserrechtlichenErlaubnis zur Beregnung mit Behandeltem Abwasser aus dem KlärwerkSteinhof für den Abwasserverband Braunschweig. Braunschweig [in German].

BFG (Biobased for growth) (2012). BFG (Biobased for growth) (2012): A PPP on biobased industries. Available from: https://www.biobasedeconomy.nl/wp-content/uploads/2012/07/Bio-Based-Industries-PPP-Vision-doc.pdf (accessed 9 October 2019).

Blair M. J., Gagnon B., Klain A. and Kulišić B. (2021). Contribution of biomass supply chains for bioenergy to sustainable development goals. *Land*, **10**(2), 181. https://doi.org/10.3390/land10020181

Bogaerts A., Neyts E., Gijbels R. and Mullen J. V. D. (2002). Gas discharge plasmas and their applications. *Spectrochim Acta*, **57**, 609–658, https://doi.org/10.1016/S0584-8547(01)00406-2

Bonechi C., Consumi M., Donati A., Leone G., Magnani A., Tamasi G. and Rossi C. (2017). Biomass: an overview. In: Bioenergy Systems for the Future: Prospects for Biofuels and Biohydrogen, F. Dalena, A. Basile and C. Rossi (eds.), Elsevier, London, pp. 3–42.

Bot A. and Benites J. (2005). The Importance of Soil Organic Matter: Key to Drought-Resistant Soil and Sustained Food Production, FAO Soils Bulletin no. 80. UN Food and Agriculture Organization, Rome.

Brands E. (2014). Prospects and challenges for sustainable sanitation in developed nations: a critical review. *Environmental Reviews*, **22**, 346–363, https://doi.org/10.1139/er-2013-0082

Bridgeman T. G., Jones J. M., Williams A. and Waldron D. J. (2010). An investigation of the grindability of two torrefied energy crops. *Fuel*, **89**(12), 3911–3918, https://doi.org/10.1016/j.fuel.2010.06.043

Bridgwater A. V. (2012). A review of fast pyrolysis of biomass and product upgrading. *Biomass & Bioenergy*, **38**, 68–94, https://doi.org/10.1016/j.biombioe.2011.01.048

Bridgwater A., Meier D. and Radlein D. (1999). An overview of fast pyrolysis of biomass. *Organic Geochemistry*, **30**(12), 1479–1493, https://doi.org/10.1016/S0146-6380(99)00120-5

Brownell H. H., Yu E. K. C. and Saddler J. N. (1986). Steam-explosion pretreatment of wood: effect of chip size, acid, moisture content and pressure drop. *Biotechnology and Bioengineering*, **28**(6), 792–801. https://doi.org/10.1002/bit.260280604

Budzinski M., Bezama A. and Thrän D. (2017). Monitoring the progress towards bioeconomy using multi-regional input-output analysis: the example of wood use in Germany. *Journal of Cleaner Production*, **161**, 1–11, https://doi.org/10.1016/j.jclepro.2017.05.090

Camacho P., Ewert W., Kopp J., Panter K., Perez-Elvira S. I. and Piat E. (2008). Combined experiences of thermal hydrolysis and anaerobic digestion – latest thinking on thermal hydrolysis of secondary sludge only for optimum dewatering and digestion. *Proceedings of the Water Environment Federation*, **2008**(15), 1964–1978. https://doi.org/10.2175/193864708788733972

Campbell H. W. (2000). Sludge management—future issues and trends. *Water Science and Technology*, **41**(8), 1–8, https://doi.org/10.2166/wst.2000.0135

Carpenter D., Westover T. L., Czernik S. and Jablonski W. (2014). Biomass feedstocks for renewable fuel production: a review of the impacts of feedstock and pretreatment on the yield and product distribution of fast pyrolysis bio-oils and vapors. *Green Chemistry*, **16**(2), 384–406, https://doi.org/10.1039/C3GC41631C

Case P. A., Wheeler M. C. and Desisto W. J. (2014). Effect of residence time and hot gas filtration on the physical and chemical properties of pyrolysis oil. *Energy and Fuels*, **28**, 3964–3969, https://doi.org/10.1021/ef500850y

CGIAR (2013). Creating wealth from waste: business plan. Available at. https://www.demos.co.uk/files/Creatingwealthfromwaste.pdf

Chan W. P. and Wang J.-Y. (2016). Comprehensive characterisation of sewage sludge for thermochemical conversion processes—based on Singapore survey. *Waste Management*, **54**, 131–142, https://doi.org/10.1016/j.wasman.2016.04.038

Chang V. S. and Holtzapple M. T. (2000). Fundamental factors affecting biomass enzymatic reactivity. In Twenty-First Symposium on Biotechnology for Fuels and Chemicals; Humana Press, Totowa, NJ, USA, **84**, pp. 5–37.

Chen H. and Qiu W. (2010). Key technologies for bioethanol production from lignocellulose. *Biotechnology Advances*, **28**(5), 556–562, https://doi.org/10.1016/j.biotechadv.2010.05.005

Chen H. and Wang L. (2016). Technologies for Biochemical Conversion of Biomass. Academic Press, The Netherlands, p. 201.

Chen G.-B., Chatelier S., Lin H.-T., Wu F.-H. and Lin T.-H. (2018). A study of sewage sludge co-combustion with Australian black coal and shiitake substrate. *Energies*, **11**, 3436, https://doi.org/10.3390/en11123436

Chew J. J. and Doshi V. (2011). Recent advances in biomass pretreatment – torrefaction fundamentals and technology. *Renewable and Sustainable Energy Reviews*, **15**(8), 4212–4222, https://doi.org/10.1016/j.rser.2011.09.017

Chiang K.-Y., Chien K.-L. and Lu C.-H. (2012). Characterization and comparison of biomass produced from various sources: suggestions for selection of pretreatment technologies in biomass-to-energy. *Applied Energy*, **100**, 164–171, https://doi.org/10.1016/j.apenergy.2012.06.063

Chohan N. A., Aruwajoye G. S., Sewsynker-Sukai Y. and Gueguim K. E. B. (2020). Valorisation of potato peel wastes for bioethanol production using simultaneous saccharification and fermentation: process optimization and kinetic assessment. *Renewable Energy*, **146**, 1031–1040, https://doi.org/10.1016/j.renene.2019.07.042

Cieslik B. M., Namiesnik J. and Konieczka P. (2015). Review of sewage sludge management: standards, regulations and analytical methods. *Journal of Cleaner Production*, **90**, 1–15, https://doi.org/10.1016/j.jclepro.2014.11.031

Cofie O. and Adamtey N. (2009). Nutrient recovery from human excreta for urban and peri-urban agriculture. WEDC International Conference, Addis Ababa, Ethiopia. Retrieved from http://publications.iwmi.org/pdf/H042722.pdf (accessed 23 August 2021).

Cofie O., Kone D., Rothenberger S., Moser D. and Zubruegg C. (2009). Co-composting of faecal sludge and organic solid waste for agriculture: process dynamics. *Water Research*, **43**, 4665–4675, https://doi.org/10.1016/j.watres.2009.07.021

Colmenares J. C., Colmenares Q. R. F. and Pieta I. S. (2016). Catalytic dry reforming for biomass-based fuels processing: progress and future perspectives. *Energy Technology*, **4**, 881–890, https://doi.org/10.1002/ente.201600195

Crombie K., Masek O., Sohi S. P., Brownsort P. and Cross A. (2013). The effect of pyrolysis conditions on Biochar stability as determined by three methods. *GCB Bioenergy*, **5**(2), 122–131, https://doi.org/10.1111/gcbb.12030

Darmawan M. A., Putra M. P. I. F. and Wiguna B. (2014). Value chain analysis for green productivity improvement in the natural rubber supply chain: a case study. *Journal of Cleaner Production*, **85**, 201–211. https://doi.org/10.1016/uj.jclepro.2014.01.098

Ddiba D. I. W. (2016). Estimating the potential for resource recovery from Productive Sanitation in Urban Areas. https://doi.org/10.13140/RG.2.1.4987.2881

Ddiba D., Andersson K., Koop S. H. A., Ekener E., Finnveden G. and Dickin S. (2020). Governing the circular economy: assessing the capacity to implement resource-oriented sanitation and waste management systems in lowand middle-income countries. *Earth System Governance*, **4**, 100063. https://doi.org/10.1016/j.esg.2020.100063

Del Porto D. and Steinfeld C. (1998). Operating and Maintaining Your Composting Toilet System. Center for Ecological Pollution Prevention, Concord, MA, USA, 234pp, ISBN: 0966678303.

Demirbas M. F. and Balat M. (2006). Recent advances on the production and utilization trends of bio-fuels: a global perspective. *Energy Conversion and Management*, **47**, 2371–2381, https://doi.org/10.1016/j.enconman.2005.11.014

Deng S., Tan H., Wang X., Yang F., Cao R., Wang Z. and Ruan R. (2017). Investigation on the fast co-pyrolysis of sewage sludge with biomass and the combustion reactivity of residual char. *Bioresource Technology*, **239**, 302–310, https://doi.org/10.1016/j.biortech.2017.04.067

Devlin D. C., Esteves S. R. R., Dinsdale R. M. and Guwy A. J. (2011). The effect of acid pretreatment on the anaerobic digestion and dewatering of waste activated sludge. *Bioresource Technology*, **102**, 4076–4082, https://doi.org/10.1016/j.biortech.2010.12.043

Dhyani V. and Bhaskar T. A. (2018). Comprehensive review on the pyrolysis of lignocellulosic biomass. *Renewable Energy*, **129**, 695–716. https://doi.org/10.1016/j.renene.2017.04.035

Díaz-Garrido E., Martín-Pena M. L. and Sanchez-Lopez J. M. (2011). Competitive priorities in operations: development of an indicator of strategic position. *International Journal of Manufacturing Science and Technology*, **4**(1), 118–125. https://doi.org/10.1016/j.cirpj.2011.02.004

Dibble C. J., Shatova T. A., Jorgenson J. L. and Stickel J. J. (2011). Particle morphology characterization and manipulation in biomass slurries and the effect on rheological properties and enzymatic conversion. *Biotechnology Progress*, **27**(6), 1751–1759. https://doi.org/10.1002/btpr.669

Diener S., Semiyaga S., Niwagaba C. B., Muspratt A. M., Gning J. B., Mbéguéré M., Ennine J. E., Zurbrugga C. and Strande L. (2014). A value proposition: resource recovery from Faecal sludge−can it be the driver for improved sanitation? *Resources, Conservation and Recycling*, **88**, 32–38. https://doi.org/10.1016/j.resconrec.2014.04.005

Dumontet S., Dinel H. and Baloda S. B. (1999). Pathogen reduction in sewage sludge by composting and other biological treatments: *a review. Biological Agriculture & Horticulture*, **16**, 409–430, https://doi.org/10.1080/01448765.1999.9755243

E4tech (UK) Ltd, Consorzio per la Ricerca e la Dimostrazione sulle Energie Rinnovabili (RECORD), Stichting Dienst Landbouwkundig Onderzoek Wageningen University and Research Centre (WUR) (2015). From the Sugar Platform to Biofuels and Biochemical, No ENER/C2/423-2012/SI2.673791, Final report for the European Commission, Directorate-General Energy. http://www.advancefuel.eu/contents/reports/d31-biomass-conversion-technologies-definitions-final.pdf (accessed 23 August 2021).

East African Science and Technology Commission (2019). Bioeconomy Strategy for Eastern Africa. Available online: https://easteco.org/2019/06/04/bioeconomy-strategy-for-eastern-africa/ (accessed on 11 January 2021).

Ellen MacArthur Foundation (2017). Urban Biocycles. Ellen MacArthur Foundation and the World Economic Forum, London, United Kingdom.

Englund M. and Strande L. (eds.) (2019). Faecal Sludge Management: Highlights and Exercises. Swiss Federal Institute of Aquatic Science and Technology, Eawag. ISBN 978-3-906484-70-9.

Eseltine D., Sankar Thanapal S., Annamalai K. and Ranjan D. (2013). Torrefaction of woody biomass (Juniper and Mesquite) using inert and non-inert gases. *Fuel*, **113**, 379–388, https://doi.org/10.1016/j.fuel.2013.04.085

European Commission (2012a). Innovating For Sustainable Growth: A Bioeconomy for Europe, Technical Report, Publications Office of the European Union. Available online: https://op.europa.eu/en/publication-detail/-/publication/1f0d8515-8dc0-4435-ba53-9570e47dbd51 (accessed 6 January 2021).

European Commission (2012b). Bioeconomy Strategy: Innovating for Sustainable Growth: A Bioeconomy for Europe. European Commission. Brussels, Belgium.

European Commission (2018). A sustainable bioeconomy for Europe: strengthening the connection between economy, society and the environment. Updated Bioeconomy Strategy.

Eze L. C. (2004). Alternative Energy Resources: With Comments on Nigeria. Macmillan Nigeria, Lagos, Nigeria. ISBN: 978-018-363-9.

Fakkaew K., Koottatep T. and Polprasert C. (2015a). Effects of hydrolysis and carbonization reactions on hydrochar production. *Bioresource Technology*, **192**(Suppl. C), 328–334. https://doi.org/10.1016/j.biortech.2015.05.091

Fakkaew K., Koottatep T., Pussayanavin T. and Polprasert C. (2015b). Hydrochar production by hydrothermal carbonization of faecal sludge. *Journal of Water Sanitation and Hygiene for Development*, **5**, 439–447. https://doi.org/10.2166/washdev.2015.017

Fan L. T., Lee Y. H. and Beardmore D. R. (1981). The influence of major structural features of cellulose on rate of enzymatic hydrolysis. *Biotechnology and Bioengineering*, **23**, 419–424, https://doi.org/10.1002/bit.260230215

FAO (Food and Agriculture Organization of the United Nations) (2019). World Fertilizer Trends and Outlook to 2022. FAO, Rome.

Feleke S., Cole S. M., Sekabira H., Djouaka R. and Manyong V. (2021). Circular bioeconomy research for development in Sub-Saharan Africa: innovations, gaps, and actions. *Sustainability*, **13**, 1926. https://doi.org/10.3390/su13041926

Fernandes T. V., Klaasse B. G. J., Zeeman G., Sanders J. P. M. and Van Lier J. B. (2009). Effects of thermo-chemical pre-treatment on anaerobic biodegradability and hydrolysis of lignocellulosic biomass. *Bioresource Technology*, **100**, 2575–2579. https://doi.org/10.1016/j.biortech.2008.12.012

Fritsche U. and Iriarte L. (2014). Sustainability criteria and indicators for the bio–based economy in Europe: state of discussion and way forward. *Energies*, **7**(11), 6825–6836. https://doi.org/10.3390/en7116825

Fromm J., Rockel B., Lautner S., Windeisen E. and Wanner G. (2003). Lignin distribution in wood cell walls determined by TEM and backscattered SEM techniques. *Journal of Structural Biology*, **143**(1), 77–84, https://doi.org/10.1016/S1047-8477(03)00119-9

Furimsky E. (2000). Catalytic hydrodeoxygenation. *Applied Catalysis A: General*, **199**, 147–190. https://doi.org/10.1016/S0926-860X(99)00555-4

Fytili D. and Zabaniotou A. (2008). Utilization of sewage sludge in EU application of old and new methods – a review. *Renewable and Sustainable Energy Reviews*, **12**(1), 116–140. https://doi.org/10.1016/j.rser.2006.05.014

Ganter C. J. (2011). Choke point: the collision between water and energy. In: Water Security: The Water–Food–Energy–Climate Nexus, D. Waughray (ed.), Island Press, Washington, DC, pp. 62–63.

Garba A. (2020). Biomass conversion technologies for bioenergy generation: an introduction. In: Biotechnological Applications of Biomass, T. P. Basso (ed.), IntechOpen Limited, London, UK. https://doi.org/10.5772/intechopen.93669; https://www.intechopen.com/books/10127 (accessed 31 August 2021).

Garrido-Baserba M., Molinos-Senante M., Abelleira-Pereira J. M., Fdez-Güelfo L. A., Poch M. and Hernandez-Sancho F. (2015). Selecting sewage sludge treatment alternatives in modern wastewater treatment plants using environmental decision support systems. *Journal of Cleaner Production*, **107**, 410–419, https://doi.org/10.1016/j.jclepro.2014.11.021

Geissdoerfer M., Savaget P., Bocken N. M. and Hultink E. J. (2017). The circular economy–a new sustainability paradigm? *Journal of Cleaner Production*, **143**, 757–768, https://doi.org/10.1016/j.jclepro.2016.12.048

Greene J. P. (2014). Introduction to sustainability. In: Sustainable Plastics, John Wiley, Hoboken, NJ, USA, pp. 1–14. ISBN: 978-1-118-89959-5.

Haas W., Krausmann F., Wiedenhofer D. and Heinz M. (2015). How circular is the global economy? An assessment of material flows, waste production, and recycling in the European Union and the World in, 2005. *Journal of Industrial Ecology*, **19**, 765–777. https://doi.org/10.1111/jiec.12244

Haller A. (1997). Wertschöpfungsrechnung. Ein Instrument zur Steigerung DerAussagefähigkeit von Unternehmensabschlüssen im Internationalen Kontext. Schäffer-Poeschel Verlag, Stuttgart [in German].

Han X., Niu M., Jiang X. and Liu J. (2012). Combustion characteristics of sewage sludge in a fluidized bed. *Industrial & Engineering Chemistry Research*, **51**, 10565–10570, https://doi.org/10.1021/ie3014988

Hellegers P., Zilberman D., Steduto P. and McCornick P. (2008). Interactions betweenwater, energy, food and environment: evolving perspectives and policy issues. *Water Policy*, **10**(S1), 1–10, https://doi.org/10.2166/wp.2008.048

Hennig C., Brosowski A. and Majer S. (2016). Sustainable feedstock potential–a limitation for the bio-based economy? *Journal of Cleaner Production*, **123**, 200–202, https://doi.org/10.1016/j.jclepro.2015.06.130

Henshall A. (2018). Process Flexibility: 4 Key Approaches and How to Use Them. https://www.process.st/process-flexibility/ (accessed 05 September 2021).

Hill G. B. and Baldwin S. A. (2012). Vermicomposting toilets, an alternative to latrine style microbial composting toilets, prove far superior in mass reduction, pathogen destruction, compost quality, and operational cost. *Waste Management*, **32**(10), 1811–1820, https://doi.org/10.1016/j.wasman.2012.04.023

Hodásová L., Jablonský M., Škulcová A. and Ház A. (2015). Lignin, potential products and their market value. *Wood Research*, **60**(6), 973–986.

Huber G. W., Iborra S. and Corma A. (2006). Synthesis of transportation fuels from biomass: chemistry, catalysts, and engineering. *Chemical Reviews*, **106**, 4044–4098. https://doi.org/10.1021/cr068360d

Hussey K. and Pittock J. (2012). The energy–water nexus: managing the links betweenenergy and water for a sustainable future. *Ecology and Society*, **17**(1), 31, https://doi.org/10.5751/ES-04641-170131

Imris I., Klenovcanova A. and Molcan P. (2005). Energy recovery from waste by plasma gasification. *Arch Thermodyn*, **26**, 3–16.

IREA (International Renewable Energy Agency) (2014). Global Bioenergy Supply and Demand Projections; A Working Paper for REmap, 2030. International Renewable Energy Agency, Abu Dhabi, UAE.

Irvine G., Lamont E. R. and Antizar-Ladislao B. (2010). Energy from waste: reuse of compost heat as a source of renewable energy. *International Journal of Chemical Engineering*, 2010, 10 pp, ID 627930. https://doi.org/10.1155/2010/627930; https://downloads.hindawi.com/journals/ijce/2010/627930.pdf (accessed 31 August 2021).

Jarvis S. M. and Samsatli S. (2018). Technologies and infrastructures underpinning future CO_2 value chains: a comprehensive review and comparative analysis. *Renewable and Sustainable Energy Reviews*, **85**, 46–68. https://doi.org/10.1016/j.rser.2018.01.007

Jazrawi C., Biller P., Ross A. B., Montoya A., Maschmeyer T. and Haynes B. S. (2013). Pilot plant testing of continuous hydrothermal liquefaction of microalgae. *Algal Research*, **2**(3), 268–277. https://doi.org/10.1016/j.algal.2013.04.006

Jenkins B. M., Bakker R. R. and Wei J. B. (1996). On the properties of washed straw. *Biomass and Bioenergy*, **10**(4), 177–200. https://doi.org/10.1016/0961-9534(95)00058-5

Jenkins B., Baxter L., Miles T. Jr. and Miles T. (1998). Combustion properties of biomass. *Fuel Processing Technology*, **54**(1), 17–46, https://doi.org/10.1016/S0378-3820(97)00059-3

Jin J., Wang M., Cao Y., Wu S., Liang P., Li Y., Zhang J., Zhang J., Wong M. H., Shan S. and Christie P. (2017). Cumulative effects of bamboo sawdust addition on pyrolysis of sewage sludge: biochar properties and environmental risk from metals. *Bioresource Technology*, **228**, 218–226, https://doi.org/10.1016/j.biortech.2016.12.103

Jindal M. K. and Jha M. K. (2016). Hydrothermal liquefaction of wood: a critical review. *Reviews in Chemical Engineering*, **32**(4), 459–488, https://doi.org/10.1515/revce-2015-0055

Jodar-Abellan A., López-Ortiz M. I. and Melgarejo-Moreno J. (2019). Wastewater treatment and water reuse in Spain. *Current Situation and Perspectives. Water*, **11**(8), 1551.

Kadam J., Dhawal P., Barve S. and Kakodkar S. (2020). Green synthesis of silver nanoparticles using cauliflower waste and their multifaceted applications in photocatalytic degradation of methylene blue dye and Hg, 2+ biosensing. *SN Applied Sciences*, **2**(4), 1–16, https://doi.org/10.1007/s42452-020-2543-4

Kalogo Y. and Monteith H. (2008). State of Science Report: Energy and Resource Recovery From Sludge. Global Water Research Coalition, Water Environment Research Foundation, Alexandria, VA, p. 238. https://edepot.wur.nl/118980 (accessed 02 September 2021).

Kalyani K. A. and Pandey K. K. (2014). Waste to energy status in India: a short review. *Renewable and Sustainable Energy Reviews*, **31**, 113–120, https://doi.org/10.1016/j.rser.2013.11.020

Kasinath A., Fudala-Ksiazek S., Szopinska M., Bylinski H., Artichowicz W., Remiszewska-Skwarek A. and Luczkiewicz A. (2021). Biomass in biogas production: pretreatment and codigestion. *Renewable and Sustainable Energy Reviews*, **150**, 111509. https://doi.org/10.1016/j.rser.2021.111509

Kaushika N. D., Reddy K. S. and Kaushik K. (2016). Biomass energy and power systems. In: Sustainable Energy and the Environment: A Clean Technology Approach, N. D. Kaushika, K. S. Reddy and K. Kaushik (eds.), Springer, Cham, pp. 123–137.

Khan S., Wang N., Reid B. J., Freddo A. and Cai C. (2013). Reduced bioaccumulation of PAHs by *Lactuca satuva* L. grown in contaminated soil amended with sewage sludge and sewage sludge derived biochar. *Environmental Pollution*, **175**, 64–68, https://doi.org/10.1016/j.envpol.2012.12.014

Khanal S. K., Grewell D., Sung S. and Van Leeuwen J. (2007). Ultrasound applications in wastewater sludge pretreatment: a review. *Critical Reviews in Environmental Science and Technology*, **37**, 277–313. https://doi.org/10.1080/10643380600860249

Kokalj F., Arbiter B. and Samec N. (2017). Sewage sludge gasification as an alternative energy storage model. *Energy Conversion and Management*, **149**, 738–747, https://doi.org/10.1016/j.enconman.2017.02.076

Koottatep T., Fakkaew K., Tajai N., Pradeep S. V. and Polprasert C. (2016). Sludge stabilization and energy recovery by hydrothermal carbonization process. *Renewable Energy*, **99**, 978–985. https://doi.org/10.1016/j.renene.2016.07.068

Koottatep T., Cookey P. E. and Polprasert C. (2019). Regenerative Sanitation: A New Paradigm For Sanitation 4.0. IWA Publishing, London. https://doi.org/10.2166/9781780409689_0209

Kruse A., Funke A. and Titirici M.-M. (2013). Hydrothermal conversion of biomass to fuels and energetic materials. *Current Opinion in Chemical Biology*, **17**(3), 515–521, https://doi.org/10.1016/j.cbpa.2013.05.004

Kujawa-Roeleveld K. and Zeeman G. (2006). Anaerobic treatment in decentralised andsource-separation-based sanitation concepts. *Reviews in Environmental Science and Bio/Technology*, **5**, 115–139, https://doi.org/10.1007/s11157-005-5789-9

Lal R. (2008). Soils and sustainable agriculture, a review. *Agronomy for Sustainable Development*, **28**(1), 57–64. https://doi.org/10.1051/agro:2007025

Lam H. L., Ng W. P. Q., Ng R. T. L., Ng E. H., Aziz M. K. A. and Ng D. K. S. (2013). Green strategy for sustainable waste-to-energy supply chain. *Energy*, **57**, 4–16. https://doi.org/10.1016/j.energy.2013.01.032

Lange J. P. (2007). Lignocellulose conversion: an introduction to chemistry, process and economics. *Biofuels Bioprod. Biorefining*, **1**(1), 39–48, https://doi.org/10.1002/bbb.7

Lebaka V. (2013). Potential bioresources as future sources of biofuels production: an overview. In: Biofuel Technology, V. Gupta and M. G. Tuohy (eds.), Springer, Berlin, pp. 223–258.

Lee H. L. (2002). Aligning supply chain strategies with product uncertainties. *California Management Review*, **44**(3), 105–119. https://doi.org/10.2307/41166135

Lee I.-S., Parameswaran P. and Rittmann B. E. (2011). Effects of solids retention time on methanogenesis in anaerobic digestion of thickened mixed sludge. *Bioresource Technology*, **102**, 10266–10272, https://doi.org/10.1016/j.biortech.2011.08.079

Lee S. Y., Sankaran R., Chew K. W., Tan C. H., Krishnamoorthy R., Chu D.-T. and Show P.-L. (2019). Waste to bioenergy: a review on the recent conversion technologies. *BMC Energy*, **1**, 4. https://doi.org/10.1186/s42500-019-0004-7

Lehmann J., Rillig M. C., Thies J., Masiello C. A., Hockaday W. C. and Crowley D. (2011). Biochar effects on soil biota – a review. *Soil Biology and Biochemistry*, **43**, 1812–1836, https://doi.org/10.1016/j.soilbio.2011.04.022

Lemmens B., Elslander H., Vanderreydt I., Peys K., Diels L., Oosterlinck M. and Joos M. (2007). Assessment of plasma gasification of high calorific waste streams. *Waste Management*, **27**, 1562–1569, https://doi.org/10.1016/j.wasman.2006.07.027

Lens P., Zeeman G. and Lettinga G. (2001). Decentralised Sanitation and Reuse. IWA Publishing, London.

Li D.-C. and Jiang H. (2017). The thermochemical conversion of non-lignocellulosic biomass to form biochar: a review on characterizations and mechanism elucidation. *Bioresource Technology*, **246**, 57–68. https://doi.org/10.1016/j.biortech.2017.07.029

Li G., Wang Z. and Zhao R. (2009). Research progress of oil making from sewage sludge by direct thermochemistry liquefaction technology. *Journal of Tianjin University of Science & Technology*, **24**, 74–78.

Li N., Tompsett G. A. and Huber G. W. (2010). Renewable high-octane gasoline by aqueous-phase hydrodeoxygenation of C5 and C6 carbohydrates over Pt/zirconium phosphate catalysts. *ChemSusChem*, **3**, 1154–1157. https://doi.org/10.1002/cssc.201000140

Li Y., Chen Y. and Wu J. (2019). Enhancement of methane production in anaerobic digestion process: a review. *Applied Energy*, **240**, 120–137, https://doi.org/10.1016/j.apenergy.2019.01.243

Littlewood K. (1977). Gasification: theory and application. *Progress in Energy and Combustion Science*, **3**, 35–71, https://doi.org/10.1016/0360-1285(77)90008-9

Liu W.-J., Jiang H. and Yu H.-Q. (2015a). Development of biochar-based functional materials: toward a sustainable platform carbon material. *Chemical Reviews*, **115**(22), 12251–12285, https://doi.org/10.1021/acs.chemrev.5b00195

Liu W.-J., Jiang H. and Yu H.-Q. (2015b). Thermochemical conversion of lignin to functional materials: a review and future directions. *Green Chemistry*, **17**, 4888–4907, https://doi.org/10.1039/C5GC01054C

Lokesh K., Ladu L. and Summerton L. (2018). Bridging the gaps for a 'Circular' bioeconomy: selection criteria, bio-based value chain and stakeholder mapping. *Sustainability*, **10**, 1695. https://doi.org/10.3390/su10061695

Lusch R. F., Vargo S. L. and Tanniru M. (2010). Service, value networks and learning. *Journal of the Academy of Marketing Science*, **38**, 19–31, https://doi.org/10.1007/s11747-008-0131-z

Ma W., Du G., Li J., Fang Y., Hou L., Chen G. and Ma D. (2017). Supercritical water pyrolysis of sewage sludge. *Waste Management*, **59**, 371–378, https://doi.org/10.1016/j.wasman.2016.10.053

Maaß O. and Grundmann P. (2016). Added-value from linking the value chains of wastewater treatment, crop production and bioenergy production: a case study on reusing wastewater and sludge in crop production in Braunschweig (Germany). *Resources, Conservation and Recycling*, **107**, 195–211, https://doi.org/10.1016/j.resconrec.2016.01.002

Mafakheri F. and Nasiri F. (2014). Modeling of biomass-to-energysupply chain operations: applications, challenges andresearch directions. *Energy Policy*, **67**, 116–126. https://doi.org/10.1016/j.enpol.2013.11.071

Magdziarz A., Dalai A. K. and Kozinski J. A. (2016). Chemical composition, character and reactivity of renewable fuel ashes. *Fuel*, **176**, 135–145, https://doi.org/10.1016/j.fuel.2016.02.069

Mahalaxmi S. and Williford C. (2014). Biochemical conversion of biomass to fuels. In: Handbook of Climate Change Mitigation and Adaptation, W. Chen, T. Suzuki and M. Lackner (eds.), Springer, New York, pp. 1–28.

Mallory A., Akrofi D., Dizon J., Mohanty S., Parker A., Vicario D. R., Prasad S., Welivita I., Brewer T., Mekala S., Bundhoo D., Lynch K., Mishra P., Willcock S. and Hutchings P. (2020). Evaluating the circular economy for sanitation: findings from a multi-case approach. *Science of the Total Environment*, **744**, 140871. https://doi.org/ 10.1016/j.scitotenv.2020.140871

Mamvura T. A. and Danha G. (2020). Biomass torrefaction as an emerging technology to aid in energy production. *Heliyon*, **6**, e03531. https://doi.org/10.1016/j.heliyon.2020.e03531

Marshall A. J. and Eng P. (2013). Commercial Application of Pyrolysis Technology in Agriculture. Ontario Federation of Agriculture, Ontario AgriCentre, Guelph, ON Canada.

Mašek O., Ronsse F. and Dickinson D. (2016). Biochar production and feedstock. In: Biochar in European Soils and Agriculture: Science and Practice, S. Shackley, G. Ruysschaert, K. Zwart and B. Glaser (eds.), Routledge, Oxon, pp. 17–40.

Maurer M., Bufardi A., Tilley E., Zurbrügg C. and Truffer B. (2012). A compatibility-based procedure designed to generate potential sanitation system alternatives. *Journal of Environmental Management*, **104**, 51–61, https://doi.org/10.1016/j.jenvman.2012.03.023

McConville J., Niwagaba C., Nordin A., Ahlström M., Namboozo V. and Kiffe M. (2020). Guide to Sanitation Resource-Recovery Products & Technologies: A Supplement to the Compendium of Sanitation Systems and Technologies, 1st edn. Swedish University of Agricultural Sciences (SLU), Department of Energy and Technology, Uppsala, Sweden. https://news.mak.ac.ug/wp-content/uploads/2021/01/McConvile-et-al.-2020-Guide-to-Sanitation-Resource-Recovery-Products-Technologies_1st_ed.pdf (accessed 11 August 2021).

McKendry P. (2002). Energy production from biomass (part, 2): conversion technologies. *Bioresource Technology*, **83**(1), 47–54, https://doi.org/10.1016/S0960-8524(01)00119-5

Miller A., Espanani R., Junker A., Hendry D., Wilkinson N., Bollinger D., Abelleira-Pereira J. M., Deshusses M. A., Inniss E. and Jacoby W. (2015). Supercritical water oxidation of a model fecal sludge without the use of a co-fuel. *Chemosphere*, **141**, 189–196, https://doi.org/10.1016/j.chemosphere.2015.06.076

Mills N., Pearce P., Farrow J., Thorpe R. B. and Kirkby N. F. (2014). Environmental & economic life cycle assessment of current & future sewage sludge to energy technologies. *Waste Management*, **34**, 185–195, https://doi.org/10.1016/j.wasman.2013.08.024

Mimmo T., Panzacchi P., Baratieri M., Davies C. A. and Tonon G. (2014). Effect of pyrolysis temperature on miscanthus (Miscanthus × giganteus) biochar physical, chemical and functional properties. *Biomass and Bioenergy*, **62**, 149–157, https://doi.org/10.1016/j.biombioe.2014.01.004

Möller K. (2006). Wertschöpfung in Netzwerken. Vahlen, München [in German].

Morales M., Pielhop T., Saliba P., Hungerbuehler K., von Rohr P. R. and Papadokonstantakis S. (2017). Sustainability assessment of glucose production technologies from highly recalcitrant softwood including scavengers. *Biofuels Bioproducts & Biorefining*, **11**, 441–453, https://doi.org/10.1002/bbb.1756

Morone P. and Imbert E. (2020). Food waste and social acceptance of a circular bioeconomy: the role of stakeholders. *Current Opinion in Green and Sustainable Chemistry*, **23**, 55–60. https://doi.org/10.1016/j.cogsc.2020.02.006

Mountouris A., Voutsas E. and Tassios D. (2008). Plasma gasification of sewage sludge: process development and energy optimization. *Energy Conversion and Management*, **49**(8), 2264–2271. https://doi.org/10.1016/j.enconman.2008.01.025

Moya B., Sakrabani R. and Parker A. (2019). Realizing the circular economy for sanitation: assessing enabling conditions and barriers to the commercialization of human excreta derived fertilizer in Haiti and Kenya. *Sustainability*, **11**, 3154. https://doi.org/10.3390/su11113154

Mulchandani A. and Westerhoff P. (2016). Recovery opportunities for metals and energy from sewage sludges. *Bioresource Technology*, **215**, 215–226, https://doi.org/10.1016/j.biortech.2016.03.075

Murray A. and Ray I. (2010). Commentary: back-end users: the unrecognized stakeholders in demand-driven sanitation. *Journal of Planning Education and Research*, **30**(1), 94–102, https://doi.org/10.1177/0739456X10369800

Narita H., Zavala M. A. L., Iwai K., Ito R. and Funamizu N. (2005). Transformation and characterization of dissolved organic matter during the thermophilic aerobic biodegradation of Faeces. *Water Research*, **39**(19), 4693–4704, https://doi.org/10.1016/j.watres.2005.09.022

Neumann P., Pesante S., Venegas M. and Vidal G. (2016). Developments in pre-treatment methods to improve anaerobic digestion of sewage sludge. *Reviews in Environmental Science and Bio/Technology*, **15**, 173–121. https://doi.org/10.1007/s11157-016-9396-8

Neyens E. and Baeyens J. A. (2003). Review of thermal sludge pre-treatment processes to improve dewaterability. *Journal of Hazardous Materials*, **98**, 51–67. https://doi.org/10.1016/S0304-3894(02)00320-5

Nielsen H. B., Thygesen A., Thomsen A. B. and Schmidt J. E. (2011). Anaerobic digestion of waste activated sludge—comparison of thermal pretreatments with thermal inter-stage treatments. *Journal of Chemical Technology and Biotechnology*, **86**, 238–245, https://doi.org/10.1002/jctb.2509

Nikiema J., Cofie O., Impraim R. and Adamtey N. (2013). Processing of fecal sludge to fertilizer pellets using a low-cost technology in Ghana. *Environment and Pollution*, 2(4), 47pp, https://doi.org/10.5539/ep.v2n4p70

Nikiema J., Cofie O. and Impraim R. (2014). Technological Options for Safe Resource Recovery From Fecal Sludge. Resource Recovery & Reuse Series, 2. International Water Management Institute (IWMI). CGIAR Research Program on Water, Land and Ecosystems (WLE), Colombo, Sri Lanka, 47p.

Nikiema J., Tanoh-Nguessan R., Abiola F. and Cofie O. O. (2020). Introducing Co-Composting to Fecal Sludge Treatment Plants in Benin and Burkina Faso: A Logistical and Financial Assessment. International Water Management Institute (IWMI). CGIAR Research Program on Water, Land and Ecosystems (WLE), Colombo, Sri Lanka, 50p. (Resource Recovery and Reuse Series, 17). https://doi.org/10.5337/2020.206

Nuagah M. B., Boakye P., Oduro-Kwarteng S. and Sokama-Neuyam Y. A. (2020). Valorization of faecal and sewage sludge via pyrolysis for application as crop organic fertilizer. *Journal of Analytical and Applied Pyrolysis*, **151**, 104903. https://doi.org/10.1016/j.jaap.2020.104903

Nunes L. J. R. (2020). A case study about biomass torrefaction on an industrial scale: solutions to problems related to self-heating, difficulties in pelletizing, and excessive wear of production equipment. *Applied Sciences*, **10**(7), 2546. https://doi.org/10.3390/app10072546

Nussbaumer T. (2003). Combustion and co-combustion of biomass: fundamentals, technologies, and primary measure for emission reduction. *Energy Fuels*, **17**(6), 1510–1521, https://doi.org/10.1021/ef030031q

Ogada T. and Werther J. (1996). Combustion characteristics of wet sludge in a fluidized bed: release and combustion of the volatiles. *Fuel*, **75**, 617–626, https://doi.org/10.1016/0016-2361(95)00280-4

Oladejo J., Shi K., Luo X., Yang G. and Wu T. (2019). A review of sludge-to-energy recovery methods. *Energies*, **12**(1), 60. https://doi.org/10.3390/en12010060

Onabanjo T., Patchigolla K., Wagland S., Fidalgo B., Kolios A., McAdam E., Parker A., Williams L., Tyrrel S. and Cartmell E. (2016). Energy recovery from human Faeces via gasification: a thermodynamic equilibrium modelling approach. *Energy Conversion and Management*, **118**, 364–376, https://doi.org/10.1016/j.enconman.2016.04.005

Otoo M. and Drechsel P. (eds.) (2018). Resource Recovery From Waste: Business Models for Energy, Nutrient and Water Reuse in Low- and Middle-Income Countries. Routledge – Earthscan, Oxon, UK, 816p.

Otoo M., Gebrezgabher S., Danso G., Amewu S. and Amirova I. (2018). Market Adoption and Diffusion of Fecal Sludge-Based Fertilizer in Developing Countries: Cross-Country Analyses. International Water Management Institute (IWMI). CGIAR Research Program on Water, Land and Ecosystems (WLE), Colombo, Sri Lanka, 68p. (Resource Recovery and Reuse Series, 12). https://doi.org/10.5337/2018.228

Pagotto M. and Halog A. (2016). Towards a circular economy in Australian Agri-food industry: an application of input-output oriented approaches for analyzing resource efficiency and competitiveness potential. *Journal of Industrial Ecology*, **20**, 1176–1186, https://doi.org/10.1111/jiec.12373

Pambudi N. A., Laukkanen T., Syamsiro M. and Gandidi I. M. (2017). Simulation of Jatropha curcas shell in gasifier for synthesis gas and hydrogen production. *Journal of the Energy Institute*, **90**, 672–679. https://doi.org/10.1016/j.joei.2016.07.010

Pan S.-Y., Du M. A., Huang I.-T., Liu I.-H., Chang E.-E. and Chiang P.-C. (2015). Strategies on implementation of waste-to-energy (WTE) supply chain for circular economy system: a review. *Journal of Cleaner Production*, **108**, 409–421, https://doi.org/10.1016/j.jclepro.2015.06.124

Panoutsou C., Singh A., Christensen T. and Pelkmans L. (2020). Competitive priorities to address optimisation in biomass value chains: the case of biomass CHP. *Global Transitions*, **2**, 60–75. https://doi.org/10.1016/j.glt.2020.04.001

Papadokonstantakis S. and Johnsson F. (2020). Biomass Conversion Technologies – Definitions D3.1 Report on Definition of Parameters for Defining Biomass Conversion Technologies. The European Union's Horizon, 2020 research and innovation programme under grant agreement N.°764799. http://www.advancefuel.eu/contents/reports/d31-biomass-conversion-technologies-definitions-final.pdf (accessed 22 August 2021).

Patinvoh R. J., Osadolor O. A., Chandolias K., Sárvári Horváth I. and Taherzadeh M. J. (2017). Innovative pretreatment strategies for biogas production. *Bioresource Technology*, **224**, 13–24. https://doi.org/10.1016/j.biortech.2016.11.083

Peppard J. and Rylander A. (2006). From value chain to value network: insights for mobile operators. *European Management Journal*, **24**(2–3), 128–141, https://doi.org/10.1016/j.emj.2006.03.003

Pilli S., Bhunia P., Yan S., LeBlanc R. J., Tyagi R. D. and Surampalli R. Y. (2011). Ultrasonic pretreatment of sludge: a review. *Ultrason Sonochem*, **18**, 1–18. https://doi.org/10.1016/j.ultsonch.2010.02.014

Pokorna E., Postelmans N., Jenicek P., Schreurs S., Carleer R. and Yperman J. (2009). Study of bio-oils and solids from flash pyrolysis of sewage sludges. *Fuel*, **88**(8), 1344–1350, https://doi.org/10.1016/j.fuel.2009.02.020

Polprasert C. and Koottatep T. (2017). Organic Waste Recycling Technology, Management and Sustainability. Published by IWA Publishing, London, UK. ISBN: 9781780408200 (Paperback).

Pongrácz E. (2007). The environmental impacts of packaging. In: Environmentally Conscious Materials and Chemicals Processing, M. Kutz (ed.), Wiley, New York, USA, Vol. **3**, pp. 237–278.

Popa V. I. (2018). Biomass for fuels and biomaterials. In: Biomass as Renewable Raw Material to Obtain Bioproducts of High-Tech Value, V. Popa and I. Volf (eds.) Elsevier, Amsterdam, The Netherlands, pp. 1–37.

Poudel J., Ohm T. I., Lee S.-H. and Oh S. C. (2015). A study on torrefaction of sewage sludge to enhance solid fuel qualities. *Waste Management*, **40**, 112–118, https://doi.org/10.1016/j.wasman.2015.03.012

Prins M. J., Ptasinski K. J. and Janssen F. J. J. G. (2006). More efficient biomass gasification via torrefaction. *Energy*, **31**(15), 3458–3470, https://doi.org/10.1016/j.energy.2006.03.008

Pulka J., Manczarski P., Koziel J. A. and Białowiec A. (2019). Torrefaction of sewage sludge: kinetics and fuel properties of biochars. *Energies*, **12**(3), 565. https://doi.org/10.3390/en12030565

Pulka J., Manczarski P., Stępień P., Styczyńska M., Koziel J. A. and Białowiec A. (2020). Waste-to-carbon: is the torrefied sewage sludge with high ash content a better fuel or fertilizer? *Materials*, **13**(4), 954. https://doi.org/10.3390/ma13040954

Puyol D., Batstone D. J., Hülsen T., Astals S., Peces M. and Krömer J. O. (2017). Resource recovery from wastewater by biological technologies: opportunities, challenges, and prospects. *Frontiers in Microbiology*, **7**, 2106. https://doi.org/10.3389/fmicb.2016.02106

Rana A. K., Frollini E. and Thakur V. K. (2021). Cellulose nanocrystals: pretreatments, preparation strategies, and surface functionalization. *International Journal of Biological Macromolecules*, **182**, 1554–1581. https://doi.org/10.1016/j.ijbiomac.2021.05.119

Raven P. H., Evert R. F. and Eichhorn S. E. (1992). Biology of Plants, 6th edn. Worth Publishers, New York, p. 115.

Remington C., Jean L., Kramer S., Boys J., and Dorea, C. (2018). Process cost analysis for the optimization of a container-based sanitation service in Haiti. Transformation Towards Sustainable and Resilient WASH Services: 41st WEDC International Conference, pp. 1–6 Available at. http://www.oursoil.org/wp-content/uploads/2018/07/Process-cost-analysis-for-the-optimization-of-a-container-based-sanitation-service-in-Haiti.pdf (accessed 12 September 2021)

Reymond P. (2014). Planning integrated faecal sludge management systems. In: Faecal Sludge Management: Systems Approach for Implementation and Operation, L. Strande, M. Ronteltap and D. Brdjanovic (eds.), pp. 363–387, IWA Publishing, London, UK.

Román S., Libra J., Berge N., Sabio E., Ro K., Li L., Ledesma B., Álvarez A. and Bae S. (2018). Hydrothermal carbonization: modeling, final properties design and applications: a review. *Energies*, **11**(1), 216. https://doi.org/10.3390/en11010216

Rong H., Wang T., Zhou M., Wang H., Hou H. and Xue Y. (2017). Combustion characteristics and slagging during co-combustion of rice husk and sewage sludge blends. *Energies*, **10**, 438, https://doi.org/10.3390/en10040438

Roth J. R. (1994). Industrial Plasma Engineering. Institute of Physics, USA.

Ruamsook K. and Thomchick E. (2014). Market Opportunity For Lignocellulosic Biomass Background Paper: Multi-tier Market Reference Framework. https://farm-energy.extension. org/wp-content/uploads/2019/04/Biomass-Market-Opportunity_Final-2014_0.pdf (accessed 01 September 2021).

Ruiz J. A., Juárez M. C., Morales M. P., Muñoz P. and Mendívil M. A. (2013). Biomass gasification for electricity generation: review of current technology barriers. *Renewable and Sustainable Energy Reviews*, **18**, 174–183, https://doi.org/10.1016/j.rser.2012.10.021

Rulkens W. (2008). Sewage sludge as a biomass resource for the production of energy: overview and assessment of the various options. *Energy Fuels*, **22**, 9–15, https://doi.org/10.1021/ ef700267m

Saarijarvi H., Kuusela H. and Spence M. T. (2012). Using the pairwise comparison method to assess competitive priorities within a supply chain. *Industrial Marketing Management*, **41**(4), 631–638. https://doi.org/10.1016/j.indmarman.2011.06.031

Sanlisoy A. and Carpinlioglu M. O. (2017). A review on plasma gasification for solid waste disposal. *International Journal of Hydrogen Energy*, **24**(2), 1361–1365. https://doi.org/10.1016/j. ijhydene.2016.06.008

Sariatli F. (2017). Linear economy versus circular economy: a comparative and analyzer study for optimization of economy for sustainability. *Visegrad Journal on Bioeconomy and Sustainable Development*, **6**, 31–34, https://doi.org/10.1515/vjbsd-2017-0005

Sarkar N., Ghosh S. K., Bannerjee S. and Aikat K. (2012). Bioethanol production from agricultural wastes: an overview. *Renewable Energy*, **37**, 19–27, https://doi.org/10.1016/j. renene.2011.06.045

Sawatdeenarunat C., Surendra K. C., Takara D., Oechsner H. and Khanal S. K. (2015). Anaerobic digestion of lignocellulosic biomass: challenges and opportunities. *Bioresource Technology*, **178**, 178–186, https://doi.org/10.1016/j.biortech.2014.09.103

Saxena R. C., Adhikari D. K. and Goyal H. B. (2009). Biomass-based energy fuel through biochemical routes: a review. *Renewable and Sustainable Energy Reviews*, **13**, 167–178, https://doi. org/10.1016/j.rser.2007.07.011

Seiple T. E., Coleman A. M. and Skaggs R. L. (2017). Municipal wastewater sludge as a sustainable bioresource in the United States. *Journal of Environmental Management*, **197**, 673–680, https://doi.org/10.1016/j.jenvman.2017.04.032

Selke S. E. M. (1990). Packaging and the Environment: Alternatives, Trends and Solutions. Technomic Publishing Co., Lancaster, PA.

Sellgren K. L., Gregory C. W., Hunt M. I., Raut A. S., Hawkins B. T., Parker C. B., Klem E. J. D., Piascik J. R. and Stoner B. R. (2017). Development of an Electrochemical Process for Blackwater Disinfection in A Freestanding, Additive-Free Toilet. RTI Press Publication No. RR-0031-1704. RTI Press, Research Triangle Park, NC. https://doi.org/10.3768/rtipress.2017. rr.0031.1704. https://www.researchgate.net/publication/317015118_Development_of_an_ Electrochemical_Process_for_Blackwater_Disinfection_in_a_Freestanding_Additive-Free_ Toilet (accessed 22 August 2021).

Sette E., Pallarès D. and Johnsson F. (2015). Influence of bulk solids cross-flow on lateral mixing of fuel in dual fluidized beds. *Fuel Processing Technology*, **140**, 245–251, https://doi.org/10.1016/j. fuproc.2015.09.017

Sialve B., Bernet N. and Bernard O. (2009). Anaerobic digestion of microalgae as a necessary step to make microalgal biodiesel sustainable. *Biotechnology Advances*, **27**, 409–416. https://doi. org/10.1016/j.biotechadv.2009.03.001

Silayoi P. and Speece M. (2005). The importance of packaging attributes: a conjoint analysis approach. *European Journal of Marketing*, **41**(11/12), 1495–1517, https://doi.org/10.1108/ 03090560710821279

Singh J., Kalamdhad A. S. and Lee B.-K. (2016). Effects of natural zeolites on bioavailability and leachability of heavy metals in the composting process of biodegradable wastes. In: Useful Minerals, Edited by Claudia Belviso, IntechOpen, London, UK, pp. 185–201. https://doi. org/10.5772/63679; https://www.intechopen.com/chapters/50844 (accessed 05 September 2021).

Singh J., Vyas P., Dubey A., Upadhyaya C., Kothari R., Tyagi V. and Kumar A. (2018). Assessment of different pretreatment technologies for efficient bioconversion of lignocellulose to ethanol. *Frontiers in Bioscience*, **10**, 350–371, https://doi.org/10.2741/s521

Sivagurunathan P., Zhen G., Kim S. and Saratale G. D. (2017). A review on bio-electrochemical systems (BESs) for the syngas and value added biochemicals production. *Chemosphere*, **177**, 84–92, https://doi.org/10.1016/j.chemosphere.2017.02.135

Siwal S. S., Zhang Q., Sun C., Thakur S., Kumar G. V. and Kumar T. V. (2019). Energy production from steam gasification processes and parameters that contemplate in biomass Gasifier- a review. *Bioresource Technology*, **2019**, 122481.

Siwal S. S., Zhang Q., Devi N., Saini A. K., Saini V., Pareek B., Gaidukovs S. and Thakur V. K. (2021). Recovery processes of sustainable energy using different biomass and wastes. *Renewable and Sustainable Energy Reviews*, **150**, 111483. https://doi.org/10.1016/j.rser.2021.111483

Smith A. D., Landoll M., Falls M. and Holtzapple M. T. (2010). Chemical production from lignocellulosic biomass: thermochemical, sugar and carboxylate platforms. *Bioalcohol Production*, 391–414.

Smith M. M. and Aber J. D. (2017). Heat Recovery from Composting: A Step-by-Step Guide to Building an Aerated Static Pile Heat Recovery Composting Facility, Research Report, University of New Hampshire Cooperative Extension, Durham, NH, 64p. https://www.researchgate.net/publication/333160048_Heat_recovery_from_composting_A_Step-by-Step_Guide_to_Building_an_Aerated_Static_Pile_Heat_Recovery_Composting_Facility (accessed 30 August 2021).

Smol M., Kulczycka J., Henclik A., Gorazda K. and Wzorek Z. (2015). The possible useof sewage sludge ash (SSA) in the construction industry as a way towards acircular economy. *Journal of Cleaner Production*, **95**, 45–54, https://doi.org/10.1016/j.jclepro.2015.02.051

Smoliński A., Howaniec N. and Bąk A. (2018). Utilization of energy crops and sewage sludge in the process of co-gasification for sustainable hydrogen production. *Energies*, **11**(4), 809. https://doi.org/10.3390/en11040809

Stegmann P., Londo M. and Junginger M. (2020). The circular bioeconomy: its elements and role in European bioeconomy clusters. *Resources, Conservation & Recycling*, **6**, 100029.

Strande L. (2021). Methods for Faecal Sludge Analysis, K. Velkushanova, L. Strande, M. Ronteltap, T. Koottatep, D. Brdjanovic and C. Buckley (eds.), Published by IWA Publishing, London, UK. ISBN: 9781780409115. https://iwaponline.com/ebooks/book/823/Methods-for-Faecal-Sludge-Analysis (accessed 22 August 2021).

Su B., Heshmati A., Geng Y. and Yu X. (2013). A review of the circular economy in China: moving from rhetoric to implementation. *Journal of Cleaner Production*, **42**, 215–227, https://doi.org/10.1016/j.jclepro.2012.11.020

Syed-Hassan S. S. A., Wang Y., Hu S., Su S. and Xiang J. (2017). Thermochemical processing of sewage sludge to energy and fuel: fundamentals, challenges and considerations. *Renewable and Sustainable Energy Reviews*, **80**, 888–913, https://doi.org/10.1016/j.rser.2017.05.262

Taherzadeh M. J. and Karimi K. (2008). Pretreatment of lignocellulosic wastes to improve ethanol and biogas production: a review. *International Journal of Molecular Sciences*, **9**, 1621–1651, https://doi.org/10.3390/ijms9091621

Tang L., Huang H., Hao H. and Zhao K. (2013). Development of plasma pyrolysis/gasification systems for energy efficient and environmentally sound waste disposal. *Journal of Electrostatics*, **71**, 839–847, https://doi.org/10.1016/j.elstat.2013.06.007

Tapia J. F. D., Samsatli S., Doliente S. S., Martinez-Hernandez E., Ghani W. A. B. W. A. K., Lime K. L., Shafrif H. Z. M. and Shaharum N. S. N. B. (2019). Design of biomass value chains that are synergistic with the food–energy–water nexus: strategies and opportunities. *Food and Bioproducts Processing*, **116**, 170–185. https://doi.org/10.1016/j.fbp.2019.05.006

TBC (Toilet Board Coalition) (2017). The circular sanitation economy new pathways to commercial and societal benefits faster at scale. Available at. http://www.toiletboard.org/media/34-The_Circular_Sanitation_Economy.pdf

Ternes T. A., Bonerz M., Herrmann N., Teiser B. and Andersen H. R. (2007). Irrigation of treated wastewater in Braunschweig, Germany: an option to remove pharmaceuticals and musk fragrances. *Chemosphere*, **66**(5), 894–904, https://doi.org/10.1016/j.chemosphere.2006.06.035

Thakur V. K., Singha A. S. and Thakur M. K. (2012). In-Air graft copolymerization of ethyl acrylate onto natural cellulosic polymers. *International Journal of Polymer Analysis and Characterization*, **17**, 48–60. https://doi.org/10.1080/1023666X.2012.638470

Thunman H., Seeman M., Vilches T. B., Maric J., Pallarès D., Ström H., Berndes G., Knutsson P., Larsson A., Breitholtz C. and Santos O. (2018). Advanced biofuel production via gasification – lessons learned from, 200 man-years of research activity with Chalmers' research gasifier and the GoBiGas demonstration plant. *Energy Science & Engineering*, **6**, 6–34, https://doi. org/10.1002/ese3.188

Toor S. S., Rosendahl L. and Rudolf A. (2011). Hydrothermal liquefaction of biomass: a review of subcritical water technologies. *Energy*, **36**(5), 2328–2342. https://doi.org/10.1016/j. energy.2011.03.013

Torjai L., Nagy J. and Bai A. (2015). Decision hierarchy, competitive priorities and indicators in large-scale 'herbaceous biomass to energy' supply chains. *Journal of Biomass and Bioenergy*, **80**, 321–329. https://doi.org/10.1016/j.biombioe.2015.06.013

Trimmer J. T. and Guest J. S. (2018). Recirculation of human-derived nutrients from cities to agriculture across six continents. *Nature Sustainability*, **1**, 427–435, https://doi.org/10.1038/ s41893-018-0118-9

Tsai W.-T., Lee M.-K., Chang J.-H., Su T.-Y. and Chang Y.-M. (2009). Characterization of bio-oil from induction-heating pyrolysis of food-processing sewage sludges using chromatographic analysis. *Bioresource Technology*, **100**(9), 2650–2654, https://doi.org/10.1016/j.biortech.2008.11.023

Tumuluru J. S., Hess J. R., Boardman R. D., Wright C. T. and Westover T. L. (2012). Formulation, pretreatment, and densification options to improve biomass specifications for co-firing high percentages with coal. *Industrial Biotechnology*, **8**(3), 113–132, https://doi.org/10.1089/ ind.2012.0004

Tursi A. (2019). A review on biomass: importance, chemistry, classification, and conversion. *Biofuel Research Journal*, **22**, 962–979. https://doi.org/10.18331/BRJ2019.6.2.3; https://www. biofueljournal.com/article_88067_bf6e89f01897e13b461e01d124fce61f.pdf (accessed 27 August 2021).

Tyagi V. K. and Lo S.-L. (2013). Sludge: a waste or renewable source for energy and resources recovery? *Renewable and Sustainable Energy Reviews*, **25**, 708–728. https://doi.org/10.1016/j. rser.2013.05.029

Urciuolo M., Solimene R., Chirone R. and Salatino P. (2012). Fluidized bed combustion and fragmentation of wet sewage sludge. *Exp. Therm. Fluid Sci*, **43**, 97–104 [CrossRef], https://doi. org/10.1016/j.expthermflusci.2012.03.019

Uzoejinwa B. B., He X., Wang S., El-Fatah Abomohra A., Hu Y. and Wang Q. (2018). Copyrolysis of biomass and waste plastics as a thermochemical conversion technology for high-grade biofuel production: recent progress and future directions elsewhere worldwide. *Energy Conversion and Management*, **163**, 468–492. https://doi.org/10.1016/j.enconman.2018.02.004

Valo A., Carrère H. and Delgenès J. P. (2004). Thermal, chemical and thermo-chemical pre-treatment of waste activated sludge for anaerobic digestion. *Journal of Chemical Technology and Biotechnology*, **79**, 1197–1203, https://doi.org/10.1002/jctb.1106

Van-Walsum G. P., Allen S., Spencer M., Laser M., Antal M. Jr. and Lynd L. (1996). Conversion of lignocellulosics pretreated with liquid hot water to ethanol. In: Seventeenth Symposium on Biotechnology for Fuels and Chemicals, C. Wyman and B. Davison (eds.), Vol. **57/58**. ABAB Symposium. Humana Press: New York, USA, pp. 157–170. https://doi. org/10.1007/978-1-4612-0223-3_14

Van Wesenbeeck S., Prins W., Ronsse F. and Antal M. J. (2014). Sewage sludge carbonization for biochar applications. Fate of heavy metals. *Energy Fuels*, **28**(8), 5318–5326, https://doi. org/10.1021/ef500875c

Vardon D. R., Sharma B. K., Scott J., Yu G., Wang Z., Schideman L., Zhang Y. and Strathmann T. J. (2011). Chemical properties of biocrude oil from the hydrothermal liquefaction of Spirulina algae, swine manure, and digested anaerobic sludge. *Bioresource Technology*, **102**(17), 8295–8303, https://doi.org/10.1016/j.biortech.2011.06.041

Vardon D. R., Sharma B. K., Blazina G. V., Rajagopalan K. and Strathmann T. J. (2012). Thermochemical conversion of raw and defatted algal biomass via hydrothermal liquefaction and slow pyrolysis. *Bioresource Technology*, **109**, 178–187, https://doi.org/10.1016/j.biortech.2012.01.008

Vavouraki A. I., Angelis E. M. and Kornaros M. (2013). Optimization of thermo-chemical hydrolysis of kitchen wastes. *Waste Management*, **33**, 740–745. https://doi.org/10.1016/j.wasman.2021.07.012

Vaxelaire J. and Cézac P. (2004). Moisture distribution in activated sludges: a review. *Water Research*, **38**, 2215–2230, https://doi.org/10.1016/j.watres.2004.02.021

Velkushanova K. (2021). Methods for Faecal Sludge Analysis, K. Velkushanova, L. Strande, M. Ronteltap, T. Koottatep, D. Brdjanovic and C. Buckley (eds.), Published by IWA Publishing,

London, UK. ISBN: 9781780409115. https://iwaponline.com/ebooks/book/823/Methods-for-Faecal-Sludge-Analysis (accessed 22 August 2021).

Verhoeff F., Arnuelos A. A., Boersma R., Pels J. R., Lensselink J., Kiel J. H. A. and Schukken H. (2011). Torrefaction Technology for the Production of Solid Bioenergy Carriers From Biomass and Waste; ECN-E-11-039. ECN, Petten, The Netherlands.

Villamayor-Tomas S., Grundmann P., Epstein G., Evans T. and Kimmich C. (2015). The water–energy–food security nexus through the lenses of the value chain and the institutional analysis and development frameworks. *Water Alternatives*, 8(1), 735–755.

Wang X., Li C., Zhang B., Lin J., Chi Q. and Wang Y. (2016). Migration and risk assessment of heavy metals in sewage sludge during hydrothermal treatment combined with pyrolysis. *Bioresource Technology*, 221, 560–567, https://doi.org/10.1016/j.biortech.2016.09.069

Wang Y., Wang D., Yang Q., Zeng G. and Li X. (2017). Wastewater opportunities for denitrifying anaerobic methane oxidation. *Trends in Biotechnology*, 35(9), 799–802. https://doi.org/10.1016/j.tibtech.2017.02.010

Wang S., Wang Y., Zang S. Q. and Lou X. W. (2020). Hierarchical hollow heterostructures for photocatalytic CO2 reduction and water splitting. *Small Methods*, 4(1), 1900586, https://doi.org/10.1002/smtd.201900586

Wannapeera J., Fungtammasan B. and Worasuwannarak N. (2011). Effects of temperature and holding time during torrefaction on the pyrolysis behaviors of woody biomass. *Journal of Analytical and Applied Pyrolysis*, 92, 99–105, https://doi.org/10.1016/j.jaap.2011.04.010

Waughray D. (2011). Water Security: The Water–Food–Energy–Climate Nexus. Island Press, Washington, DC.

Weiss N. D., Farmer J. D. and Schell D. J. (2010). Impact of corn stover composition on hemicellulose conversion during dilute acid pretreatment and enzymatic cellulose digestibility of the pretreated solids. *Bioresource Technology*, 101(2), 674–678. https://doi.org/10.1016/j.biortech.2009.08.082

Werther J. and Ogada T. (1999). Sewage sludge combustion. *Progress in Energy and Combustion Science*, 25(1), 55–116. https://doi.org/10.1016/S0360-1285(98)00020-3

Wielemaker R. C., Weijma J. and Zeeman G. (2018). Harvest to harvest: recovering nutrients with New Sanitation systems for reuse in Urban Agriculture. *Resources, Conservation and Recycling*, 128, 426–437. https://doi.org/10.1016/j.resconrec.2016.09.015

Wikipedia contributors (2021). Oleochemistry. In: Wikipedia, The Free Encyclopedia. Retrieved 08:44, June 17, 2022, from https://en.wikipedia.org/w/index.php?title=Oleochemistry&oldid=1064136984 (accessed 22 August 2021)

Williams C. L., Emerson R. M. and Tumuluru J. S. (2017). Biomass Compositional Analysis for Conversion to Renewable Fuels and Chemicals. https://doi.org/10.5772/65777; https://www.intechopen.com/chapters/52751 (accessed 25 August 2021)

Wilson C. A. and Novak J. T. (2009). Hydrolysis of macromolecular components of primary and secondary wastewater sludge by thermal hydrolytic pretreatment. *Water Research*, 43, 4489–4498, https://doi.org/10.1016/j.watres.2009.07.022

Winkler M. K. H., Bennenbroek M. H., Horstink F. H., van Loosdrecht M. C. M. and van de Pol G. J. (2013). The biodrying concept: an innovative technology creating energy from sewage sludge. *Bioresource Technology*, 147, 124–129, https://doi.org/10.1016/j.biortech.2013.07.138

World Bank (2019). Evaluating the Potential of Container-Based Sanitation: Sanergy in Nairobi, Kenya. World Bank, Washington, DC. Available at: http://documents.worldbank.org/curated/en/661201550180019891/Evaluating-the-Potential-of-Container-Based-Sanitation-Sanergy-in-Nairobi-Kenya (accessed 25 August 2021)

WWAP (United Nations World Water Assessment Programme) (2017). The United Nations World Water Development Report, 2017: Wastewater, The Untapped Resource, UNESCO, Paris. https://unesdoc.unesco.org/ark:/48223/pf0000247153 (accessed 04 September 2021).

Xie Q., Peng P., Liu S., Min M., Cheng Y., Wan Y., Li Y., Lin X., Liu Y., Chen P. and Ruan R. (2014). Fast microwave-assisted catalytic pyrolysis of sewage sludge for bio-oil production. *Bioresource Technology*, 172, 162–168, https://doi.org/10.1016/j.biortech.2014.09.006

Xu C., Chen W. and Hong J. (2014). Life-cycle environmental and economic assessment of sewage sludge treatment in China. *Journal of Cleaner Production*, 67, 79–87, https://doi.org/10.1016/j.jclepro.2013.12.002

Yadav K. D., Tare V. and Ahammed M. M. (2010). Vermicomposting of source-separated human faeces for nutrient recycling. *Waste Management*, **30**, 50–56, https://doi.org/10.1016/j.wasman.2009.09.034

Yermán L., Hadden R. M., Carrascal J., Fabris I., Cormier D., Torero J. L., Gerhard J. I., Krajcovic M., Pironi P. and Cheng Y.-L. (2015). Smouldering combustion as a treatment technology for faeces: exploring the parameter space. *Fuel*, **147**, 108–116, https://doi.org/10.1016/j.fuel.2015.01.055

Yoshida T. and Antal M. J. (2009). Sewage sludge carbonization for terra preta applications. *Energy and Fuels*, **23**(11), 5454–5459, https://doi.org/10.1021/ef900610k

Young S. (2002). Packaging design, consumer research, and business strategy: the march toward accountability. *Design Management Journal*, **13**(4), 10–14.

Zarei M. (2020a). Wastewater resources management for energy recovery from circular economy perspective. *Water-Energy Nexus*, **3**, 170–185. https://doi.org/10.1016/j.wen.2020.11.001

Zarei M. (2020b). Sensitive visible light-driven photoelectrochemical aptasensor for detection of tetracycline using ZrO_2/g-C_3N_4 nanocomposite. *Sensors International*, **1**, 1–8. https://doi.org/10.1016/j.sintl.2020.100029

Zarei M. (2020c). The water-energy-food nexus: a holistic approach for resource security in Iran, Iraq, and Turkey. *Water-Energy Nexus*, **3**, 81–94. https://doi.org/10.1016/j.wen.2020.05.004

Zarei M. and Aalaie J. (2019). Profiling of nanoparticle–protein interactions by electrophoresis techniques. *Analytical and Bioanalytical Chemistry*, **411**(1), 79–96, https://doi.org/10.1007/s00216-018-1401-3

Zeeman G. (2012). New Sanitation: Bridging Cities and Agriculture. Wageningen University, Wageningen, Netherlands.

Zhang L., Xu C. C. and Champagne P. (2010). Overview of recent advances in thermo-chemical conversion of biomass. *Energy Conversion and Management*, **51**, 969–982, https://doi.org/10.1016/j.enconman.2009.11.038

Zhang X., Wang T., Ma L., Zhang Q. and Jiang T. (2013). Hydrotreatment of bio-oil over Ni-based catalyst. *Bioresource Technology*, **127**, 306–311, https://doi.org/10.1016/j.biortech.2012.07.119

Zhang W., Yuan C., Xu J. and Yang X. (2015). Beneficial synergetic effect on gas production during co-pyrolysis of sewage sludge and biomass in a vacuum reactor. *Bioresource Technology*, **183**, 255–258, https://doi.org/10.1016/j.biortech.2015.01.113

Zhang W., Alvarez-Gaitan J. P., Dastyar W., Saint C. P., Zhao M. and Short M. D. (2018). Value-added products derived from waste activated sludge: a biorefinery perspective. *Water*, **10**, 545. https://doi.org/10.3390/w10050545

Zhen G., Lu X., Kato H., Zhao Y. and Li Y. Y. (2017). Overview of pretreatment strategies for enhancing sewage sludge disintegration and subsequent anaerobic digestion: current advances, full-scale application and future perspectives. *Renewable and Sustainable Energy Reviews*, **69**, 559–577. https://doi.org/10.1016/j.rser.2016.11.187

Zielinska D., Rydzkowski T., Thakur V. K. and Borysiak S. (2021). Enzymatic engineering of nanometric cellulose for sustainable polypropylene nanocomposites. *Industrial Crops and Products*, **161**, 113188. https://doi.org/10.1016/j.indcrop.2020.113188

doi: 10.2166/9781789061840_0181

Chapter 7

Marketplace and sales

Walter Gibson, Peter Emmanuel Cookey, Thammarat Koottatep and Chongrak Polprasert

---

Chapter objectives

The aim of this chapter is to guide the reader in assessing sanitation end-market conditions that determine production and services in the IFSVC as well as explore some of the characteristics of this market, why its potential has not yet been realized, and some of the ventures that are bringing about change.

7.1 INTRODUCTION

It is only relatively recently that sanitation began to be considered to a marketplace where sales of products and services can be made. Historically, in many countries it was considered to be a public good which should be provided by the state. While governments do still have an important role to play, it is now widely accepted that the private sector also has a key role in meeting sanitation needs and can do so profitably and sustainably. The emerging sanitation marketplace is large, diverse, complex and underdeveloped, but represents a huge opportunity for business, for improved public health, for resource recovery, and for reducing human impact on the environment. In this chapter we will explore some of the characteristics of this market, why its potential has not yet been realized, and some of the ventures that are bringing about change. Although global estimates are not readily available, the commercial case for the sanitation market has now been made in several countries (Toilet Board Coalition, 2020a, 2020b, 2020c) and for new technologies such as the Janicki Omniprocessor (Gates Foundation, 2017, 2018). With 2.3 billion people, about 30% of the world's population, still to access decent sanitation (JMP, 2017) the potential for market growth in all sanitation-related goods and services must be very large. Furthermore, there is a huge potential market for resources – water, energy and nutrients – which can be recovered from the 'waste' collected from sewers (wastewater) and on-site sanitation systems (faecal sludge), and for the associated plant and equipment. Globally there is a marked lack of capacity to manage and treat both types of waste, particularly in developing countries (Harada *et al.*, 2016; WWAP, 2017). Thus, the potential for recovery and reuse is very high, as is the potential to reduce harmful pollution and disease caused by direct discharge of untreated waste into the environment.

The future development and impact of the sanitation market and of the Integrated Functional Sanitation Value Chain (IFSVC) are thus closely intertwined. The IFSVC can be viewed as a series of market-based transactions taking place within and between the different stages, and involving different customers and suppliers, different products and services, and differing motivations for purchase and sale. For the concept to be realized, value should be added at each stage and accumulated along the chain, with every part of the chain aligned to and serving the overall goal of reuse and recovery. Equally, revenues should flow in the opposite direction, providing the incentives for value creation and covering the costs of the transactions involved (Figure 7.1).

The current reality falls short of this ideal for a number of reasons that will be explored in this chapter, although recent innovations in market development, wastewater reuse, container-based sanitation and reuse product technology are bringing it closer. However difficult, achieving an efficient IFSVC is important for several reasons:

(i)   economically, it should reduce the cost burden of sanitation;
(ii) environmentally, it should reduce harmful pollution caused by dumping waste, enhance resource reutilization and reduce pressure on scarce resources; and
(iii) socially, it should drive increased access to decent sanitation, and the associated health benefits, for the underserved.

Making markets work effectively and sustainably is critical to achieving this vision. This necessitates private-sector involvement and there has been a growing recognition over the past decade of the essential role that the private sector can play in delivering the benefits of sanitation and meeting the Sustainable Development Goals (SDGs). However, significant barriers and constraints remain, and different innovations are being explored to overcome them and make markets work more effectively. In this chapter we examine progress, what has been learnt, and opportunities for the future. The focus will be on attempts to drive universal access to safely managed sanitation through market-based sanitation and at the same time realize the vision for the IFSVC.

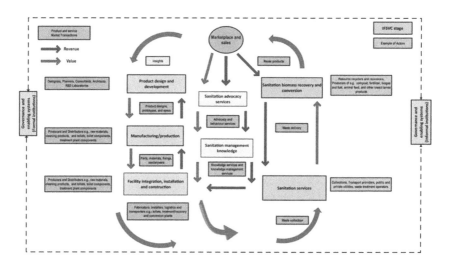

Figure 7.1 IFSVC: revenue and value flow, market transactions and actors. (Diagram by author, after Koottatep *et al.* 2019).

7.2 MARKET LANDSCAPE

7.2.1 Sanitation market

At its simplest the sanitation market is where demand meets supply, where buyers or customers meet suppliers and a transaction related to sanitation provision takes place. This is conducted by the sanitation enterprise (or in some cases a public utility) which arranges for promotion, production, distribution, sale, and delivery of the goods or services through its operations. There are many different sub-markets, characterized in terms of size (number of customers or products sold), value (revenue from sales), segments (different groups of customers with similar characteristics who share the same desire for a particular product), and the available products or services and their prices and perceived value. As an example, the 'toilet' market is one major segment, but it can be further sub-divided by location (e.g., urban/rural), by type (e.g., household, communal or public), by product type (e.g., container-based vs on-site treatment) and so on. Other major segments include sewered and non-sewered, 38% and 62% of the world's population respectively (JMP, 2017), and the sanitaryware market, estimated to be worth \$32 bn and growing, driven by developing countries (Business Wire, 2020).

The sewered market predominates in countries and regions with more mature developed economies (e.g., EU, USA): the value of the market is difficult to estimate as tariffs typically cover combined water and sanitation services, but the figure of €37bn for the total production value in the EU is indicative of its size (Schouten & Van Dijk, 2007). In developing countries, where non-sewered (on-site) sanitation is the dominant type, the market value is also very large: detailed analyses of India, Kenya and Nigeria by the Toilet Board Coalition estimate the 2021 values of the sanitation economy in those countries alone to be \$97.4bn, \$3.2bn and \$15.5bn respectively (Toilet Board Coalition 2020a, 2020b, 2020c). The toilet economy, which includes household, public and community toilets, operations and maintenance and auxiliary products, is by far the greatest sector currently. The circular economy, comprising the market for reuse products, is relatively small at present. To some extent this reflects the dominant paradigm in the sanitation sector for the past 20 years, which has been toilet provision. The concept of reuse, essential for the IFSVC, is relatively recent and the associated technologies, markets and business models are still being explored.

The sanitation market is poised for significant growth, much of it in developing and emerging economies, and much of it driven by growing wealth and the ambitions of the SDGs to provide universal access to adequate sanitation: as noted above around 2.3bn people still do not have access to decent sanitation (JMP, 2017). This should enhance sales of toilets and on-site waste collection and treatment systems, as well as cleaning products. The gap in wastewater and faecal sludge capacity is being recognized in countries such as India and will drive investment in new technology and plants. Production and sales of reuse products will increase, driven by improved technology, market forces (e.g., resource scarcity) and government legislation. Product innovations will open new and valuable markets: the Bill & Melinda Gates Foundation estimates the market potential for their portfolio of 'Reinvented Toilets' to be around \$6bn (Gates Foundation, 2017, 2018). It is striking however that as yet there is no actual market for raw faecal sludge or other 'toilet resources'; the emergence of significant global demand for this as a feedstock, with a concomitant value in dollars/tonne, will signal the integration, full functionality and widespread operation of the IFSVC. If the value is high enough it could lead to a new paradigm in sanitation, one where instead of being a cost burden, it becomes value creating, offering the prospect of lower charges and better products and services for customers providing the raw material for subsequent transformation.

7.2.2 Sanitation marketplace and the IFSVC

How does this marketplace of sanitation goods and services map onto the IFSVC? Conceptually, within the IFSVC, individual marketplace transactions should be connected and add value at each stage to achieve the overall purpose of optimal reuse and recovery of sanitation resources. The model implies an alignment of market transactions from stage to stage with each one adding value in exchange for revenue. The customer for one product or service becomes the supplier to the next, with the reuse customer being the ultimate driver of demand. However, at present the customer for an individual product or service is rarely buying into a joined-up, safely managed sanitation value chain. Indeed, the reuse customer is presently not the dominant driver of market transactions further back in the chain. Other customers, often sitting outside the chain, and not necessarily connected to later stages, can interrupt value and revenue flows as shown in Figure 7.2.

This reflects the fact that, for good reasons, the focus of many organisations, both private and public, is and has been the provision of toilets, with less thought being given to completing the chain and adding value at every step. However, the purchase or provision of a new toilet does not mean that the new owner will then have access to related sanitation services or markets further along the chain. Nor does it mean that the value of the waste it collects will be recovered at a later stage. As we will see later, for the domestic customer it can appear as a very fragmented marketplace. At present, sanitation product and service providers rarely span the whole chain and operate it as an IFSVC. More typically each stage of the IFSVC is made up of different types of enterprise that may or may not supply customers in a later stage. The wide range of different enterprises active in the IFSVC has been comprehensively summarized by Koottatep *et al.* (2019) and is mapped onto the IFSVC conceptually in Figure 7.1.

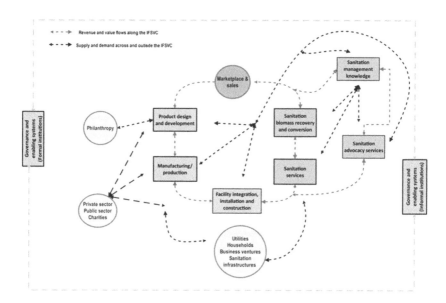

Figure 7.2 Sanitation marketplace and the IFSVC: internal and external demand. (Diagram by author, after Koottatep *et al.* 2019).

Although Figure 7.1 recognizes that public utilities have an important role to play and can be considered as enterprises if they are charging market-based fees, it is to be expected that private enterprises will dominate within the IFSVC and its associated market transactions. Although there is a view that sanitation is a public good and there is evidence that private-sector involvement will not necessarily improve access to services for the poor (De Oliveira, 2018), there is a growing acceptance that the private sector does have important roles to play and that its contribution will be necessary to meet the SDGs (Mason *et al.*, 2015). Sparkman and Sturzenegger (2016) also recognized this trend and highlighted some of the benefits of a market-based approach involving private-sector providers, which include:

(i) greater sense of ownership by the end customer if they are investing their own resources;
(ii) continuous improvement in quality of goods and services due to greater accountability of, and competition between, providers;
(iii) greater focus on what people actually want and are willing to pay for;
(iv) greater potential for sustainability due to reduced dependence on outside aid; and
(v) greater potential for self-scalability due to incentives to seek growth and new customers.

Businesses can therefore contribute not just value to the IFSVC but also to the sustainability of service delivery. For the IFSVC to function optimally, each enterprise must be able to achieve its full market potential, that is, efficiently create and meet demand, and have incentives to participate in value creation. However, they sit within a complex system, that, along with market reality and market forces, will influence how far they can go to achieving that goal.

7.2.3 Market system

Marketplaces and enterprises do not operate in isolation: they sit within a broader market system which supports and influences it through a broader context and range of actors such as government institutions, investors and so forth. Market rules will apply which also affect business operation. Figure 7.3 illustrates how IFSVC enterprises are influenced by the wider environment and external enablers and supporters (based on Agarwal *et al.*, 2018, and Koottatep *et al.* 2019). For market potential to be realized effectively an entrepreneur must identify the target customer, create and produce a product they want to buy, activate demand through sales and marketing, and develop a model to deliver the product to the customer. As shown in Figure 7.3 these activities take place within the broader market system, which includes enablers, supporters and the external environment. Here the latter is taken to include factors that influence businesses directly and the broader context in which they have to operate. For sanitation this system is quite complex and involves a range of different actors including NGOs, utilities, regulators and other government bodies, policy makers, builders, manufacturers, designers, philanthropists, donors and investors. Some of the constraints to its effective operation are considered later in this chapter in the section on market failures.

NGOs have historically played a major role in sanitation provision; their role, as we will see later, is now changing in some cases towards facilitating market development. Non-sanitation businesses such as property developers, architects and builders may influence the design and choice of sanitation system used in homes, factories and offices. Academics study market dynamics and opportunities and support new policy development and market development. Scientists and technologists develop improved products and services, some of which may enable new markets for existing or start-up enterprises. Finance providers help with purchases and help entrepreneurs to grow their

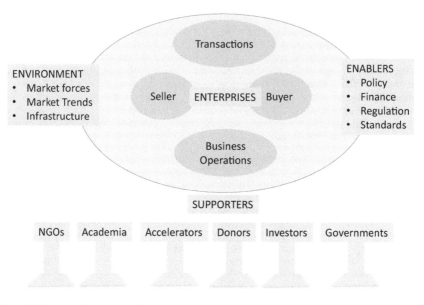

Figure 7.3 The market system: wider influences on IFSVC enterprises.

businesses. Philanthropy has a role to play too in supporting new ventures, developing and introducing better products and services, and learning how to make the marketplace work. The actions of each of these players are not necessarily coordinated, leading to market complexity and the potential that they will work against, rather than with, each other.

Governments play a particularly important role. Sanitation has historically been viewed as a public good although this does not mean that governments have taken complete responsibility for an end-to-end provision of sanitation as a social service. It does mean, however, that in most countries it is viewed as a government responsibility to ensure that sanitation is safely managed for all its citizens. How this is discharged, for example through policy, regulation and direct interventions such as subsidies, can have a positive or negative effect on the market.

The market and business viability are also influenced by the environment in which they operate. This can include broader factors such as infrastructure (affecting, e.g., movement of goods), as well as others that affect businesses more directly, such as finance availability, laws and regulations, and availability and process of raw materials (Agarwal *et al.*, 2018).

7.2.4 Sanitation market status

The market transactions and enterprises at the heart of the IFSVC are thus part of a much larger system and influenced by it. This market-led approach is however a relatively recent trend, especially in developing countries, and is not yet well established. So what is the current status? The position in developed countries is exemplified by the UK where sanitation is largely delivered by connection to sewers and waste treatment in large, centralized plants. Since privatization in 1989 these are maintained and operated by what are effectively private monopoly companies (derived from previous regional public utilities) within a market framework designed and regulated by an independent government-established regulator (Ofwat, 2020). The cost is borne largely by fees charged to individual households, offices and other premises connected to the network. In the past the focus of sewage treatment has largely been on safely treating it and disposing of residual sludge either by landfill or at sea. More recently the value of sludge for use in agriculture or in energy production has been recognized and there has been a shift towards reuse and recycling, encouraged by the regulator (Ofwat, 2015). External factors, such as demand for clean energy, and scarcity of water and other natural resources such as phosphorus are also driving greater recovery and reuse of energy and nutrients from wastewater and sludge treatment. Recovery and reuse potential from wastewater is highest in urban areas (WWAP, 2017). The market for biogas globally is predicted to grow from \$25.5bn in 2019 to \$31.69bn by 2027; currently the main biogas installations are in the US, Germany and the UK (Fortune Business Insights, 2020). Water is being recycled directly or indirectly for drinking or for irrigation, and phosphorus recovery is now getting closer to being economically viable (WWAP, 2017).

Unlike many other markets, consumer choice is somewhat limited for the majority of users in developed countries where sanitation is sewered: there is likely to be only one local provider. Those consumers who are off-grid have somewhat more choice in terms of the range of septic tanks available for purchase and there is also competition between private service providers for emptying tanks and transport to treatment plants. The market for sanitaryware is however highly developed and competitive: although the basic functionality of toilets has changed little (except in Japan, where very innovative features have emerged in the last 30 years), designs are constantly changing to keep pace with fashion and trends in interior design. By contrast, in developing countries, the focus in sanitation has been ending open defecation. In regions such as Sub-Saharan Africa and Central and South Asia only a minority of end users are connected to sewers, mostly in urban rather than rural areas (JMP, 2017). The majority of rural households and low-income urban households are off-grid and have no real prospect of being connected to sewers. Installation of sewer lines in rapidly growing cities is simply too expensive and difficult, so even wealthier families may not be able to get a connection and have to rely on septic tanks. Connection to a sewer does not of course guarantee safely managed sanitation if there is not adequate provision for waste treatment, as is the case in many countries. Likewise, there is nothing wrong with on-site sanitation so long as excreta are disposed of safely *in situ* or safely removed and treated off-site. The latest JMP figures, while noting a lack of data on excreta management, suggest there is a long way to go to achieve universal safely managed sanitation (JMP, 2017).

Typically, sanitation markets aimed at low-income customers are focused on provision of toilets – generally septic tanks and latrines along with superstructure and fittings – and emptying and transport services by small-scale entrepreneurs. Public services may also be available although utilities tend to focus on the sewer network and operation of treatment plants. Management of faecal sludge and septage has not kept pace with increased coverage. Regulations may have been adopted from developed countries and not aligned with the dominant local non-sewered form of sanitation. There is low treatment capacity and reuse is still on a very small scale: in India alone something like 72 000 tonnes of faecal sludge are dumped in the environment every day without treatment or value recovery (Sivaramakrishnan, 2019). This is a key issue for the IFSVC: a flourishing and high-value market for reuse products is critical to drive the chain.

Sanitation markets in such countries are fairly embryonic, limited and fragmented relative to their potential. A report in 2015 concluded that 'sanitation markets that meet the needs of poor rural consumers are virtually non-existent' (Dumpert & Perez, 2015). This may reflect the fact that the off-grid market was viewed as an interim step on the way to fully sewered sanitation, instead of a viable and acceptable alternative in its own right. However, recognition is growing that it is precisely these markets which need to be developed if universal access to safe sanitation is to be achieved. This is beginning to happen as the underlying market failures and barriers to growth are better understood and new business models and interventions are explored. Standards for product and service performance will also contribute to the development of the market as they will provide reassurance to customers of the quality of their purchase, build trust and encourage product use, which is crucial to achieving the health impact of sanitation.

7.3 MARKET FAILURES

At present, sanitation markets in developing countries, particularly those serving the majority with on-site facilities, lack capacity: that is, goods and services are not being provided in sufficient quality and quantity. This can be broadly attributed to market failures due to external events, information asymmetry or imperfect competition (Tremolet, 2012). Looking along the value chain and across the market system in more detail, specific failures and constraints can be identified using a number of different approaches, including economic analysis (Tremolet, 2012), qualitative research (Murta et al., 2018; Scott et al., 2018; Sy et al., 2014), mixed methods (Muhkerjee et al., 2019; Tsinda et al., 2015), expert interview/literature review (Sparkman & Sturzenegger, 2016), workshops(Mulumba et al., 2014) and case studies (Agarwal et al., 2018). The conclusions from these studies are summarized in Table 7.1 using the market system framework of Agarwal et al. (2018) where the environment is broken down into the business environment and the broader context; it is not an exhaustive list but will exemplify the wide range of issues in the market.

Many of these failures fall under the heading of 'information asymmetry', a category which applies in other sectors such as climate change (Andrew, 2008). As concluded by Dumpert and Perez (2015) information flow to market actors is often the missing link in the value chain, due to poor communication and misperceptions: consumers are unaware of products because suppliers don't promote them, and suppliers think there is no demand from consumers. Others relate to the intrinsic market attractiveness in terms of business viability, competition, market rules and access to finance. Another critical factor is coordination and alignment of public- and private-sector players and interventions.

In addition, there seems to be a damaging combination of weak demand coupled with a fragmented supply chain. The lack of demand seems to be more related to lack of focus on products that people want to buy as opposed to the products being unaffordable. Indeed, there is evidence that if the products are designed to meet key wants and needs, and are of good quality, then people are willing to invest in improved sanitation. Willingness-to-pay studies are a key part of evaluating new products and can be a guide to likely demand, as well as indicating the trade-offs that customers are willing to make in terms of price vs product attributes. Dumpert and Perez (2015) in a comprehensive review of the sanitation market in countries across the world noted that low-income consumers – contrary to common beliefs – value quality and are willing to pay for high-quality products even if lower-cost alternatives are available. In this sense, in some cases the challenge may be not so much about increasing demand per se but about finding ways to enable such consumers to purchase products designed for them and to which they aspire, for example, through financing. Agarwal et al. (2018) also noted the importance of addressing *ability* as well as *willingness* to pay and in addition to microfinance mention the use of 'market-compatible' subsidies such as incentives to credit providers. Perhaps the single greatest factor influencing the IFSVC is the relative lack of a significant market for reuse products, which is influenced by the relative lack of treatment capacity, weak demand for reuse products, lack of awareness of its potential and regulatory factors. Currently the flow of revenue along the chain originates largely

Table 7.1 Market system failures along the IFSVC.

Failure/Constraint	Causes
Core Market	
Weak demand	Not providing products people want and offering products they do not want
	Lack of market segmentation
	Lack of promotional activities and low awareness of products and services
	Limited access to finance
Limited availability products/ services	Weak, fragmented supply system
Lack of entrepreneurs	Lack of access to finance and training
	Lack of market attractiveness
Inadequate market structure	Lack of skills and training of service providers, presence of illegal operations, no economies of scale
Poor revenues for small-scale operators	Lack of awareness of reuse opportunities
Lack of attractiveness of reuse markets	High transport costs, consumer perception, limited financial value of reuse products,
Business Environment	
Barriers to entry/ entrepreneur participation	Public utilities may hold monopoly but be unwilling to provide services to poorer customers
	Lack of access to capital
	Lack of transparent regulatory framework
	Lack of representation for start-ups to influence policy
Distortion by subsidies	Ineffective targeting or over-liberal use
Governments invest in expensive solutions	Lack of awareness of cheaper solutions
Financial value of reuse products fluctuations	Changes in external factors, for example, energy prices
Inhibitory market rules	Lack of alignment with market functions
Broader Context	
Poor infrastructure	Insufficient cross-sectoral action due to poor planning
Challenging physical environments	Difficult terrain, physically dispersed population

from household customers for sanitation goods and services (Figure 7.2). For the IFSVC to flourish this needs to be matched or exceeded by a flow of revenue from customers in other markets for reuse products.

7.4 MARKET DEVELOPMENT INNOVATIONS

The challenges outlined above are almost certainly incomplete but are not insurmountable. Recent work has shown that there is considerable scope for many of the actors in the market system to intervene to improve its efficiency and develop the core market for sanitation goods and services. Innovation across the IFSVC, not just in products but in delivery models, marketing approaches, financial interventions, is vital for improved service delivery (Dumpert & Perez, 2015) and indeed for the market to develop into a viable, sustainable means of delivering sanitation at scale. Because the private sector has only recently become engaged with sanitation and many of the players are SMEs, often lacking the necessary business skills, market development has been slow and few market-based sanitation enterprises have achieved significant scale (Agarwal *et al.*, 2018). Help is at hand however as many of the other actors in the sanitation market system are starting to innovate and play leading roles in its development.

7.4.1 Role of NGOs
Historically NGOs in the sanitation sector have been implementers, acting as providers of goods and services for underserved populations. There is a growing realization and trend for them now to act as facilitators of market systems, helping them to develop and become self-sustaining (Text Boxes 7.1 and 7.2 below).

Box 7.1 WaterSHED's Market Systems Approach in rural Cambodia

WaterSHED have taken a market systems approach to delivering sanitation to rural customers in Cambodia, deliberately playing a facilitation role with the ultimate aim of exiting once the market has become self-sustaining – the 'hands off' strategy (Jenkins *et al.*, 2019). Over a period of roughly 10 years, they went from research and development, to scale-up, and finally to strategic phase-out – while enabling well over 200 000 rural families to purchase a toilet. The approach involved multiple interventions across the sanitation market system, engaging with consumers, local businesses, local NGOs, lenders, community leaders and higher levels of government across the market system. WaterSHED worked closely with academics and another NGO, Lien Aid, to design their strategy (Pedi *et al.*, 2011); a key initial intervention was to carry out very detailed market research into demand for toilets in rural areas and in parallel understand the supply chain in the same districts. User-centred design was employed to develop an affordable and desirable latrine product. Demand data were used to encourage local suppliers with appropriate capacity to take up production of the new design. Market activations were then carried out village by village with the help of local community-based independent promoters and village leaders to create sales which were then fulfilled by the local entrepreneurs. In effect they created a replicable micro-market system to encourage local suppliers to deliver sanitation products and meet locally generated demand. The transition to a self-sustaining market involved ongoing monitoring, identifying success factors and gaps in the market system, expansion to new areas and continued focus on making WaterSHED's role obsolete (Jenkins *et al.*, 2019). Initially WaterSHED acted through local market facilitators who brokered relationships between key actors in the system. This transitioned to a more business consultancy relationship, which kept the risk of creating dependency low. A key challenge was to find a more sustainable way to activate demand. One solution was the *civic champions* leadership development program, based on the recognition of the potential role commune officials could play in demand generation: this program trained and motivated local government officials, in part through peer-to-peer learning. Later iterations were led by the Ministry of Interior. In 2018 WaterSHED moved into its final phase to support provincial government staff and the Ministry of Interior to take over its roles in market system facilitation and work towards universal access. Complete exit was planned for 2020 (Jenkins *et al.*, 2019).

Box 7.2 PSI's Market development approach in rural India: the 3SI project

Like WaterSHED, PSI and the 3SI project team began by analysing the sanitation market landscape in rural Bihar, which at the start of the project (2012) had a high rate of open defecation. This enabled them to identify four key market failures which they sought to tackle through the 3SI (Supporting Sustainable Sanitation Improvements) project. They also played the role of facilitator with the aim of creating a self-sustaining market system: it required the team to think as a viable business has to think, but work through actors who could sustain it into the future. The process involved iterative market testing and evaluation of possible solutions, then strengthening the wider market infrastructure, particularly in terms of finance and supply chain (Singh *et al.*, 2017). The four key market failures were the lack of an affordable and aspirational product, a fragmented supply chain which was difficult and costly for consumers to navigate, a lack of finance for customers and entrepreneurs, and a prevailing acceptance of open defecation. Each of these was addressed through the 3SI project and integrated within a scalable business model. The final product addressed key customer concerns about frequency of emptying, quality and price. It offered a standard, basic model of good quality at an affordable price with a pit which would not fill up quickly and a superstructure which could be customized depending on the available budget. The fragmented supply chain – there were as many as 30 different actors in Bihar across the value chain – was a major factor deterring potential customers from purchasing a toilet. Customers also wanted to deal with certain market actors they could trust. This was tackled through an 'aggregation' approach: selected entrepreneurs dealt with all the individual suppliers necessary to build a toilet and became a single point of contact (Turnkey Solution Provider) offering a complete installation service for would-be purchasers. While attractive on paper, this did not work out as well as hoped, for either the customers or the entrepreneurs. A further iteration took account of the reality that customers wanted to oversee construction themselves and were prepared to source some of the materials: this blend of DIY and one-stop shop was much more successful and was focused on entrepreneurs (concrete ring manufacturers) for whom sanitation was a core market. As with WaterSHED, working out a market activation strategy was important: demand existed but was latent. Several models were tested and that of sales facilitator (promoting awareness and generating sales interest) worked better than sales agent (actually taking orders). Alongside salaried sales promoters, local champions ('toilet motivators') played a key role in generating community interest. While not explicitly sales agents they played a role very similar to sales agents in convincing families to make a purchase and they did facilitate toilet sales and received a commission on each toilet sold. They were key in delivering the behavior change campaign. Figure 7.4 illustrates sales activity in action. Bringing in a sustainable source of finance was crucial to oil the wheels of the market system: the project was able to demonstrate successfully to MFIs that loans would be repaid and to establish methods for assessing the creditworthiness of entrepreneurs. The project acted as a demonstrator of what works, building a solid foundation in terms of sales (around 220 000 by 2017), trained entrepreneurs (759) and loans provided (37 175 to households and 251 to enterprises). The transition to sustainability was through helping entrepreneurs to strengthen their own supply chains, encouraging them to take over payment and responsibility for demand creation via sales facilitators and toilet motivators, and through helping to mobilize further credit.

Figure 7.4 The 3SI Project – sales activity in progress (image credit, Kiran Thejaswi).

What these examples show is that sanitation market systems can be made to work even for low-income customers but need patience, detailed market analysis, creative and targeted interventions, multiple iterations, and external soft funding to allow proof of concept of key elements of the mix.

7.4.2 Role of start-ups

In the above examples NGOs acted as market facilitators with an endgame of exiting and handing over to entrepreneurs and other actors in the market system to sustain product and service delivery. Entrepreneurs themselves, through innovative start-ups, also have a key role to play by experimenting with delivery models and new ways to capture value along the IFSVC. In particular, in the past 10 years a number of enterprises have been exploring the potential of container-based sanitation (CBS), which represents an attempt to marry improved service provision with collection and transport of the raw material and its conversion to reuse products that increase business revenue. Their main target is the urban poor, who are poorly served by other types of service, and their activities span the IFSVC. Different approaches in terms of waste collection, revenue models and reuse products are being explored (World Bank, 2019). While such businesses do not currently recover all their costs (Remington *et al.*, 2018; Russel *et al.*, 2019; World Bank, 2019) and require external funding to sustain their operations, there are different approaches to accommodating or reducing this requirement. Some level of ongoing subsidy may be acceptable and justified from public funds: however, there is also scope to increase service charges as the businesses become better known and established, and to seek cost reductions and efficiencies (World Bank, 2019). In addition, were the values in terms of environmental, health and economic impacts of waste to value businesses taken into account, this would make a material difference (Parker *et al.*, 2020).

Revenue can also be increased through innovation in value recovery technologies (Diener *et al.*, 2014) and finding new markets for reuse products where demand is high but demonstrating their potential at scale is important. One such innovative technology, included in the analysis of Diener *et al.* (2014), is the use of BSF larvae to process faecal waste (Banks *et al.*, 2013) and convert it into protein and oil, which can be further converted into animal feed (e.g., Sanergy's KuzaPro, see Chapter 2) and a variety of products, for example, biodiesel (Nguyen *et al.*, 2018) respectively. This technology has been the subject of an at-scale study in Durban, South Africa, to explore its business and technical viability (Grau & Alcock, 2019). The study brought together in a public–private partnership the eThekwini municipality, who wished to explore the potential of this reuse technology to deal with toilet waste, and The BioCycle, who had done much of the early proof-of-concept work on nutrient recovery from faecal waste using BSF larvae and wished to test its potential on a bigger scale. KwaZulu Natal University was also involved and the implementing agency was Khanyisa Projects. With funding for capital expenditure from the Bill & Melinda Gates Foundation, a plant was designed, built and operated at a scale of 2–3 tons/day for several months. Food waste was included as well as faecal waste in the feedstock.

This study provided valuable data on key parameters such as conversion efficiency, allowing for a more reliable projection of capital and operating costs within a business model for the existing plant. Assuming an upgrade of the facility and operation to a cycle of 12 tons per day of 50:50 faecal sludge:food waste, the projected loss based on sales of protein and oil was found to be $9 per ton (Grau & Alcock, 2019). This is coming quite close to achieving profitability and the authors identified several routes to enhancing profitability. The learning from the study will enable the eThekwini municipality to make more informed decisions about future waste management processes. The BioCycle conclusion was that to be commercially viable, scale would be a prerequisite, specifically for the farming of the fly (biological capacity) and that simple cross-subsidization would not be sufficient (Lewis, 2022, personal communication). Another key insight was about the patience required to take such new technologies to the point where they can viably operate at scale: the period from the initial lab work (Banks *et al.*, 2013) to the Durban study (Grau & Alcock, 2019) spanned some 9 years (2010–2019): perhaps another 5–10 years will be required.

The learning from this and other research, and how it is taken forward and applied, is vital to increase the impact of value recovery processes on sanitation.

A further idea of the size of the gap that remains is provided by a detailed economic and financial analysis of a trial of Sanivation's fuel briquette process in a refugee camp linked to a CBS service (Parker *et al.*, 2020). This is outlined in Box 7.3 and Figure 7.5.

Since that study, Sanivation have taken several important steps towards addressing these challenges and bridging the gap. Firstly, they have scaled their technology by developing a product called 'superlogs' which is suitable for large industrial customers such as tea producers, flower farms and dairy producers with a large firewood furnace. They are currently operating at >200 tonnes/month. Secondly, costs of production are less because the process is more automated, at a bigger scale, and uses a cheaper more abundant co-ingredient (sawdust). Thirdly, it is a more favourable market: because the earlier 'superballs' were mostly sold at a household level, distribution costs were too high at the scale they could achieve. Finally, Sanivation have recently learnt that their products reduce greenhouse gas emissions in comparison with the fuels they replace and may thus be eligible for carbon credits of $12–70 per tonne of product.

This type of market and product innovation by entrepreneurs will drive integration along the IFSVC: as demand for the end product grows and capacity to produce it has to increase, so will demand grow for the feedstock (i.e., faecal waste) and the products and services needed to collect and transport it safely. As further and even more economically attractive biomass conversion technologies become available (see Chapter 2), the incentives to operate a fully integrated IFSVC will increase. However as highlighted above, action from other actors can help to create an enabling environment for the

Box 7.3 The benefits and challenges of waste to value innovation

Sanivation's mission is to increase access to safe and cost-effective sanitation services in urbanizing communities and refugee camps. Their core innovation is a waste treatment process which takes faecal sludge, combines it with other organic waste, and turns it into briquettes which can be burnt as fuel. The briquettes are safe to handle as pathogens are destroyed during the manufacturing process. Both the first-generation briquettes ('superballs') and the second generation 'superlogs' have significant customer advantages over the fuels they are designed to replace, that is, charcoal and firewood respectively (Parker *et al.*, 2020; Sanivation, 2020). At Kakuma refugee camp in Kenya, Sanivation piloted a CBS service linked to production and sale of fuel briquettes: 500 toilets were in use and being serviced and peak sales of 11 tonnes/month of briquettes achieved, in a context where a certain amount of free solid fuel was distributed to camp residents. An economic and financial analysis was undertaken to understand the underlying cost structure and revenue potential, as well as to study cost-effectiveness in comparison with two other sanitation options. Overall, the revenue from briquette sales did not cover the operational costs and there was a gap of some $122 per toilet per year, similar to the net cost of burying the waste in landfill without treatment. However, this does not take account of the non-monetized benefits in terms of environmental, health and employment impacts. Placing an actual value on these is complex but not impossible and perhaps could be part of a wider sustainability assessment when making choices about sanitation provision. The cost-effectiveness analysis showed that the Sanivation model was less cost-effective than pit latrines and urine-diverting double vault (UDDT) toilets where the waste went to landfill. However, either a 25% saving in costs or a 67% increase in revenues brings the Sanivation model in line with UDDT. Can this gap be closed? Savings could undoubtedly be made through process and material improvements, and scale of operations. Revenue could be increased if distribution challenges were overcome. More difficult to achieve are realizing the full value of all the benefits of the approach and establishing a more level playing field with respect to competition – for example through regulation to ensure that all sludge is made safe prior to disposal and that firewood and charcoal are sustainably produced. So, the challenges of delivering the vision for the IFSVC are not purely economic or technical and within the control of the entrepreneur.

markets to flourish. Recently a number of CBS businesses have joined forces to promote this approach and influence the enabling policy environment (Russel *et al.*, 2019).

7.4.3 Role of governments

Governments have very significant roles to play in encouraging and enabling greater private-sector involvement and market development (Dumpert & Perez, 2015; Mukherjee *et al.*, 2019; Sparkman & Sturzenegger, 2016; Tremolet, 2012). Among the innovations being advocated and explored in different countries are:

(i) more targeted subsidies, aimed at enabling access for the poorest households;
(ii) framing and enforcing regulations to encourage safe disposal and reuse;
(iii) providing technical assistance on options available;

Figure 7.5 Sanivation's Superlogs being used in an industrial boiler at a flower farm in Naivasha, Kenya (image credit: Sanivation).

(iv) framing and enforcing regulations to support fair competition and adherence to service standards;
(v) supporting small-scale entrepreneurs;
(vi) improving infrastructure, including treatment facilities and roads; and
(vii) targeted social marketing campaigns to create demand.

While regulations are needed, analysis in India (Mukherjee *et al.*, 2019) suggests that over-regulation can negatively impact business viability. Likewise, Dumpert and Perez (2015) concluded from a global survey that government policies can have a positive influence on the market, but that direct interference should be avoided. Governments cannot act in isolation: there needs to be an interaction between legislation, available markets and products of suitable quality, as highlighted by Christodoulou and Stamatelatou (2016) in an overview of how legislation is being used to encourage more sustainable forms of sewage sludge management in several different countries. Japan is one of the pioneers in this respect, with a series of legislative and innovation efforts aimed at minimizing sludge production and maximizing the recovery of energy and nutrients, such as phosphorus (Christodoulou & Stamatelatou, 2016). Legislation was passed in 2015 to require sewage operators to use recovered biosolids as a carbon-neutral source of energy (UNESCO, 2017) in an effort to double the amount of energy recovered. Legislation is not the only tool being deployed: financial incentives in the form of feed-in tariffs and measures to encourage investment in new technologies for energy recovery were also introduced (Christodoulou & Stamatelatou, 2016; WWAP, 2017).

Encouragingly, Christodoulou and Stamatelatou (2016) found a general shift in legislation favouring realization of the value of energy and nutrients in sewage sludge, and they have set out some guidelines for such frameworks which balance human health and safety with the desire to achieve economic, environmental and social impacts. Governments can also play a key role in helping new enterprises prove, scale and commercialize their innovations in the sanitation market. In Australia, the government agency ARENA (Australian Renewable Energy Agency) has provided significant funding to the Hazer Group to support the construction and operation of a new facility to demonstrate its process for converting sewage-derived biogas into hydrogen and graphite at a large (100 tonne per annum) scale. This is in line with Australia's increasing focus on hydrogen as an alternative fuel (ARENA, 2019).

7.4.4 Role of philanthropy
Once a role for the private sector is established then multiple actors, not only NGOs and governments, can play a role to develop the markets. Philanthropic organisations such as the Bill & Melinda Gates Foundation have a deliberate strategy to encourage market development through market-making grants. This has involved patient support for initiatives such as 3SI and learning the lessons from them, which included that reaching the poor through markets is possible, understanding their wants and aspirations is critical, and working within the prevailing local market system is essential (Rosenboom, 2017). As mentioned above, philanthropy also plays a key role in developing new technologies to the point where they are proven at scale: commercial investors are likely to be reluctant to bear this kind of risk and since scale is a key part of achieving commercial viability, it is vital that such risks continue to be addressed by philanthropic grants. In addition to supporting technical development of their portfolio of 'Reinvent the Toilet' technologies, the Bill & Melinda Gates Foundation also conducted market analyses to identify key areas of opportunity for them in a number of key market segments: this also provides encouragement to potential commercial partners to become engaged. These analyses can be viewed on the Bill & Melinda Gates Foundation Open Research Gateway.

7.4.5 Role of multi-national corporations
Large companies bring resources, knowledge and capacity of working at very large scale, understanding of consumer-based markets and technical and business model innovation expertise. There is growing interest in promoting the role that the private sector can play in the development of the sanitation market among a number of multi-nationals, as evidenced by the establishment and membership of the Toilet Board Coalition, which includes three major corporations with global reach, Kimberly–Clark, Unilever and LIXIL.

Alongside their mentorship role within the Toilet Board Coalition's Accelerator Program for start-ups, such companies are also conducting their own direct exploration of new markets, often through partnerships with other organisations. It was announced in 2018 that LIXIL would partner with the Bill & Melinda Gates Foundation to apply such commercial skills to bear on preparing for a market test of a household 'reinvented toilet' (3BLMedia, 2018). LIXIL had already become directly involved in new products and business model development through the design, development and marketing of the SATO pan, which has had a significant impact worldwide (see Chapter 2 for more information).

Unilever, one of the world's leading manufacturers of toilet cleaners, is partnering with UNICEF to enable more people to have access to toilets: so far they have helped over 16 million such individuals. They have also launched a toilet cleaner aimed at low-income households in India, which represents a growing market as a result of the Swachh Bharat Mission (Unilever, 2017). Unilever has also contributed to the development of new business models such as Clean Team in Ghana (Narracott & Norman, 2011) and has set up the

Transform fund with the Department for International Development in the UK to provide finance and mentoring to innovative businesses. Major utility and waste management companies also have significant roles to play, particularly in leading the development of the recovery and reuse markets. Veolia is one such company: perhaps responding to demand created by stricter legislation in many countries regarding waste disposal and reuse, it offers customers in local authorities and industry globally the plant and equipment for biogas production from sewage sludge (Veolia, n.d.). This reduces the environmental impact of the operators and provides an additional potential income stream.

7.4.6 Need for coordination
It is thus becoming clearer how each individual actor in the system can best work and deploy their skills and resources to support market development. What then becomes vital is that their actions, interventions and incentives are coordinated with the common aim of enabling the market to grow and become self-sustaining. NGOs can work as market makers, supported by philanthropy and other funders, demonstrating viable business models to the point where private enterprises can take over and run them; innovative entrepreneurs can test new products and service delivery models in the market financed by grant-making organisations until they can attract commercial investment; governments can review and design policy and regulations to encourage and support the private sector; investors and finance providers can provide resources to households and enterprises to support individual purchases and improvements in capacity.

7.5 TOWARDS AN IDEAL MARKET
The sanitation market system is far from ideal at present, but much has been learnt about what is wrong and many initiatives are underway to drive improvements. It is possible to think in terms of designing the ideal market. After all, this is the function of the regulator in countries like the UK. It will not be a blank sheet of paper and will have to start with the system imperfect as it is, but it is possible to envision what an ideal market system would look like and then consider strategies to get there. In functional terms the characteristics of this ideal market system would align with the aims of the IFSVC, namely:

(i)   Every customer along the chain is able to purchase the products and services they want (and aspire to) at an affordable price.
(ii)   Each step of the chain/market transaction is performed to high standards of public and environmental safety.
(iii)   Each step of the chain/market transaction realizes the maximum potential for value creation.
(iv)   Each step of the chain/market transaction facilitates optimum recovery of resources and minimizes waste.
(v)   Each step of the chain/market transaction is supported by the business environment and wider context, including policy and regulatory factors.
(vi)   Each actor in the chain recognizes their role and contribution and is incentivized to do so.
(vii)   Effective partnerships between key players are built and new possibilities created.
(viii) The overall chain is working at the required scale for commercial viability and optimal health and environmental impact.

Two ways to make this happen could be considered. The first is that current efforts by different actors individually and through working together will continue to address key issues and opportunities for improvement. While valuable, these are likely to be incremental, unless there is a concerted effort by all the players at a large enough scale, such as a major city, to create, evaluate and demonstrate a viable market system. The second is that some kind of disruptive innovation will occur which dramatically alters the economic incentives at one or more stages of the chain. An example of this would be a novel high-value end product for which market demand and value was such that the need for faecal waste as a raw material drove backwards vertical integration along the IFSVC. This integration would involve all elements of the chain: toilets would be redesigned not just for consumer appeal, but also to preserve the raw material ingredients; collection and transport would be optimized for the same purpose; and treatment facilities would be redesigned to make the necessary conversions.

Such a strong market pull does not exist at present, but there are already signs of what it could look like as a result of valuable marketplace experiments and learning. Further research in economics and science and technology would be welcome to help frame the next round of business exploration and discovery. Economically it should be possible, using the data already in existence from different business models on the costs of operation along the IFSVC, to estimate the market value needed to drive backwards vertical integration sustainably. This will set the challenge and the target for science and technology, for example through the continued exploration of the potential of bioconversion (Chapter 2).

7.6 Take action

(I)   Explore the sanitation marketplace in your city and show actors, activities and interactions within the market.

(II)   Visit your chamber of commerce to find out the existing sanitation businesses and the chain linkages between them.

7.7 Journal entry

(I)   Determine the sanitation market status in your locality and country and the economic value addition at local and national levels.

(II)   What are the sewered and non-sewered sanitation markets? Where is each predominant?

7.8 Reflection

(I)   Consider the enterprises and actors involved in the marketing, sales and distribution/supply of sanitation products and services.

(II)   With the aid of a diagram illustrate the linkages and interactions between the players that participate in the marketplace and sales.

7.9 Guiding questions

(I)   What are the three main reasons for achieving an efficient IFSVC?
(II) What is a sanitation market and sanitation market system?
(III) What are the estimates of the sanitation economy in India, Kenya and Nigeria?
(IV) What is the implication of these estimates for sanitation markets in these countries?
(V)   What is sanitation market failure and what implication does it have for the IFSVC?
(VI) Why are marketing and sales considered to be a primary duty of the IFSVC?

7.10 Tools

Tools have been developed to explore the market attractiveness of different reuse options and tested in five different cities (Andriessen *et al.*, 2017). A tool is also available to estimating and valuing the resources which could be recovered from municipal waste streams (Ddiba *et al.*, 2016).

ACKNOWLEDGEMENTS

The author is grateful to David Drew, Marc Lewis, John Sauer, Geoff Revell and Jim Lane for their helpful advice and image provision.

REFERENCES

3BLMedia (2018). LIXIL to pilot household reinvented toilets in partnership with the Gates Foundation. https://www.3blmedia.com/News/LIXIL-Pilot-Household-Reinvented-Toilets -Partnership-Gates-Foundation (accessed 2021)

Agarwal R., Chennuri S. and Mihaly A. (2018). Scaling Market-Based Sanitation: Desk Review on Market-Based Rural Sanitation Development Programs. USAID (WASHPaLS), Washington, DC.

Andrew B. (2008). Market failure, government failure and externalities in climate change mitigation: the case for a carbon tax. *Public Administration and Development*, 28, 393–401, https://doi.org/10.1002/pad.517

Andriessen N., Schoebitz L., Bassan M., Bollier S. and Strande L. (2017). Market driven approach for faecal sludge treatment products. In: Local Action with International Cooperation to Improve and Sustain Water, Sanitation and Hygiene (WASH) Services, R. J. Shaw (ed.), Proceedings of the 40th WEDC International Conference, Water Engineering and development Centre, Loughborough University, pp. 1–6, 24–28 July 2017. Available at: https://repository.lboro.ac.uk/articles/conference_contribution/Market_driven_approach_for_faecal_sludge_treatment_products/9589361/1 (accessed 2020)

ARENA (2019). World-first project to turn biogas from sewage into hydrogen gas. https://arena.gov.au/news/world-first-project-to-turn-biogas-from-sewage-into-hydrogen-and-graphite/ (accessed 2021)

Banks I. J., Gibson W. T. and Cameron M. M. (2013). Growth rates of black soldier fly larvae fed on fresh human faeces and their implication for improving sanitation. *Tropical Medicine & International Health*, **19**, 14–22, https://doi.org/10.1111/tmi.12228 (accessed 2020)

Business Wire (2020). Global sanitary ware market (2020 to 2025) – technological advancements present lucrative opportunities. https://www.businesswire.com/news/home/20200615005245/en/Global-Sanitary-Ware-Market-2020-2025-- (accessed 2020)

Christodoulou A. and Stamatelatou K. (2016). Overview of legislation on sewage sludge management in developed countries worldwide. *Water Science and Technology*, **73**, 453–462, https://doi.org/10.2166/wst.2015.521

Ddiba D. I. W., Andersson K. and Rosemarin A. (2016). Resource Value Mapping (REVAMP): A Tool for Evaluating the Resource Recovery Potential of Urban Waste Streams. Discussion Brief, Stockholm Environment Institute, Stockholm.

De Oliveira A. (2018). Market solutions and inequalities in sanitation services access in Brazilian cities. *Theoretical and Empirical Researches in Urban Management*, **13**(4), 28–42.

Diener S., Semiyaga S., Niwagaba C., Muspratt A. M., Gning J. B., Mbeguere M., Ennin J. E., Zurbrugg C. and Strande L. (2014). A value proposition: resource recovery from faecal sludge – can it be the driver for improved sanitation? *Resources, Conservation and Recycling*, **88**, 32–38, https://doi.org/10.1016/j.resconrec.2014.04.005

Dumpert J. and Perez E. (2015). Going beyond mason training: enabling, facilitating and engaging rural sanitation markets for the base of the pyramid. *Waterlines*, **34**, 210–226, https://doi.org/10.3362/1756-3488.2015.021

Fortune Business Insights (2020). Biogas Market Size, Share and COVID-19 Impact Analysis, Report ID: FBI100910. https://www.fortunebusinessinsights.com/industry-reports/biogas-market-100910 (accessed 2021)

Gates Foundation (2017). Bill & Melinda Gates Foundation Open Research Gateway, J-OP market landscape – India. https://gatesopenresearch.org/documents/3-1698 (accessed 2021)

Gates Foundation (2018). Bill & Melinda Gates Foundation reinvented toilet expo overview. https://www.fsmtoolbox.com/assets/pdf/Reinvented_Toilet_Expo_Overview_2018.pdf (accessed 2021)

Grau M. and Alcock N. (2019). Viability of A Black Soldier Fly Plant for Processing Urine Diversion Toilet Faecal Waste. Gates Open Research. https://gatesopenresearch.org/documents/3-1665 (accessed 2021)

Harada H., Strande L. and Fujii S. (2016). Challenges and opportunities for faecal sludge management for global sanitation. In: Towards Future Earth: Challenges and Progress of Global Environmental Studies, T. Katsumi and S. Hashimoto (eds.), Kaisei Publishing, Tokyo, pp. 81–100.

JMP (2017). Progress on Drinking Water, Sanitation and Hygiene: 2017 Update and SDG Baselines, WHO & UNICEF, Geneva. https://washdata.org/sites/default/files/documents/reports/2019-05/JMP-2017-report-final.pdf (accessed 2020)

Jenkins M. W., McLennan L., Revell G. and Salinger A. (2019). Strengthening the sanitation market system: WaterSHED's hands-off experience. All Systems Go Symposium, IRC, 2019. http://watershedasia.org/wp-content/uploads/Strengthening-the-sanitation-market-system_WaterSHED%E2%80%99s-Hands-Off-experience.pdf (accessed 2020)

Koottatep T., Cookey P. E. and Polprasert C. (2019). Regenerative Sanitation: A New Paradigm for Sanitation 4.0. IWA Publishing, London, UK, chapter 6.

Mason N., Matoso M. and Smith W. (2015). Private Sector and Water Supply, Hygiene and Sanitation, ODI Report, Overseas Development Institute, London, UK. 74pp.

Mukherjee A., Arya P., Desgupta S. and Chabbra S. S. (2019). Bridging the Gap: Opportunities for Private Sector Participation in Faecal Sludge and Septage Management. Centre for Policy Research, New Delhi, India, https://doi.org/10.13140/RG.2.2.22926.72006

Mulumba J. N., Nothomb C., Potter A. and Snel M. (2014). Striking the balance: what is the role of the public sector in sanitation as a service and a business? *Waterlines*, **33**, 195–210, https://doi.org/10.3362/1756-3488.2014.021

Murta J. C. D., Willetts J. R. M. and Triwahyudi W. (2018). Sanitation entrepreneurship in rural Indonesia: a closer look. *Environment, Development and Sustainability*, **20**, 343–359, https://doi.org/10.1007/s10668-016-9883-7

Narracott A. and Norman G. (2011). Clean Team, a human-centred approach to sanitation: initial trials in Ghana. WSUP Practice Note. https://www.wsup.com/content/uploads/2017/08/PN008-ENGLISH-CleanTeam.pdf (accessed 2021)

Nguyen H. C., Liang S.-H., Li S.-Y., Su C.-H., Chien C.-C., Chen Y.-J. and Huong D. T. M. (2018). Direct transesterification of black soldier fly larvae (*Hermetia illucens*) for biodiesel production. *Journal of Taiwan Institute of Chemical Engineers*, **85**, 165–169, https://doi.org/10.1016/j.jtice.2018.01.035

Ofwat (2015). Water 2020 Regulatory Framework for Wholesale Markets and the 2019 Price Review. Appendix 1. Sludge Treatment, Transport and Disposal: Supporting Evidence and Design Options. Ofwat, London, UK. https://www.ofwat.gov.uk/wp-content/uploads/2015/12/pap_tec20151210water2020app1.pdf (accessed 2020)

Ofwat (2020). Water Sector Overview. Ofwat, London, UK. https://www.ofwat.gov.uk/regulated-companies/ofwat-industry-overview/ (accessed 2020)

Parker J., Hakspiel D., Foote A. and Woods E. (2020). Waste to Value Sanitation in Kakuma Refugee Camp. UNHCR, Geneva. https://wash.unhcr.org/download/waste-to-value-sanitation-in-kakuma-refugee-camp/ (accessed 2020)

Pedi D., Jenkins M., Aun H., McLennan L. and Revell G. (2011). The 'hands–off' sanitation marketing model: emerging lessons from rural Cambodia. In: The Future of Water, Sanitation and Hygiene in Low-Income Countries – Innovation, Adaptation and Engagement in a Changing World, R. J. Shaw (ed.). Proceedings of 35th WEDC International Conference, Water Engineering and Development Centre, Loughborough University, Loughborough https://wedc-knowledge.lboro.ac.uk/index.html 6–8 July 2011. http://wedc-knowledge.lboro.ac.uk/index.html (accessed 2020)

Remington C., Jean L., Kramer S., Boys J. and Dorea C. (2018). Process cost analysis for the optimization of a container-based sanitation service in Haiti. Proceedings of 41st WEDC International Conference, Nakuru, Kenya. https://www.oursoil.org/wp-content/uploads/2018/07/Process-cost-analysis-for-the-optimization-of-a-container-based-sanitation-service-in-Haiti.pdf (accessed 2020)

Rosenboom J. W. (2017). Developing markets for sanitation. SuSanA and BEAM Exchange Webinar 12. https://www.youtube.com/watch?v=ReZTD8XCOGU&feature=youtu.be (accessed 2020)

Russel K., Hughes K., Roach M., Auerbach D., Foote A., Kramer S. and Briceno R. (2019). Taking container–based sanitation to scale: opportunities and challenges. *Frontiers in Environmental Science*, **7**, 00190. https://doi.org/10.3389/fenvs.2019.00190 (accessed 2020)

Sanivation (2020). Superlogs Overview. https://sanivation.com/approach (accessed 2020)

Schouten M. and van Dijk M. P. (2007). The European water supply and sanitation markets. In: Water and Liberalisation: European Water Scenarios', M. Finger, J. Allouche and P. Luis-Manso (eds.), IWA Publishing, London, pp. 11–33.

Scott P., Forte J. and Mazeau A. (2018). Barriers and opportunities for SMEs: a study of the wider market system in Ghana. https://www.issuelab.org/resources/28424/28424.pdf (accessed 2020)

Singh S., Singh A., Sharma V. and Sinha B. (2017). PSI White Paper Building a sustainable market for toilets: lessons learned from rural Bihar, India. https://www.psi.org/wp-content/uploads/2020/07/India-Sanitation-Report_FINAL.pdf

Sivaramakrishnan S. (2019). 120,000 Tonnes of Faecal Sludge: Why India Needs A Market for Human Waste. World Economic Forum. https://www.weforum.org/agenda/2019/09/how-to-improve-sanitation-in-india/ (accessed 2020)

Sparkman D. and Sturzenegger G. (2016). Fostering water and sanitation markets in Latin America and the Caribbean, Inter-American Development Bank Report, Washington DC.

Sy J., Warner R. and Jamieson J. (2014). Tapping the Markets: Opportunities for Domestic Investment in Water and Sanitation for the Poor. World Bank. http://doi.org/10.1596/978-1-4648-0134-1

Toilet Board Coalition (2020a). The sanitation economy in India. https://www.toiletboard.org/media/65-Sanitation-Economy-Markets-India-2020.pdf (accessed 2021)

Toilet Board Coalition (2020b). The sanitation economy in Kenya. https://www.toiletboard.org/media/64-Sanitation-Economy-Markets-Kenya_2020.pdf (accessed 2021)

Toilet Board Coalition (2020c). The sanitation economy in Nigeria. https://www.toiletboard.org/media/63-Sanitation-Economy-Markets-Nigeria-2020.pdf (accessed 2021)

Tremolet S. (2012). Sanitation markets. Using economics to improve the delivery of services along the sanitation value chain. https://sswm.info/sites/default/files/reference_attachments/TREMOLET%202012%20Sanitation%20Markets.%20Using%20economics%20to%20 improve%20the%20delivery%20of%20services%20along%20the%20sanitation%20 value%20chain.pdf (accessed 2020)

Tsinda A., Abbot P. and Chenoweth J. (2015). Sanitation markets in urban informal markets of East Africa. *Habitat International*, **49**, 21–29, https://doi.org/10.1016/j.habitatint.2015.05.005

Unilever. (2017). Two Domestos innovations that will help tackle the sanitation crisis. https://www. unilever.com/news/news-and-features/Feature-article/2017/two-domestos-innovations-that-will-help-tackle-the-sanitation-crisis.html

Veolia (n.d.). When sewage sludge becomes a source of green energy. https://www.veolia.com/en/ solution/sewage-sludge-green-energy-biogas-wastewater (accessed 2021)

World Bank (2019). Evaluating the potential of container-based sanitation. https://www.worldbank. org/en/topic/water/publication/evaluating-the-potential-of-container-based-sanitation (accessed 2020)

WWAP (UN World Water Assessment Programme) (2017). The United Nations World Water Development Report 2017. Wastewater: The Untapped Resource, UNESCO, Paris, 198pp. https://unesdoc.unesco.org/ark:/48223/pf0000247153/PDF/247153eng.pdf.multi (accessed 2021)

doi: 10.2166/9781789061840_0203

Chapter 8

Sanitation advocacy services

Jack Sim and Peter Cookey

Chapter objectives

The aim of this chapter is to help the reader understand the sanitation advocacy value chain (SAVC) and its contributions in sanitation and hygiene improvement, and also provide the opportunity to situate organizations, actors and professionals in the sanitation advocacy sub-sector within a sanitation economy. It will also show how advocacy-aided value chain business models build a critical mass of people to support a common sanitation cause by creating high social impact with a sizable market for sanitation products and services. Furthermore, this chapter will provide more insights on how the SAVC strategy enables advocacy organizations (enterprises) to evaluate their operations and processes so that they can provide the greatest opportunities that reduce operational costs and optimise their efforts in improving access to safely managed sanitation and hygiene.

8.1 INTRODUCTION

Influencing sanitation and hygiene public policy change can be very difficult and complex, particularly for those with limited power and resources, which is constructed through complex interactions and negotiations amongst a range of stakeholders, including politicians, professionals, interest groups, advisers, bureaucrats, and a range of other actors (Bridgman & Davis, 2004; Clavier & De Leeuw, 2013; Cullerton *et al.*, 2018), and could be policy, market and/or behaviour change driven. The term 'advocacy' suggests systematic efforts (as opposed to sporadic outburst) by actors that seek to further specific policy goals (Prakash & Gugerty, 2010). In other words, the process of undertaking active interventions with explicit goals of influencing government policies is known as advocacy (Cullerton *et al.*, 2018; Onyx *et al.*, 2010) and in the case of sanitation, primarily directed at achieving policy practice, social and/or political change for the implementation of the SDG6. Advocacy activities can include public education and influencing public opinion; research for interpreting problems and suggesting preferred solutions; constituents' actions and public mobilizations; agenda setting and policy design; lobbying; policy implementation, monitoring, and feedback; and election-related activities (Reid, 2001). Advocacy has grown from being focused on service delivery, which is often felt to have limited impact, to engaging with and influencing key policies and decisionmakers at different levels (Arensman, 2020). Advocacy pursues outcomes as structural changes in social, political and organizational systems while challenging existing power structures (Hudson, 2001; Keck & Sikkink, 1998). As an interaction between government, society, and enterprises it can be viewed through three lenses, that is:

(i)   *global-level advocacy* where it is internationally referred to as advocacy among organizations and their networks in civil societies, international institutions and national governments;

(ii) *national-level advocacy* which involves larger, more formal organizations, structures, and practices; and

(iii) *grassroots-level advocacy*, which takes place at the level of states/provinces/ municipalities/districts (Reid, 2001); these organizational networks and practices are less formal at the local level.

Advocacy activities are embedded in distinct organizational models, setting boundaries around the practice of advocacy and participation in the political and social processes that improve sanitation and hygiene service delivery by insiders and outsiders alike (Minkoff, 1999). They include interest groups, political organizations, mobilizing groups, public interest groups, citizen organizations, multi-issue organizations, social movement organizations, and other types of non-profit organizations. Thus, advocacy as participation for sanitation and hygiene improvement addresses the ways organizations stimulate public actions, create opportunities for people to express their concern in social and political arenas, and build the resources and skills necessary for effective actions (Verba *et al.*, 1995).

Subsequently, in order to engage in an advocacy campaign, it is important to understand the various levels of the value chain involved in the process. The value chain as a sanitation advocacy management tool (SAMaT) enables the manager to visualize (in a systematic and integrated manner) the operations and set of processes that exist in the organization (Monteiro *et al.*, 2017), especially the comprehension of cause and effect relations, interfaces and overlaps, as well as the results and impacts that contribute to the efficiency of the sanitation advocacy (SA) and identification of the value and benefits aggregated in the process, particularly with the identification of processes that add value to SA products and services. The design of the value chain can help advocacy campaign managers/facilitators prioritise the improvement of their processes and activities to achieve an increase in service delivery as well as coverage amongst vulnerable groups (Monteiro *et al.*, 2017). In this regard, the advocacy value chain is a presentation of the activities performed to design, produce, deliver and sustain a sanitation advocacy campaign (SAC) for policy change and societal change that encourage improved budgetary allocation for sanitation and hygiene (Monteiro *et al.*, 2017; Porter, 1989) and sanitation behaviour. Dicken (2007) defines it as a sequence of activities in which each activity adds value to the sanitation advocacy campaign.

Many advocacy campaigns around the world start from various areas of the value chain and realize they need other areas of the value chain in order to claim success (Elens-Edeh, 2017). The sanitation value chain consists of a variety of activities within the advocacy organization that are undertaken to provide valuable sanitation products or services to the markets (Chofreh *et al.*, 2019; McGuffog, 2016). It provides opportunities for sanitation advocacy organizations (SDOs) to evaluate their processes and activities to leverage their influence on policy processes and societal/community engagements for the achievement of the SDG 6 targets on sanitation and hygiene. It is important to understand the various levels of the sanitation value chain (SVC), especially the sanitation advocacy value chain (SACV). Advocacy value chain consists of citizens acting individually with different professional titles and/or as a collective (Reid, 2001) often represented by non-profit or not-for-profit organizations and/or social-preneurs or enterprises (Reid, 2006). There is no agreement on which activities constitute advocacy, and no singular source gives a full account of the many kinds of activities and strategic groups used to leverage influence in the processes of change in sanitation and hygiene services (Gen & Wright, 2013).

8.2 SANITATION ADVOCACY

Advocacy as participation addresses the ways organizations stimulate public action, create opportunities for people to express their concerns in social and political arenas, and build the resources and skills necessary for effective actions (Verba *et al.*, 1995). Advocacy as organizational representation often referred to as 'direct advocacy' includes lobbying and other appearances before key decisionmakers on behalf of others (McCarthy & Castelli, 1996) while 'indirect advocacy' describes participatory aspects of non-profit advocacy, particularly the capacity of groups to stimulate individual citizens to take actions on their own behalf (McCarthy & Castelli, 1996; Reid, 2001). However, advocacy describes a wide range of individual and collective expressions or actions on a cause, idea, or policy, and may also refer to specific activities or organizations. Sometimes a distinction is made between advocacy on behalf of others and grassroots advocacy or civic or political participation. Words associated with 'advocacy' include defending, influencing, sensitizing, interviewing, change, decision-making, persuasion, selling an idea, exposure, lobbying, communication and attracting attention (AALEP, 2013). 'Programmatic' (or issue) advocacy is when an organization takes a position on a public policy that affects their work (AALEP, 2013; Chofreh *et al.*, 2019). Specific advocacy approaches described by (Hopkins, 1993) include:

- legislative advocacy, or lobbying of legislators;
- political campaign advocacy to support or oppose political candidates whose agenda do not support improve sanitation and hygiene programmes;
- demonstrations and rallying public support around an issue or policy;
- boycotts to encourage or discourage businesses with a targeted entity;
- litigation or using legal action to advance a cause (Chofreh *et al.*, 2019);
- grassroot advocacy, or engaging individual citizens in advocacy effort;
- change advocacy (e.g. behaviour and perception change advocacy) (Cookey *et al.*, 2020); and
- capacity building for policy improvement (Chofreh *et al.*, 2019; Morariu *et al.*, 2009).

Advocacy for sanitation and hygiene can be described as a combination of individual and social actions that are expected to achieve collaborated and coordinated sustained public information campaign to change sanitation and hygiene policies and improve service delivery at local, national and international level (de Jong, 2003; Uzochukwu *et al.*, 2020) as well as behaviour/perception change (Cookey *et al.*, 2020). This can include many activities that organizations undertake such as media campaigns, public speaking, commissioning and publishing research, capacity building, and relationships development. The types and kinds of advocacy may include (Carlisle, 2000; Uzochukwu *et al.*, 2020):

- facilitation advocacy focused on the empowerment of the disadvantaged;
- representational advocacy designed to ensure that the systems support and protect the vulnerable;

Box 8.1: Some global sanitation and hygiene advocacy organizations with grassroots spreads presented according to their year of formation

Water Supply and Sanitation Collaborative Council (WSSCC) – a global, multi-stakeholder membership and partnership organization hosted by the United Nations that works with poor people, organizations, governments and local entrepreneurs to improve sanitation and hygiene at scale. Established in 1990, WSSCC advocates for improved sanitation and hygiene, paying attention to the needs of women, girls and people in vulnerable situations. To reach the SDG 6.2 target of safely managed sanitation, there is an urgent need to globally prioritize sanitation, hygiene and menstrual health. That is why WSSCC evolved into the *Sanitation and Hygiene Fund* in 2021.

World Toilet Organization (WTO) – a global non-profit organization committed to improving toilet and sanitation conditions worldwide. Founded on 19 November 2001, World Toilet Organization (WTO) empowers individuals through education, training and building local marketplace opportunities to advocate for clean and safe sanitation facilities in their communities. WTO established World Toilet Day and the World Toilet Summit in 2001; this was followed by the World Toilet College in 2005. On 24 July 2013, WTO achieved a key milestone for the global sanitation movement when 122 countries co-sponsored a UN resolution tabled by the Singapore government to designate 19 November, World Toilet Day, as an official UN day. World Toilet Organization was granted consultative status with the United Nations Economic and Social Council in 2013.

End Water Poverty (EWP)– a global civil society coalition, campaigning to end the water and sanitation crisis. Established in 2007 with a coalition of 270 CSOs working to end water and sanitation crisis and in more than 90 countries around the world, working in all regions across Africa, Asia, Latin America, Europe and North America. EWP vision is to ensure that everyone everywhere has access to safely managed water and sanitation services and good hygiene. To achieve EWP we work at different levels of advocacy to leverage sustainable change such as globally, regionally and nationally).

Sanitation and Water for All (SWA) – a multi-stakeholder partnership of governments and their partners from civil society, the private sector, UN agencies, research and learning institutions and the philanthropic community. Established in 2008 to create a Global Framework for Action on Sanitation and Water Supply (GF4A), which was launched at a side-event during the UN MDG High-Level Event. Partners share the belief that government-led, collaborative and multi-stakeholder decision-making leads to more effective and sustainable solutions. More specifically, SWA's work focuses on encouraging and motivating partners to increase political prioritization of water, sanitation and hygiene; ensure adequate financing; and build better governance structures and institutions to achieve SDG6 by the year 2030. SWA's communications team was hosted by the Water Supply and Sanitation Collaborative Council (WSSCC) until the end of 2019. From 2020 onwards, it is being hosted by UNICEF.

The Toilet Board Coalition (TBC) – is a global alliance of corporations, government agencies, multilateral institutions, sanitation experts and non-profit organisations that aims to bring sanitation to millions of families by catalysing and accelerating scalable market-based initiatives that bring together the resources and skills of corporations, the know-how of the development sector, and the expertise of the non-profit sector. The motivations of the corporates involved in the TBC range from developing new markets for products, equipment and services, to collaborating and learning from others, exploring innovative business models and BoP solutions, attracting young talent, providing content for their communications, or even contributing the transformation of their organisations. The TBC was officially launched in November 2014.

- social justice advocacy – seeking to organize communities and citizens to come together and speak out about their concerns;
- administrative advocacy – focused on the implementation phase of the policy process when rules and regulations are promulgated and service delivery systems designed and put in place, sometimes with feedback from citizen groups (Reid, 1998);
- programme advocacy – describing the everyday work of organizations carrying out their charitable missions or providing services, as long as the activities are not outside the realm of related speech; does not refer to specific legislation and does not become partisan activity (Hopkins, 1993);
- society-related advocacy – where non-profits have an important role to play outside government in shaping public opinion, setting priorities for the public agenda, and mobilizing civic voices and action, social change, or social movements; and
- behaviour change intervention advocacy, seeking to improve access to safely managed sanitation and hygiene facilities, perceptions of sanitation and related materials, as well as for proper use of provided infrastructure.

For advocacy to be sustained over time, it should include multiple tracks of activities and messages directed towards targeted audiences at all levels and this requires better understanding of the various advocacy goals, added value and the motivation for their use (Data Harvest, 2009). The global advocacy sanitation value could be described as relatively nascent, but because of the activities in recent years, it is off to a good start. Therefore, the effectiveness in the implementation of the advocacy value-chain over the years has contributed to raising and sustaining general awareness about sanitation and hygiene across the development sector, especially advocacy around the establishment of the annual UN World Toilet Day by the World Toilet Organization (WTO), the International Year of Sanitation (IYS), Global WASH Campaign of the Water and Sanitation Collaborative Council (WSSCC) now Sanitation and Hygiene Fund (SHF), Global Handwashing Day (UNICEF), eThekwini Declaration/ AfricaSan Action Plan of 2008, German Toilet Organization's 'Where Would You Hide Campaign?' and much work by WaterAid.

A value chain strategy enables advocacy organizations (enterprises) to evaluate their operations and processes so that they can provide the greatest opportunities that reduce operational costs, optimise efforts in improving access to safely managed sanitation and hygiene through public education and influencing public opinion, research for interpreting sanitation problems and suggesting preferred solutions, constituent actions and public mobilizations, agenda setting and policy decisions, lobbying, policy implementation, monitoring and feedback, and so on. Therefore, the aim of sanitation advocacy is to change policy (Reisman *et al.*, 2007) or the policy-making process, generally to make it more accessible and transparent to the public to participate in sanitation and hygiene improvement of their community and also to change the perception and behaviour that are inimical to safe sanitation and reuse of recovered materials (Cookey *et al.*, 2020).

8.3 ADVOCACY-DRIVEN-SANITATION VALUE CHAIN

Sanitation advocacy value chain provides a way of understanding the significant contributions of the role of advocacy in sanitation and hygiene improvement and offers the opportunity to situate organizations, actors and professionals in the sanitation advocacy sub-sector within a sanitation economy. Advocacy-aided value chain business models will build a critical mass of people to support a common cause by creating high social impact with a sizable market for sanitation products and services as well as discouraging negative impacts of 'hero-preneur' self-defeating tendencies as they grow to become obstacles to their own missions. The value chain approach removes redundancies, duplications, bureaucracies, centralized hierarchies, intermediaries, expensive consultancies, and any cost of general distrust, to unleash an integrated delivery model at a fraction of current costs. Sanitation can add value to farmers' and artisans' products up the agricultural value chain through the process of biomass recovery and products transformation (see Chapter 6). Sanitation can mimic the value-added processes of agricultural products where chilli peppers can become chilli pepper sauce at 50× higher value; coffee beans to cafe drinks are 100× value-added; straw hats and designer hats maybe 10× more expensive. Sanitation's examples are compost, bricks and so on. To top that up, business models can be deployed to uplift the income of the poor by using digitized platform cooperatives – cooperatively owned, democratically governed businesses that establish a computing platform and make use of a website, mobile applications or a protocol to facilitate the sales of goods and services. The involvement of Mondragon Team Academy (MTA World), a global network of social innovation ecosystem laboratory using Finnish Educational methods operating in Spain, and the New School University in New York City (NYC) teaches these value-added innovative business models.

The Base-of-Pyramid (BoP) population, that is the largest and poorest socio-economic groups that earn less than $2.50 a day (2.7 billion), need to have access to knowledge/training, access to customers/markets, access to finance, access to logistics, and access to technologies for safe sanitation. One way is to map all assets, match them into alignments, facilitate pathfinding, and then motivate each of the stakeholders individually and collectively. These can be done via a combination of algorithms and human interventions. With an ecosystem approach, sanitation products and services can be delivered faster, cheaper, better, easier, and sustainably. This will attract investments to entrepreneurs and create jobs which in turn generates income. This income becomes expenditure and the velocity of money creates the multiplier effects needed for economic growth and then supports poverty reduction and promotes self-reliance, as well as needed sanitation and hygiene infrastructure. A sanitation economy is then in view through the mechanism of advocacy-driven sanitation value chain systemic strategies. The Government can now tax these new middle-class demographics and invest in public goods so that the people get the quality of life they deserve. Changing the world is not so complicated; the status quo is much more complicated. And that's why we need to change the status quo and at the end everyone gets safely managed sanitation, and finally toilets. This is the theory of change that will motivate each stakeholder to act for their own self-interests while simultaneously delivering the common good for the sanitation mission. The lesson is that when we want to solve a problem as a movement, it is cheaper, faster, better, and easier and best done through the lens of value-added sanitation advocacy. Box 8.2 illustrate some global advocacy initiatives.

Advocacy-driven-sanitation value chain provides the effectiveness needed to end open defecation and bring everyone good sanitation by designing customized value-added incentives for all the stakeholders involved in each of the 17 SDGs because of the interlinkages of the SDG 6 and this could encourage others to join in the ecosystem to improve and sustain sanitation and hygiene services. Thus, unlocking the spirit

---

Box 8.2: Some global advocacy initiatives

International Year of Sanitation (IYS) – the year 2008 was declared the International Year of Sanitation by the United Nation. The goal of IYS was to help raise awareness of the sanitation crisis and to accelerate progress towards reaching the UN's Millennium Development Goals (now replaced by the UN's Sustainable Development Goals (SDGs)) and cutting the number of people without access to basic sanitation and hygiene in half by the year 2015. IYS was considered to be one of the biggest international advocacy initiatives in sanitation/hygiene and demonstrated significant outcomes.

Global Handwashing Day – an initiative of the Global Handwashing Partnership showcasing a successful public–private partnership is an annual global advocacy day dedicated for handwashing with soap as an easy, effective, and affordable way to prevent diseases and save lives. It is an opportunity to design, test, and replicate creative ways to encourage people to wash their hands with soap at critical times. Global Handwashing Day is celebrated every year on 15 October. The first Global Handwashing Day was held in 2008, when over 120 million children around the world washed their hands with soap in more than 70 countries. Since 2008, community and national leaders have used Global Handwashing Day to spread the word about handwashing, build sinks and tippy taps, and demonstrate the simplicity and value of clean hands. Global Handwashing Day is endorsed by governments, schools, international institutions, civil society organizations, NGOs, private companies, individuals, and more.

World Toilet Day (WTD) observed annually on 19 November, was established by the UN General Assembly in 2013 as an advocacy initiative of Jack Sim, President of the World Toilet Organization (WTO). The goal of WTD is to recognize the importance of sanitation for development and how it impacts on the environment. Safe access to clean toilets is also vital to achieving target 6.2 of the Sustainable Development Goal 6 (SDG6), on adequate and equitable sanitation and hygiene for all. It calls for an end to open defecation, paying special attention to the needs of women and girls and those in vulnerable situations, by 2030.

---

of enterprise and good work ethic of the poor, with value added to their labour with technologies and downstream production, they will have money to buy toilets, clean water, housing, education, healthcare and all the quality of life beyond sanitation.

Consider the fact that each year about $150bn is donated into the development sector, yet these monies have hardly moved the needle forward. Instead of uniting the players, funders seem to be asking NGOs and Social Entrepreneurs to compete for their monies and dividing the sector into a competitive community suspicious of each other. This is not helping the cause for improved and safe sanitation and hygiene. It is essential that funders and investors embrace the ecosystem approach rather than competitive silo funding. An advocacy-driven-sanitation value chain gives each stakeholder what they want, allows different expertise at the table at the same time and, through an alignment of incentives, gives them the opportunity to take ownership and aspire for different rewards such as:

- creating powerful stories and soundbites which could interested the media to sell to large readerships with increased advertising revenue and income;

- making sanitation agenda a viable election ticket for the politicians who are interested to get into the media by promising improved sanitation services to win popularity;
- mobilizing bureaucrats and policymakers to work in collaboration with their politician bosses so they allocate budget for sanitation and hygiene;
- helping donors for sanitation and hygiene projects to realized that prevention is cheaper than cure and gives them a bigger impact for their funding support;
- providing academia with research materials for publications to validate the advocacy hypothesis and strategy with a cost–benefit analysis to position sanitation as the cheapest preventive medicine;
- leveraging the corporate social responsibility of supply chain organizations with impact on sanitation and hygiene of the vulnerable population in the community in which they operate to create shared value programs with win–win strategies;
- creating the right narratives for sanitation and hygiene programme donors/aid agencies/funders to support advocacy projects that influence public policies and integrate the strategy with blended capital from public–private partnerships (PPP);
- mobilizing non-governmental organizations (NGOs) and social entrepreneurs to attract funding for building toilets as well as undertake public awareness campaigns on educating communities on the importance of sanitation and hygiene;
- spicing it up with celebrities to join in for publicity and goodwill to create visibility for sanitation as well as for themselves;
- generating impact investments by social entrepreneurs that create sustainable business models by scaling good practices across sectors and geographies;
- making toilets sexy and fashionable by making the users and local communities take ownership of its upkeep and maintenance, thereby making ownership of toilets a status symbol – and using jealousy to change their habits into showing off their toilets; and
- providing education and training to develop skilled and knowledgeable manpower for sanitation and hygiene as well as related sectors.

8.4 SANITATION ADVOCACY VALUE CHAIN (SAVC) MAPPING

Advocacy activities are embedded in distinct organizational models, setting boundaries around the practice of advocacy and participation in process to improve sanitation and hygiene of the vulnerable groups of the society by insiders and outsiders alike (Minkoff, 1999). Interest groups, political organizations, mobilizing groups, public interest groups, citizen organizations, multi-issue organizations, social movement organizations and other descriptions of non-profit organizations as policy actors adopting different activities and strategies (Berry, 1999).

Sanitation advocacy value chain mapping is the structural description of value-added activities of professionals and organizations engaging in sanitation advocacy. The chain map not only provides an overview of the system identifying the position of the value chain actors, but also helps to visualize many aspects of the advocacy value chain analysis by structuring the information according to the functions and stages of the chain (Springer-Heinze, 2018). Sanitation advocacy value chain (SAVC) mapping is a process that will identify the main activities associated with an advocacy campaign organization's product line that is often used in sanitation advocacy organization's (SAOs) corporate strategy in order to identify performance improvement opportunities designed to achieve desired objectives. Understanding of SAVC is made possible by mapping its value chain, which describes the activities required for sanitation advocacy, from conception, passing through different stages of delivery to target populations and/

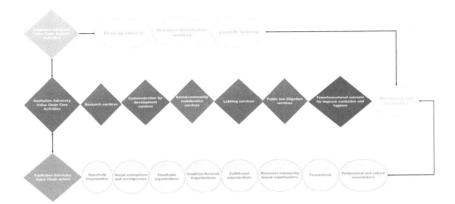

Figure 8.1 Sanitation advocacy value chain mapping (Source: Authors).

or groups (Monteiro *et al.* 2017). The mapping process identifies opportunities in the chain for the sector and social enterprises that grow and expand support for sanitation actors' engagement on how to improve or upgrade the value chain. The main sanitation advocacy value chain activities start from: research, public awareness/campaign, mass mobilization/demonstration/protest, lobbying, litigation, resource mobilization, planning, implementation of the advocacy, education and training, monitoring and evaluation (see Figure 8.1).

(I)   ***Research services*** provide facts and credibility to reach out to target populations (Elens-Edeh, 2017). Researchers add value to advocacy by helping to gain a clear understanding of the causes and effects related to sanitation and hygiene issues from the perspective of identifying practical and feasible solutions that make it possible to build a consensus in favour of change. It is impossible to argue logically and coherently for change without a strong understanding of the insight that research provides (Elens-Edeh, 2017; World Animal Net, 2017). Research is the foundation for successful advocacy, and it is important for both an effective advocacy strategy by enabling thorough strategic analysis and successful advocacy work and then providing authoritative and accurate evidence to support advocacy (Elens-Edeh, 2017; World Animal Net, 2017).

(II) **Social/community mobilization services** bring together all societal and personal influences to raise awareness and demand for better sanitation and hygiene infrastructure, usage, management and service delivery (WHO, 2014). This service adds value by empowering local communities/vulnerable groups by combining awareness, creation, self-organization and action through dialogue and collaboration to facilitate change through an interdisciplinary approach coupled with organizing face-to-face interpersonal communications, group discussions/community dialogues, community outreach and road shows. It is influenced in part by the spread of social media and new information technology (Mostafa, 2020; WHO, 2014). Social mobilization professionals are trained to assess community needs, issues, and resources around sanitation and hygiene, design a social mobilization strategy, identify and partner with other local organizations, design, test and produce social mobilization sanitation and hygiene advocacy materials, and implement and monitor sanitation and hygiene social mobilization activities.

(III) **Communication for development services** adds value across the advocacy value chain by creating a favourable ecology for sanitation and hygiene development programmes through re-thinking and facilitating interactions between economically, politically and culturally disconnected groups and ideas – between indigenous knowledge and science, elite national policymakers and rural communities, donor agencies and local NGOs, industry and academia (e.g., products/services), providers and users, men and women, and didactic pedagogy and participation (Inagaki, 2007) as well as it empowers people, enables expressions and dialogues, raises awareness of socio-structural problems, and fosters self-reflection among marginalized and disadvantaged populations, and then promotes participation and social change using methods and instruments of interpersonal communication, community media and modern information technologies (Inagaki, 2007; Melkote, 1991).

(IV) **Lobbying services** is an accepted and legal process which ensures that all the voices of citizen groups, associations, industry, local leaders and others are heard in the political arena and by policymakers. It is a specialized form of advocacy which is strategically planned and an informal way of influencing decision-makers (Berg, 2009). It adds value as a communication function and closely resembles the work of public affairs that builds and maintain relations with government primarily for the purpose of influencing legislation and regulations (Berg, 2009; Toth, 1986). Lobbying is a very powerful service, which, if it is well used and performed in a professional manner, can help improve sanitation and hygiene policies that will ensure sustainable service delivery.

Advocacy value chain supporting services such as planning, resource mobilization, capacity building and monitoring and evaluation provides an enabling environment for effective advocacy works. *Planning* is the first step in starting advocacy. It helps to avoid surprises, ineffectiveness, clumsiness and incompetency as well as identify issues of interest and the objectives of the sanitation and hygiene advocacies, groups or individuals affected by such issues. It should also define the goals of the advocacy and possible solutions, and also build coalitions and networks, identify target decision makers and the right strategies to be adopted (UN-WATER, 2009). Advocacy activities are resource intensive and as such requires investments of funds, staff time and materials. Thus, *resource mobilization* serves as an agenda for advocacy. The job of resource mobilizers will require influencing donors and institutions to fund certain issues, encouraging individual supporters to give to advocacy, accepting funds from private sector and sharing resources in alliance and coalitions. Resource mobilization adds value for effective implementation of advocacy strategies by ensuring that funds continue to flow for the work (Gosling & Cohen, 2007; PARIS 21, 2010; Sprechmann & Pelton, 2001).

Advocacy capacity intervention helps to identify under-capacity problems and best context specific solutions and also defines effectiveness in advocacy contexts that enhances performance. Capacity building interventions can be targeted at *individual levels and* address issues of skills and abilities; at *project and programme levels*, it addresses issues of single issue campaigns and broader advocacy programmes; and at *organizational levels*, the interventions are designed to address issues relating to organizational structure, processes and resources management, and governance issues; at *external linkages*, capacity interventions look at extent and quality of coordination between organizations, links between organizations and the groups and communities they are supporting and representing; at the *level of enabling environment*, capacity interventions addresses issues of political and policy contexts within which advocacy processes take place; while *multiple levels* enhance the connections between each of the levels stated above and how they work together to enhance advocacy capacity (Stalker & Sandberg, 2011). In addition, **Monitoring and evaluation** shape and transform advocacy strategy and help ensure results have the maximum effect. It is generally conducted to establish accountability, to determine whether a case can be made that advocacy effort produced its intended results and to ensure performance, inform decision-making, providing data that will inform and strengthen advocacy efforts, and then encourage learning experiences (UNICEF, 2010).

The **sanitation advocacy value chain actors, networks and/or organizations** are value chain operators, providers of operational services and support services that enhance delivery of advocacy products and services as well as share benefits that accrue when championing the values of their constituents (Prakash & Gugerty, 2010). In most cases, the value chain actors and their organizations are conceived as being formed largely for the purpose of improvement in sanitation and hygiene policies, frameworks, products and to increase access to more equitable sanitation services delivery to all. These actors could be groups of individuals, enterprises, labour and professional organizations and public agencies which are commonly referred to as advocacy networks/organizations and can operate transnationally, regionally and/or domestically (Keck & Sikkink, 1999). These advocacy value chain actors/organizations could also be professional organizations consisting of salaried employees with established organizational infrastructures or volunteer organizations representing sustained collective actions by non-salaried actors (Prakash & Gugerty, 2010). The major actors in advocacy networks acting in the sanitation value chain include: (i) international and domestic NGOs; (ii) research and advocacy organizations; (iii) local social movements; (iv) foundations; (v) the media; (vi) churches, trade unions, consumers organizations, intellectuals; (vii) parts of regional and international intergovernmental organizations; and (viii) parts of the executive and/ or parliamentary branches of government (Keck & Sikkink, 1999; Prakash & Gugerty, 2010). Groups in the sanitation advocacy network value chain share values and frequent exchange of information among the actors both formal and informal. The movement of funds and services is notable and also some actors within the value chain provide services such as training and personnel who also circulate within and among networks (Keck & Sikkink, 1999). Thus, sanitation advocacy value chain actors, networks and/ or organizations are special types of firms functioning in policy market. These markets vary in terms of entry and exit barriers as well as levels of competition, all of which provide organizations with opportunity to supply distinct products to well defined constituencies. The structure of sanitation policy markets provides opportunities for competition and collaboration (Prakash & Gugerty, 2010).

8.5 ADVOCACY CASE STUDY

8.5.1 Advocacy for the creation of the World Toilet Day

The sanitation agenda often (unsuccessfully) competes for media visibility with other global and industry agendas, especially in development arena. News outlets are even less likely to discuss sanitation concerns over sports, movies, games, food, scandals, technologies and such likes. This was the strategy that Jack Sim, co-founder of the World Toilet Organization (WTO) deployed to break the ceiling of the sanitation challenge in his home country, Singapore. Once forced silence turned to public discourse, enlightenment and willingness followed and then came the designed change gradually. The WTO movement started as a global organization with a presence in several countries across the continent. Subsequently, the WTO proposed having a day globally that focuses on toilets and other related sanitation concerns. This required another level of advocacy using lobbying, networking and the media to push for a UN World Toilet Day.

The WTO team started by lobbying the Singapore government to table a UN resolution for the creation of a UN World Toilet Day. The value of lobbying, contact, networking and public-speaking experience added to the success of this aspect. Then came laborious task of winning over other countries' representatives to lend their support to the resolution. This involved meeting with the diplomatic corps of nations and key international NGOs over dinners, lunches, coffee and conversations at opportune moments and events organized by WTO – to convince them to join the crusade. The added value of social interactions, persuasion and collaboration provided a strong mobilization mechanism for this advocacy drive.

The unique blend of humour, empirical facts and value-added advocacy created headlines and news story with powerful soundbites and amusing photographs that were either shocking or funny. The element of humour captured the imagination of the general public who have always wanted to speak about the toilet subject but felt that they did not have the permission to do so. By calling things as they were, it was possible to unlock the mental blockage and release the freedom of expression in public discourse. Eventually, even though 19 November is Monaco's National Day and Indira Gandhi's birthday in India, after a whole year of hard work in 2013, all 193 countries' governments at the UN General Assembly unanimously adopted the founding day of the World Toilet Organization, which is 19 of November, as the official UN World Toilet Day. This legitimacy has contributed to some countries' politicians using it as an election agenda to win votes, popularity and visibility.

Sanitation products and services cut across all works of life and different markets within homes, workplaces, schools, tourism and hospitality, transportation, health facilities, public spaces and all places where people spend enough time to need the restroom. These artefacts range from low cost and affordable, simple and basic and then luxury and sophisticated, and levels of treatment and recovery plants/systems. Therefore, the sanitation value chain consists of several actors from public, private and civil society sectors and so any sanitation advocacy campaign should emphasise integrated collaboration for it to be successful. In essence, the competition for donor funds amongst different organizational and individual actors and the suspicion played out within the sanitation advocacy sphere should be de-emphasised as each player recognizes the value they and other players bring to overall global pursuit. The WTO avoids this pitfall by avoiding the competition in the playing field and any rigorous fund-raising approach. WTO advocacy adopts an ecosystem approach that recognises the value of other actors, thereby facilitating everybody with the legitimacy of sanitation as a fundable subject, which in turn allows everyone to attract more support, funding, and investments as well as political support and even change public policies and ensure implementation.

8.5.2 Clean India campaign constructed 110 million toilets
Prime Minister Modi's ambitious sanitation campaign often referred to as 'Clean India' help provide access to safely managed sanitation to almost half of India's 1.3 billion population in five years. The advocacy campaign for 'open defecation free India' started when India's President Abdul Kalam opened the World Toilet Summit in Delhi in 2007. Prime Minister Modi in his remark said that everyone in the country contributed to the success of the sanitation campaign. Public participation in the 'Clean India Campaign' was key, for example featuring of Bollywood stars and filmmakers helped bring the issue to screens, and they were well publicized. Since that year, Indian politicians have realized that they could gain a lot of popularity with sanitation, which eventually led to Union Minister Jairam Ramesh building 32 million toilets. The trend of sanitation

awareness in India is now very high, but because the country's population is very large with multiple cultures, we still need to do a lot of work in changing attitudes, and driving demand for sanitation from the ground. Even if we build toilets for them, if they don't use it, it would still not be a success.

8.5.3 China Toilet Revolution

In China, President Xi Jing Ping has become a Toilet champion as well. After he created the China Toilet Revolution in the last five years, there is a complete change in the cleanliness level of tourism toilets. But the seed of the tourism toilet revolution was planted in 2004, at the World Toilet Summit hosted by Beijing Tourism Bureau. They were preparing for the 2008 Olympics, and we worked with them to renovated 4000 public toilet blocks in Beijing's tourism areas. Since then, the transformation of tourism pilot culture in China improved year by year as all tourism bureaus realized that toilets are a profitable feature for facilitating tourism growth. And now, if you go to the Chinese first-tier, second-tier and third-tier cities you will find their tourism toilets are as clean as in a developed country like Japan or Singapore. However, there is still a lot of work to be done, for rural sanitation and rural school sanitation in China. But this momentum is growing very well.

8.5.4 Brazil institutional framework for private sector participation in sewage management

In Brazil, during the World Toilet Summit 2019, an advocacy campaign day was launched to provide a law that allows public–private partnership (PPP) investment in government sanitation treatment companies. After series of debates at the summit, the Senate held several hearings where the negative impacts on health and tourism from the 50% of untreated sewage dumped into the country's rivers and water bodies were highlighted. In the end the Senate was persuaded and in November 2019 many congressmen and senators from Brazil met with global experts during the World Toilet Summit, which then culminated in enough votes and support for a change of law. Seven months after the Summit, the law was passed and now private investors can invest in government sewage treatment companies to enable the country to address the challenges of sewage management in Brazil.

8.6 CONCLUSION

Value chain mapping and analysis help to provide answers to what exactly sanitation advocacy organizations do. They combine norms, information, strategies, mass communication channels, and coalitions to produce a consensus about sanitation and hygiene issues, to influence policy or implementation, or to change behaviour (Lecy *et al.*, 2010). To achieve these, they primarily rely on communicative power – the ability to persuade or influence key decision-makers or general public. Advocacy is a marketplace of ideas whereby participants exchange competing ideas in public in ways similar to a transaction. In this way, the primary activity of an advocacy group is to produce or reproduce norms shaping the public sphere, or campaigns targeting institutional and policy changes by challenging the status quo through ideas, persuasion, education and lobbying (Lecy *et al.*, 2010). Also, advocacy organizations carefully manage their brand as a key factor for their funding strategies because name recognition translates to legitimacy in the eyes of donors and membership fees from individuals. Such legitimacy translates into income through the commodification of the brand. The larger and better known an organization is, the more opportunities will arise to markets its brand (Lecy *et al.*, 2010; Prakash & Gugerty, 2010). Working through the value chain strategies of advocacy organizations means that they rarely act alone to further their agenda. Critically, they bring value contributions together through a collection of individuals and organizations and gives norms a voice, aligns interests and strategies, develop campaigns, and create emotional links between membership/staff and the mandate of the different causes. This produces strong incentive to enter networks and partnerships in order to utilize shaming strategies, exchange information and resources, amplify their voices and/or, or extend their reach in the large-scale mobilization of public opinion (Lecy *et al.*, 2010; Prakash & Gugerty, 2010).

8.7 Take action

(I)   Visit sanitation advocacy organizations to gain a first-hand experience of their operations and business model.
(II) Pick anyone of the stages of the IFSVC and design a business plan/ feasibility study for an advocacy organization towards the development of that sub-value chain within the overall IFSVC.

8.8 Journal entry

(I)   Draw an overview map of the sanitation advocacy value chain of your area.
(II) Write a brief review of the following advocacy organizations paying special attention to business models, operations, strategies, funding and donor base and indicate the difference and similarities: World Toilet Organization (WTO) and Toilet Board Coalition (TBC).

8.9 Reflection

(I)   What roles do advocacy organizations like World Toilet Organization (WTO), Toilet Borad Coalition (TBC), and Sustainable Sanitation Alliance (SUSANA) play in the sanitation advocacy value chain?
(II) What is advocacy and what are the components of advocacy activities?

8.10 Guiding questions

(I)   With the aid of a diagramme, describe the advocacy value chain.
(II)   Describe advocacy approaches with sanitation examples.
(III) Explain sanitation advocacy value chain mapping with examples?
(IV)   Why do you think advocacy-driven-sanitation value chain could provide the effectiveness needed to end open defecation?
(V)   Describe advocacy for sanitation and hygiene.
(VI)   How does the SAVC fit into the IFSVC and why is it a crucial component of the IFSVC?

REFERENCES

AALEP (Association of Accredited Public Policy Advocates to the European Union). (2013). Understanding Advocacy: Context and Use. http://www.aalep.eu/understanding-advocacy-context-and-use (accessed 06 November 2020)

Arensman B. (2020). Advocacy outcomes are not self-evident: the quest for outcome identification. *American Journal of Evaluation*, **41**(2), 216–233. https://doi.org/10.1177/1098214019855137

Berg K. T. (2009). Finding connections between lobbying, public relations and advocacy. *Public Relations Journal*, **3**(3), 1–19.

Berry J. (1999). New Liberalism: The Rising Power of Citizen Groups. Brookings Institution, Washington, D.C.

Bridgman P. and Davis G. (2004). The Australian Policy Handbook. Allen & Unwin, Australia.

Carlisle S. (2000). Health promotion, advocacy and health inequalities: a conceptual framework. *Health Promotion International*, **15**(4) 369–376. https://doi.org/10.1093/heapro/15.4.369

Clavier C. and De Leeuw E. (2013). Health Promotion and the Policy Process. Oxford University Press, United Kingdom.

Chofreh A. G., Goni F. A., Zeinalnezhad M., Navidar S., Shayestehzadeh H. and Klemeš J. J. (2019). Value chain mapping of the water and sewage treatment to contribute to sustainability. *Journal of Environmental Management*, **239**, 38–47. https://doi.org/10.1016/j.jenvman.2019.03.023

Cookey P. E., Kugedera Z., Alamgir M. and Brdjanovic D. (2020). Perception management of non-sewered sanitation systems towards scheduled faecal sludge emptying behaviour change intervention. *Humanities and Social Science Communications*, **7**, 183. https://doi.org/10.1057/s41599-020-00662-0

Cullerton K., Donnet T., Lee A. and Gallegos D. (2018). Effective advocacy strategies for influencing government nutrition policy: a conceptual model. *International Journal of Behavioral Nutrition and Physical Activity*, **15**, 83. https://doi.org/10.1186/s12966-018-0716-y (accessed 29 November 2020)

Data Harvest. (2009). Sanitation Advocacy Research Project. Global Development Water, Sanitation and Hygiene (WSH) initiative of the Bill & Melinda Gates Foundation. http://www.ipcinfo.org/fileadmin/user_upload/unwater_ext/documents/pdf/TF_Sanitation/Sanitation_Advocacy_Research_Project_-_Final_Report.pdf (accessed 21 November 2020)

de Jong D. (2003). Advocacy for Water, Environmental Sanitation and Hygiene: Thematic Overview Paper. IRC International Water and Sanitation Centre, The Netherlands. https://www.susana.org/_resources/documents/default/2-439-jong-de-et-al-2003-advocacy-water-sanitation-health-irc-en.pdf (accessed 04 November 2020)

Dicken P. (2007). Global Shift: Mapping the Changing Contours of the World Economy. 4th edn. Sage, London.

Elens-Edeh E. (2017). The Six Steps of the Advocacy Value Chain (published 13 June 2017). https://www.linkedin.com/pulse/6-steps-advocacy-value-chain-efua-edeh (accessed 04 November 2020)

Gen S. and Wright A. C. (2013). Policy advocacy organizations: a framework linking theory and practice, *Journal of Policy Practice*, **12**(3), 163–193. https://doi.org/10.1080/15588742.2013.795477

Gosling L. and Cohen D. (2007). Participant's Manual: Advocacy Matters – Helping Children Change Their World. International Save the Children Alliance, London, UK. https://resourcecentre.savethechildren.net/node/1979/pdf/1979.pdf (accessed 06 December 2020)

Hopkins B. (1993). Charity, Advocacy and the Law. Wiley, New York.

Hudson A. (2001). NGOs' transnational advocacy networks: from 'legitimacy' to 'political responsibility?' *Global Networks*, **1**, 331–352. https://doi.org/10.1111/1471-0374.00019

Inagaki N. (2007). Communicating the Impact of Communication for Development – Recent Trends in Empirical Research. World Bank Working Paper NO. 120. https://doi.org/10.1596/978-0-8213-7167-1. eISBN: 978-0-8213-7168-8, Washington, D.C. 20433, U.S.A. https://openknowledge.worldbank.org/bitstream/handle/10986/6728/405430Communic180821371 67101PUBLIC1.pdf?sequence=1&isAllowed=y (accessed 28 November 2020)

Keck M. and Sikkink K. (1998). Activists Beyond Borders: Advocacy Networks in International Politics. Cornell University Press, Ithaca, NY.

Keck M. and Sikkink K. (1999). Transnational advocacy networks in international and regional politics. *International Social Science Journal*, **51**, 89–101. https://doi.org/10.1111/1468-2451.00179

Lecy J. D., George E. M. and Hans P. S. (2010). Advocacy organizations, networks and the firm analogy. In: Advocacy Organizations and Collective Action, A. Prakash and M. K. Gugerty (eds.), Cambridge University Press, Cambridge, pp. 229–251.

McCarthy J. D. and Castelli J. (1996). 'Studying Advocacy in the Nonprofit Sector: Refocusing the Agenda.' Draft paper prepared for the Independent Sector Spring Forum.

McGuffog T. (2016). Building Effective Value Chains: Value and its Management. Kogan Page Publishers, London, UK.

Melkote S. R. (1991). Communication for Development in the Third World: Theory and Practice. Sage Publications, New Delhi, India.

Minkoff D. C. (1999). 'Organizational Barriers to Advocacy.' Paper presented to the Nonprofit Sector Research Fund Strategy Group, Wye River, Maryland.

Monteiro S., Pereira M., Branco I. and Reis A. C. (2017). Value Chain Mapping Methodology: A proposal for a process mapping project. International Joint Conference – ICIEOM-ADINGOR-IISE-AIM-ASEM (IJC 2017), 6–7 July 2017, Valencia, Spain.

Morariu J., Reed E., Brennan K., Stamp A., Parrish S., Pankaj V. and Zandniapour L. (2009). Pathfinder evaluation edition: A practical guide to advocacy evaluation. Innovation Network, Inc. Retrieved from http://www.innonet.org/index.php?section_id=3&content_id=601

Mostafa M. (2020). Youth mobilization for civic rights. In: Gender Equality. Encyclopedia of the UN Sustainable Development Goals, W. Leal Filho, A. M. Azul, L. Brandli, A. Lange Salvia and T. Wall (eds.), Springer, Cham, pp. 1–6. http://doi-org-443.webvpn.fjmu.edu.cn/10.1007/978-3-319-70060-1_118-1

Onyx J., Armitage L., Dalton B., Melville R., Casey J. and Banks R. (2010). Advocacy with gloves on: the 'manners' of strategy used by some third sector organizations undertaking advocacy in NSW and Queensland. *VOLUNTAS: International Journal of Voluntary and Nonprofit Organizations*, **21**(1), 41–61. https://doi.org/10.1007/S11266-009-9106-Z

PARIS 21. (2010). Advocating for the National Strategy for the Development of Statistics, Country-level Toolkit. https://www.paris21.org/sites/default/files/advocacytoolkit.pdf (accessed 06 December 2020)

Porter M. E. (1989). Vantagem Competitiva das Nações: Criando E Sustentando um DesempenhoSuperior. Campus, Rio de Janeiro.

Prakash A. and Gugerty M. K. (eds.) (2010). Advocacy organizations and collective actions: an introduction. In: Advocacy Organizations and Collective Action, Cambridge University Press, London, UK, pp. 1–29. https://doi.org/10.1017/CBO9780511762635.002

Reid E. J. (1998). Nonprofit advocacy and political participation. In: Nonprofit and Government: Collaboration and Conflict, E. T. Boris and C. E. Steuerle (eds.), The Urban Institute Press, Washington, D.C., pp. 291–325.

Reid E. J., E. T. Boris and C. E. Steuerle (eds.) (2001). Understanding the word 'advocacy': context and use. In: Nonprofit Advocacy and the Policy Process: Structuring the Inquiry Into Advocacy, The Urban Institute, Washington, DC, vol. **1**, pp. 1–7.

Reid E. (2006). Nonprofit advocacy and political participation. In: Nonprofit and Government: Collaboration and Conflict, E. T. Boris and C. E. Steuerle (eds.), The Urban Institute Press, Washington, D.C., pp. 343–372.

Reisman J., Gienapp A. and Stachowiak S. (2007). A guide to measuring advocacy and policy. Prepared by Organizational Research Services for Annie E. Casey Foundation.

Sprechmann S. and Pelton E. (2001). Advocacy Tools and Guidelines Promoting Policy Change. CARE. https://onthinktanks.org/wp-content/uploads/2016/01/CARE_Advocacy_Guidelines.pdf (accessed 06 December 2020)

Springer-Heinze A. (2018). ValueLinks 2.0. Manual on Sustainable Value Chain Development, Vol 1: Volume 1 Value Chain Analysis, Strategy and Implementation. GIZ, Eschborn, Germany. Page: 356. https://beamexchange.org/uploads/filer_public/f3/31/f331d6ec-74da-4857-bea1-ca1e4e5a43e5/valuelinks-manual-20-vol-1-january-2018_compressed.pdf (accessed 06 December 2020).

Stalker C. and Sandberg D. (2011). Praxis Paper 25: Capacity Building for Advocacy. https://www.intrac.org/wpcms/wp-content/uploads/2016/09/Praxis-Paper-25-Capacity-building-for-advocacy-Chris-Stalker-with-Dale-Sandberg.pdf (accessed 06 December 2020)

Toth E. L. (1986). Broadening research in public relations. *Public Relations Review*, **13**, 27–36.

UNICEF. (2010). Advocacy Toolkit: a guide to influencing decisions that improve children's lives. https://www.unicef.org/evaluation/files/Advocacy_Toolkit.pdf (accessed 06 December 2020)

UN-WATER. (2009). Advocacy – Influencing Leaders (WD), Factsheet. https://sswm.info/water-nutrient-cycle/water-distribution/softwares/awareness-raising/advocacy---influencing-leaders-%28wd%29 (accessed 06 December 2020)

Uzochukwu B., Onyedinma C., Okeke C., Onwujekwe O., Manzano A., Ebenso B., Etiaba E., Ezuma K. and Mirzoev T. (2020). What makes advocacy work? Stakeholders' voices and insights from prioritisation of maternal and child health programme in Nigeria. *BMC Health Services Research*, **20**, 884. https://doi.org/10.1186/s12913-020-05734-0

Verba S., Kay L. S. and Henry B. (1995). Voice and Equality: Civic Voluntarism in American Politics. Harvard University Press, Cambridge.

World Animal Net (WAN). (2017). http://worldanimal.net/our-programs/strategic-advocacy-course-new/module-3/advocacy-research-and-analysis/advocacy-research-and-its-importance#:~:text=Research%20is%20the%20foundation%20for,accurate%20evidence%20to%20support%20advocacy (accessed 26 November 2020)

WHO. (2014). Health in All Policies (HiAP) Framework for Country Action. https://www.who.int/cardiovascular_diseases/140120HPRHiAPFramework.pdf?ua=1 (accessed 26 November 2020)

doi: 10.2166/9781789061840_0221

Chapter 9

Sanitation management knowledge value chain

Mayowa Abiodun Peter-Cookey, Peter Cookey, Thammarat Koottatep and Chongrak Polprasert

Chapter objectives

The aim of this chapter is to help the reader to understand the working of the Sanitation Management Knowledge Value Chain, which shows the linkages between knowledge enterprises that source, acquire, create, distribute and utilize sanitation knowledge to produce effective, functional and sustainable novel and innovative solutions. Other issues to be covered under this chapter include but are not limited to sanitation knowledge, sanitation management knowledge and learning, sanitation knowledge management processes, and sanitation management knowledge marketplace.

9.1 INTRODUCTION

The knowledge of sanitation management and about sanitation is embedded in various activities, several organisations and individuals with different focus, disciplines and in different sectors as sanitation has become everybody's concern (TBC, 2017). Sanitation is no longer just a development concern, but increasingly an integral aspect of enterprises' operations because providing access to safely managed sanitation has shown spillover benefits for occupational health and safety (OHS), environmental sustainability, socioeconomics and cultural sustainability, corporate social responsibility (CSR) and even business success (ADBI, 2019; TBC, 2019a). Creating and using knowledge is now central to sanitation management for both public and private sector (Simard, 2006) and is probably why knowledge and learning in sanitation management has seen increased focus in the last two decades. When sanitation knowledge is properly managed and disseminated then quality and effective innovations will ensue (Darroch, 2005). The SDG 6 and its targets for sanitation and the related links to other SDGs like end poverty, health, education, sustainable cities, clean energy, gender equality and collaborative partnerships among others have opened up an urgent need to create and share knowledge across sectors, disciplines and regions for deeper understanding and innovative and contextual solutions (TNUSSP, 2018). The demand for increased access, improved service and inclusivity has stirred up a drive for innovation that requires sanitation management professionals to operate with up-to-date knowledge and opportunities to learn what is required to deliver quality and innovative products and services.

However, the dispersed nature of sanitation management knowledge, whereby content required to build on the quest for safely managed accessible and inclusive sanitation is scattered among diverse groups with different perspectives and no authoritative sources of knowledge (Becker, 2001; Dew *et al.*, 2004), creates uncertainties in the sanitation sector and too many actors with different ideas and practices. But, when the myriad of data, information and existing knowledge about sanitation and its knowledge are tied together as one coordinated systemic whole, that is gathered or collated and organised in order, then a clear and comprehensive picture of the complex and dynamic nature, problems and solutions of the sanitation phenomena can be addressed (Becker, 2001). This is where knowledge and learning management service providers find their place in the Integrated Functional Sanitation Value Chain (IFSVC) as they create, use and distribute relevant knowledge content, products and services that support innovation and management in sanitation. They operate in the Knowledge Economy (KE) as support mechanisms for sanitation management enterprises and organisations as well as those in related industry that need sanitation knowledge (Sani-K). In fact, the IFSVC (proposed in this book) cannot function effectively and sustainably without a sanitation knowledge market (Sani-KMart) where there is a demand and supply of knowledge about sanitation and its management that will enhance the creation, manufacture and delivery of related content, products and services. In essence, the sanitation sector will be strongly dependent on the acquisition, creation, distribution and utilization of knowledge to produce effective, functional and sustainable novel and innovative solutions towards the progress of the SDG 6 related Targets (Kefela, 2010) with knowledge as both input and output.

As has been pointed out by many scholars, the global economy has transited to a Knowledge Economy (or digital) (Powell & Snellman, 2004) that depends primarily on knowledge as the key asset for knowledge-intensive activities, which creates added value to advancements in innovations that are heavily reliant on human and intellectual capital (Powell & Snellman, 2004; Pluta-Olearnik, 2013; World Bank, 2007). Knowledge has quickly and easily become an important tool for value creation with ideas as the ingredients and intellectual property as the merchandise that fuels the drive for change in the face of society's demands for smart solutions that offer more convenience and affordability (Bryan, 2004). The knowledge economy (KE) is directly based on the acquisition, creation, distribution and utilisation of knowledge more effectively for novel innovations toward progress in society (Kefela, 2010) – with knowledge as both significant input and output. This implies that for the sanitation sector to meet the demands and expectations of clients and users, it must rest strongly on knowledge drawn from information and data, and also the experience and expertise of highly skilled workers, as well as the increasing need for readily accessed knowledge input and output sourced for, produced and used by private and public entities across industrial sectors (Bryan, 2004; World Bank, 2007) to sustainably and effectively manage sanitation at all levels; enterprises and organisations that anchor their competitive advantage on knowledge-based innovations and solutions will drive the sector (EMCC, 2005; Miles, 2005, 2007; Miles *et al.*, 1995; World Bank, 2007). It becomes imperative that knowledge-intensive activities and services become essential to the complex and dynamic innovations and inventions that will move the sanitation economy to the next level. Sanitation management enterprises and organisations that produce content, products and services for industries, governments, businesses, communities and households will have to shift to knowledge-intensive interactions, internally and externally (e.g., research, customer engagement, training, education, etc.) to activate new knowledge and use available knowledge in conjunction with new knowledge to enhance their deliverables, customer management and to meet the SDG Targets.

As safe sanitation management demands for innovations and novelty in product manufacture and service delivery rises, in turn, the demand for knowledge content, products and services has become an urgent necessity. Considering that knowledge is regenerative and can keep reproducing itself or another version of itself or even a new knowledge entirely in a different area, and can be delivered as a product or service that is available for distribution based on demand and supply (i.e., there are those who will provide and those who will pay for them) (Simard, 2006), the sanitation knowledge market (Sani-KMart) is crucial to all stakeholders in the IFSVC. In fact, this is big business as the art and act of turning knowledge into products and services is a key competitive advantage for sanitation enterprises, organisations, professionals and also for economies and societies of the future. The Sani-KMart creates a circular platform for sharing and/ or exchanging information as well as distributing knowledge sources between users and suppliers – either for a fee (e.g., IWA) or free (e.g., Susana) (Simard, 2006). They are made up of problem-solving, innovations, civil/social, business and research activities, and are transactional systems that trade on contents, agents' experiences and relevant interactions determined specifically through the dynamic properties of intellectual capital creation and exchange (Carrillo, 2016; Pluta-Olearnik, 2013). Sani-KMart are the conduits from which transactions in knowledge products and services are conducted to provide content, support, guidance, and other merchandise towards the demands of consumers (OECD, 2013, 2012; St Clair & Reich, 2002). There are four ways to deliver Sani-KMart as a service (Simard, 2006): generate content, develop products, provide assistance and share solutions. Sanitation knowledge services (Sani-KServ) could then include, education, training, research/development, ICT, design, media content, databases, repositories, legal, finance, marketing, and other professional services while sanitation knowledge products could be reports, manuals, publications, agreements, contracts, and so on. They could be produced and/or provided by knowledge-intensive firms/organisations (KIFs/KIOs) by employees and/or outsourced to knowledge-intensive business services firms (KIBSFs) (Alvesson, 2004; Den Hertog, 2000; EMCC, 2005; Khadir-Poggi & Keating, 2013; Muller & Doloreux, 2009; Swart & Kinnie, 2003) in the sanitation economy.

The characteristics of sanitation knowledge (Sani-K) consumers are diverse as are their expectations, but one thing is common, supply of solutions that meet the knowledge demand are equally wide and far-reaching. This indicates that the Sani-KMart is not linear, but exists within the circular economy whereby relevant knowledge is also desired by users outside the core sanitation sector (e.g., transport, healthcare, construction, tourism and hospitality, events management, etc.) and by the producers of Sani-K. But, surviving in the knowledge marketplace is highly dependent on seamless and high access to knowledge and the ability to create and use knowledge faster than others (Amidon, 1997; Davenport & Prusak, 1998; Nonaka & Konno, 1998; Simard, 2006) as this could enhance efficiency, novelty, innovation and competitiveness (World Bank, 2007). Thus, knowledge should drive the IFSVC as the push towards better sanitation management fuels the need for innovations in developed and developing countries. This value chain rests strongly on knowledge drawn from information and data, history and the repository of experience and expertise of highly skilled workers, and upon the increasing need for ready access to knowledge input and output by private and public entities across industrial sectors (World Bank, 2007). It operates as actors in the sector acquire, create, disseminate and apply knowledge that facilitates sustainable growth and innovative progress in accessibility, inclusivity, functionality, affordability and profitability. The increasing need for safe, inclusive, sustainable, practical and affordable sanitation facilities that are accessible to all and acceptable in different contexts highlight the importance of a strong Sani-KMart that will rest heavily on knowledge production, sharing and workers (Davenport, 2005). Sanitation enterprises and organisations can trade on (i.e., buy and sell) knowledge for innovative solutions for products and services (World Bank, 2007). In fact, the Bill and Melinda Gates Foundation's (BMGF) 'Reinventing the Toilet' program has created a knowledge-driven paradigm in sanitation management by funding research to develop novel and affordable toilet options.

Contemporary societies and economies are knowledge-driven and the creation of value and innovative progress is often dependent on knowledge utilization and/or new knowledge (Landry *et al.*, 2006; Venkatraman & Venkatraman, 2018) as the need to exchange knowledge-based products and services continue to increase (Simard, 2006). Managing knowledge sourcing, acquisition, creation, transformation, dissemination and usage is key to developing innovations and competitive advantage (Holsapple & Singh, 2003; Lee, 2016) where Sani-K is not just a resource, but a product or service, that when value is added to or created by available knowledge capital/assets, could produce improved performance, capabilities and competences in individuals, organisations and industrial sectors (Alawneh *et al.*, 2009; Lee, 2016; Malik *et al.*, 2010; Marr *et al.*, 2003). But while knowledge can be considered a resource in and of itself, the manner in which it is used and managed will determine the quality of whatever it produces (Darroch, 2005). In essence, when knowledge resources (KRs) are gathered and coordinated, they can be used to build skills, abilities and capacities of individuals, organisations, governments and communities of practice (CoPs). However, this depends on the knowledge capital, that is, the sourced, acquired and stored knowledge available in the enterprise, government and/or community of practice. This means that Sani-K capital and the management thereof is vital to productivity and quality performance at any level of the IFSVC (Bernet *et al.*, 2005; Lee, 2016; Lowitt *et al.*, 2015; Saliola & Zanfei, 2009). The knowledge management services of the IFSVC consists of individuals, enterprises (business and social) and government organisations that manage, produce and deliver knowledge products and services whether internally as knowledge workers or externally as expert consultants, contractors or businesses. This is chapter addresses the sub-value chain of the IFSVC referred to as the Sanitation Management Knowledge Value Chain that explores the value creation processes within the sanitation knowledge management (Sani-KM) of enterprises and organisations and even governments in domestic, regional and global levels. It considers a knowledge value chain (KVC) for operations within sanitation KIFs/KIOs and KIBSFs and then proposes a creative concept map for Sanitation Management Knowledge Value Chain (SaniM-KVC) in the sanitation industry. These proposed concepts are not yet tested but could be a guide for research into the KVC for sanitation management. To proceed, it is important to comprehend the concepts of sanitation knowledge (Sani-K) and learning, sanitation management knowledge (SaniM-K), sanitation knowledge management (Sani-KM) processes and the sanitation knowledge market place (Sani-Kmart).

9.1.1 Sanitation knowledge (Sani-K), sanitation management knowledge (Sani-KM) and learning

Translating Sani-K into valuable resource transits from an understanding and identification of what constitutes Sani-K and the ability to manage this resource (i.e., Sani-KM); and is crucial for making up a sanitation management knowledge value chain (SaniM-KVC) that will support the IFSVC. Although sanitation is such a prevalent topic for discourse in development quarters, according to Revilla *et al.* (2021), only 51 (out of 18,329) academic papers in the top development journals globally focused on sanitation and related issues. Sanitation is the different ways that excreta and urine (i.e., faecal sludge) as well as menstrual blood (UNICEF/WHO, 2020) and wastewater is collected and treated to prevent human exposure and environmental contamination (Naughton & Mihelcic, 2017; UNICEF/WHO, 2020) while **sanitation management (SaniM)** is the process (which could include facilities, products, services, and systems) for safely managing the collection, transportation, disposal, treatment and conversion of sanitation matter to protect the socioecological integrity of contextual locations and **sanitation management knowledge (SaniM-K)** is the knowledge about all of these. Sani-K is the body of information, data, wisdom, expertise, skill and experiences embedded in individuals, firms and organisations in either tacit and explicit (or both) about what makes up faecal sludge and the processes that are involved within the sanitation service chain (SSC); and **sanitation knowledge management (Sani-KM)** is how Sani-K can be and is used to design, develop, build and provide Sani-K content, products and services that ensure the safe management of the SSC and the activities that lead to the production of faecal sludge and could be used to engender innovative solutions in products, services, processes and/or governance towards SDG 6 Targets and how it relates to other SDGs as well as growing and strengthening the IFSVC.

This Sani-K/Sani-KM is embedded in people, processes and best practices of firms, organisations and governments and also constitutes a potentially regenerative resource (Davenport & Prusak, 1998; Nonaka & Takeuchi, 1995) that enhances competitive advantage and economic value generations, and is strongly dependent on the ability to learn, innovate and change (Khadir-Poggi & Keating, 2013). Learning is acquiring and imbibing knowledge assets/capital and skills as resources to use, exploit or create for problem-solving – and/or decision-making. It can be received through study, experience, or instruction; with the ultimate aim to improve performance (whether know-how, what, when, where or why). Learning requires a balanced interaction between people, organisations, knowledge-providers and technology – culture, governance. The purpose of learning is to improve the knowledge base and competences (Bereiter, 2002; Maclellan & Soden, 2007) and to allow the sustainable utilisation of knowledge (King, 2009), and the acquisition of knowledge that replaces old existing knowledge with new content, behaviours and skills, which in turn adds value to the overall (Ermine, 2013; Nonak, 1994). The process of learning improves the individual/group's knowledge base and performance, which then culminates in knowledge-building (different from just arbitrarily learning (Maclellan & Soden, 2007)), creates and articulates solutions and/or new knowledge with added value (Bereiter, 2002).

Sani-K/Sani-KM facilitates learning as individuals, groups and teams build up knowledge in a way that enables consistent and effective continuous improvement (Argyris, 1999; King, 2009; Pedler *et al.*, 1997). Knowledge-building produces critical and transformative learning, and so learners of sanitation management need to understand how to integrate old and new knowledge to create solutions and/or even more new knowledge (Maclellan & Soden, 2007). Knowledge and learning fuels competency, capability and strategic abilities (Grant, 1991; Sveiby, 2001) that increases innovative capacities of individuals, groups, teams, enterprises, organisations, and CoPs. It will also improve domestic activities, economies and global relationships and interactions (King, 2009). Without learning, knowledge-transfer/sharing/dissemination remains at the abstract level and cannot be translated into any value addition or creation for innovation and inventions. Therefore, knowledge-building in the sanitation management sector is key to innovation that adds and creates value; and this makes Sani-KM essential to any SaniM-KVC. It should lead to informed actions and competence (Ermine, 2013), and when used appropriately, it should improve performance, decision-making, problem-solving and competitive advantage.

Table 9.1 Knowledge characteristics (King, 2009; Lundvall & Johnson, 1994; Powell & Snellman, 2004; Pluta-Olearnik, 2013).

1	Know what	knowing the facts as to what actions to take in given circumstances, for example, knowing which toilet system is appropriate for peculiar contexts
2	Know how	knowing how to respond in any given situation, that is, the appropriate, experience, skills and expertise, for example, determining the best treatment system as per sewage versus faecal sludge
3	Know why	having a deep understanding of theoretical basis, causal relationships, interactive effects and uncertainties
4	Know-who	having the ability to reach out to key persons or groups, that is, experts that possess the appropriate knowledge needed
5	Know-when	having the ability to comprehend and predict times, variations, seasons, and so on.
6	Know-where	knowing how to determine and comprehend peculiar contexts, locations, geographies, culture, and so on.

Determining what is sanitation knowledge (see Table 9.1) and the processes of learning will help to identify and collate knowledge capital and/or assets. Sani-K involves learning what, how, where, when and why; whether it is from stored, codified and formal data/information (explicit knowledge) and personal experiences, perceptions, intuition and insights (tacit knowledge).

Sani-K can be explicit knowledge, which is tangible, searchable and can be easily found in books, documents, repositories, libraries, and so on.; can be recorded and expressed in texts, numbers, codes, formulas, programmes, and so on., making transfer easier (King, 2009; Lee, 2016; Nonaka & Takeuchi, 1995), for example, policy documents and research findings. People acquire explicit knowledge through formal training, education, dialogue, reading, viewing and listening to codified knowledge content (Bryan, 2004). On the other hand, it could be tacit knowledge that is intangible and locked in the individual mind and not easily transformed into tangible forms. It is built from experiences on the job or life, from experts and/or peers, lectures/classes, norms, cultures, traditions, and so on. It makes up about 95 percent of all knowledge and is mostly transferred from one person to another through personal interactions like conversations (training, discussions, stories, etc.) and/or practical tasks (supervision, coaching, mentoring and apprenticeship/internship, etc.) (IRC, 2006; King, 2009; Lee, 2016; Nonaka & Takeuchi, 1995; Sandelin *et al.*, 2019). However, tacit knowledge can be presented as explicit knowledge when it is codified and formally communicated in a way that can be captured, stored and disseminated (Allee, 2003; Nonaka & Takeuchi, 1995), while Wilson (2002) argues that knowledge can be implicit when tacit knowledge that is expressible is not expressed. Sani-K is greatly reliant on indigenous knowledge, which is an example of tacit knowledge, in particular for contextual cases for rural communities and specific communities in urban centres. This tacit knowledge is generally embedded in the minds of residents and is developed over time (IRC, 2006; Nonaka & Takeuchi, 1995) and to transfer such knowledge will require a process of translating into explicit knowledge by collating, organising, documenting and archiving/storing or data-listing (King, 2009).

Progress and sustainability in the sector, thus, rest upon the quality and quantity of knowledge that is available and accessible and how they are applied. This implies that Sani-K should be used and managed as a value-adding resource that extensively and expansively contributes to society in and of itself as well as other activities, systems and processes (Landry *et al.*, 2006). Sani-K is multi- and trans-disciplinary and sometimes cross-disciplinary giving it a unique dynamic and complex learning curve that requires skills and content from a myriad of disciplines. Ultimately, creating knowledge innovations (KI) that fuel the progress and success of the SDG 6 and other agenda towards sustainability and continuous improvement in sanitation management across all levels as well as in the Sani-KMart (Kostas & John, 2006). These KIs are made up of processes that create, evolve, exchange and apply new ideas towards commercialised situations that boost the bottom line of sanitation enterprises, sanitation industry and economy as well as societies in general (Amidon, 1997).

Making sense of existing knowledge (and knowledge waiting to be discovered) in the universe as well as those available in tacit and explicit forms will require human capital and ICT infrastructure to source, discover, create (and recreate), store, transfer/share and use (Bhagwath, 2014; Gunday *et al.*, 2011; Lee, 2016; Marr *et al.*, 2003; Venkatraman & Venkatraman, 2018); as well as knowledge agents, individuals (i.e., workers, students, experts, etc.), groups (teams, units, departments, etc.), networks (professional associations, communities of practice, etc.), and entities (enterprises, organisations, state and non-state actors, industry, academic and research institutions, etc.) that manage the diffusion of these knowledge to create innovations at different levels (Lee, 2016; IRC, 2006; Landry *et al.*, 2006; Venkatraman & Venkatraman, 2018). This is the Sani-KM processes within companies and organisations that make up the SMKVC.

9.2 SANITATION KNOWLEDGE MANAGEMENT (SANI-KM) PROCESSES

Sanitation Knowledge management (Sani-KM) involves the planning, organising, motivating and controlling of people, processes and systems to improve knowledge assets and effectively use them. O'Dell and Hubert (2011) describe Knowledge Management (KM) as a systematic approach for presenting knowledge that will 'grow, flow and create value' through processes that provide appropriate knowledge to where it is needed so that it can aid actions that improve performance. It is concerned with the generation, capture, storage and sharing of knowledge with an intent to take timely actions for increasing an organisation's competitive advantage (Venkatraman & Venkatraman, 2018), and relates to activities such as learning and innovation, benchmarking and best practices, strategy, culture and performance measurement (Nonaka & Takeuchi, 1995). Some have argued that knowledge cannot be managed, and that even the 'knower' can only know imperfectly (Wilson, 2002); what can be managed is the way knowledge is 'created, discovered, captured, shared, distilled, validated, transferred, adopted, adapted and applied' (Collison & Parcell, 2004). The ultimate goal of KM, then, is to leverage and improve organisations', government agencies' and/or enterprises' knowledge assets in order to strengthen and upgrade existing competitive advantage, knowledge workers abilities, and overall performance (King, 2009). Thus, KM services are very important for Sani-K creation and transfer to stakeholders and clients at different levels and for varied purposes (Bratianu, 2015; King, 2009; Sandelin *et al.*, 2019).

In this knowledge-driven economy, competitive advantage is dependent on what is known (individual or organisation), how the 'known' is used and how fast the 'known' can be transformed into valuable assets (Prusak, 1996). These knowledge assets could be used to create value that translates to innovations (Sveiby, 2001) and include intangible resources of intellectual capital (IC) (Kok, 2007) derived from tacit and explicit knowledge. The process of managing knowledge assets in such a way that motivates knowledge sharing, creation, acquisition, storage and dissemination will lead to improved productivity, performance, problem-solving, innovation and decision-making

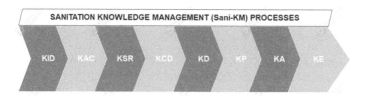

Figure 9.1 Sanitation Knowledge Management Processes. (Source: Authors)

(Bhagwath, 2014; Dei & van der Walt, 2020; Kok, 2007; Lev, 2001; Marr *et al.*, 2003; Tsuneo, 2001) for those enterprises and organisations that operate in the sanitation management sector. Based on literature, the Sani-KM processes include the following (see Figure 9.1):

(1) ***Knowledge Identification/Sourcing:*** This is where the relevant and related knowledge required for safe sanitation management product manufacture and service delivery is determined. The process of searching and discovering necessary information, data and knowledge from across boundaries (sourcing) and then selecting and classifying such content (identification) to determine what is available and what is required to create a knowledge inventory that guides what knowledge resources (KRs) exist as assets and also what KRs need to be acquired (Ermine, 2013; Landry *et al.*, 2006; Probst, 1998; Wang & Ahmed, 2005; Weggeman, 1997, 2000). This could reduce the multiplication of knowledge across different organisations with differing interpretations. Knowledge products and services include codification, learning, research, analysis, collation, organise, publications, and so on.; while players could involve experts/specialists; knowledge workers; higher education institutions; research institutions; knowledge service providers; media; media; publishers; state and non-state actors; enterprises; networks and CoP.

(2) ***Knowledge Acquisition/Capturing (KAD):*** This is the process of locating, discovering and capturing relevant knowledge assets or resources from different sources (individuals, groups, organisations, stakeholders, etc.) and continuously updating knowledge capital in the sanitation sector, organisation, economy or Sani-K expert/CoP. Knowledge could be acquired internally or externally through personal interactions and/or physical artefacts like books, articles, repositories, and so on. it involves learning via training, research, education, and other intuitive forms (Carrillo, 2016; Darroch, 2003; King, 2009; Landry *et al.*, 2006; Lee & Yang, 2000; Nonaka & Takeuchi, 1995; Probst, 1998; Venkatraman & Venkatraman, 2018). Activities, products and services comprise personal development and one-on-one interactions (tacit knowledge); search engines; repositories and inventories (codified info); libraries and archival systems; intellectual property and patents; knowledge sharing platforms; document management systems; expert network systems; digital products (e.g., software, apps, etc.); data and information management systems; media (audio-visual, print, social); education, training and research; publications; industry/sector reports; and so on.; and players are state and non-state actors; enterprises; networks and CoP; experts/specialists; knowledge workers; higher education institutions; research institutions; knowledge service providers; media; media; publishers; primary/secondary schools; consumers and other stakeholders in the community/society; professional service providers (e.g., legal, accounting, financing, marketing, design, management, etc.); educators and trainers, and so on.

(3) *Knowledge Storage and Retrieval (KSR):* This process involves activities that store and retrieve acquired knowledge for future use. It includes building knowledge capital that could be stored as individual (i.e., tacit), organisational, institutional and industry/sector memory (i.e., explicit) from resources acquired and retained in individuals, groups/teams, organisations and institutions (tacit) from processes, products, services, systems, activities, best practices, routines and/or socioeconomic interactions (e.g., producer/user, buyer/seller, etc.) (Cross & Baird, 2000; de Jesus Ginja Attunes & Pinheiro, 2020; Irani *et al.*, 2009; King, 2009; Lee & Yang, 2000; Nonaka & Takeuchi, 1995; Venkatraman & Venkatraman, 2018; Walsh & Ungson, 1991). They comprise personal development and one-on-one interactions (tacit knowledge); digital repositories and inventories (codified info); libraries and archival systems; intellectual property and patents; knowledge sharing platforms; document management systems; expert network systems; digital products (e.g. software, apps, etc.); data and information management systems; media (audio-visual, print, social, web); education, training and research; workshops/seminars, and so on.; publications; organisational/institutional/sector memory (e.g., industry/sector reports); and so on.; and players are enterprises; networks and CoP; experts/specialists; knowledge workers/providers; experts and peers; higher education institutions; research institutions; knowledge service providers; media; media; publishers; professional service providers (e.g., legal, accounting, financing, marketing, design, management, etc.); educators and trainers, and so on.

(4) *Knowledge Creation and Development (KCD):* This is the process of creating new sanitation knowledge resources (Sani-KRs) with available and acquired Sani-K that have been transformed, refined, combined and integrated at both individual and collective levels, internally (organisation, enterprises, institutions, industry/sector, profession, discipline, communities of practice, etc.) and externally (inter- and transdisciplinary, sectors, organisations, etc.). It involves filling knowledge gaps via learning and knowledge-building to integrate tacit and explicit from individual and collective intuitions and shared experiences to generate new knowledge content with added value for consumer use, optimal performance and competitive advantage. This is also referred to as knowledge innovation (Lee, 2016) whereby combined and integrated knowledge are used to develop commercialised sanitation merchandise (products/services), which could contribute to the viable growth of enterprises, industry/sector, economies and societies. This accrues from the capacity to transform knowledge into actions, decisions, products, services, and even policies by integrating knowledge assets from different sources (individuals/collective and internally/externally) and transdisciplinary interactions. Knowledge is created by individuals, groups and organisations. Research can create knowledge innovations (KI) and intellectual capital (IC) that produces new (and/or upgraded) products and services, applications, processes, policies, and so on. – which could be commercialised for sale to users or for further research (Amidon, 1997; Giebels *et al.*, 2020; Landry *et al.*, 2006; Lee, 2016; Lee & Yang, 2000; King, 2009; Nonaka & Takeuchi, 1995; Probst, 1998; Simard, 2006; Venkatraman & Venkatraman, 2018; Weggeman, 2000).

(5) *Knowledge Dissemination (KD):* This is the process of distributing KRs between individuals (sharing) and groups (transfer) across boundaries. It is the transmission of new and valuable information, data, expertise, ideas and knowledge from different sources – that is making SaniK available to those who need it (free or for a fee) in organisations, industry, sectors, governments,

enterprises, institutions, societies, and so on. It drives the creation of new knowledge from the gaps in existing knowledge to develop innovations. Knowledge dissemination includes transfer, sharing, diffusion, donation and convection depending on the process of transmission. Knowledge-sharing is the process by which an individual imparts knowledge to others (e.g., expertise, insight, understanding, etc.) whether tacit or explicit (Ford & Staples, 2010) and is the most important mode of knowledge-dissemination (Bratianu & Bejinaru, 2017). As a key component of KM and a driver of innovation, it reaches a broad and generic audience to make relevant knowledge available to others to support value creation, problem-solving and decision-making; and disseminated information for appropriate use (Bartol & Srivastva, 2002; Chyi Lee & Yang, 2000; Liu & Cheng, 2007). Knowledge transfer is focused and purposeful and is diffused when knowledge is made available far and wide across borders. Knowledge convection is when knowledge-holders move with their knowledge (cognitive, emotional, spiritual, etc.) from one place to another and then transfer or share such knowledge with others in a different place (Bratianu & Bejinaru, 2017; Bartol & Srivastva, 2002; Berends *et al.*, 2011; Braunerhjelm *et al.*, 2010; Cowan & Nicolas, 2004; King, 2009; Landry *et al.*, 2006; Lee, 2016; Lee & Yang, 2000; Liu & Cheng, 2007; Venkatraman & Venkatraman, 2018).

(6) ***Knowledge Protection (KP):*** This is the process of preserving KRs within a system and also guarding knowledge assets, mostly tacit knowledge – for example expert employees. It involves legal mechanisms for protecting intellectual capital (i.e. intellectual property), patents, copyrights, trademarks, brands and also specific know-how of processes for innovations, inventions and such likes (Chyi Lee & Yang, 2000; Probst, 1998).

(7) ***Knowledge Application (KA):*** This is the process of using available SaniK to perform tasks, create new knowledge and innovations, make decisions and also respond to acquired and available knowledge with appropriate actions and interactions (e.g., response to customer feedback). It involves exploiting and exploring existing knowledge capital and/or memory to perform tasks, make changes, updates and upgrades, reach decisions and innovations (including new knowledge). The productive deployment or application of knowledge resources for developing safe sanitation management products, services, and best practices could lead to new knowledge, new processes and systems, new business/enterprise ideas and practices, new merchandise, new academic fields, new programmes, new markets, new policies and a host of other results from applying Sani-K for different purposes (Darroch, 2003; Probst, 1998; Rowley, 2001; Venkatraman & Venkatraman, 2018).

(8) ***Knowledge Evaluation: (KE):*** Knowledge evaluation or measurement is the ultimate purpose of knowledge management. It seeks to determine if KM has made any inputs on productive and market performance and if knowledge assets are worth the investments. Measuring the SaniKRs gained and available, and growth impacts (on organisations, institutions, enterprises, industry/sector, economies, governance, societies and individuals) against specific mission, vision, goals and strategies to determine future actions and investments (King, 2009; Landry *et al.*, 2006; Probst, 1998; Simard, 2006) will provide new knowledge for decision-making, strategic actions, and even content, product and service design.

These processes create a demand for knowledge workers and experts in Sani-KIFs/ KIOs and the services of KIBSFs in the Sani-KMart that create a SaniM-KVC in the IFSVC.

9.3 SANITATION MANAGEMENT KNOWLEDGE MARKETPLACE

In this section, the term Sani-K and SaniM-K will be used interchangeably. An overview of the Sanitation Knowledge marketplace will seek to identify the operations, activities and interactions that take place within the Sani-K (at domestic and global levels) and explore how they relate within the overall value chain (Miles, 2005; Springer-Heinz, 2018a, 2018b; TBC, 2019a, 2019b). It is anchored on activities within the knowledge management processes, and actors, enterprises, organisations and operations that make up the sanitation knowledge marketplace. The sanitation knowledge sector has been mostly operated by donor and charity funds and some profit-making ventures; thus, this value chain will consider both streams of income as they both contribute to the IFSVC. Actors are primarily made up of knowledge workers, knowledge experts, knowledge brokers, knowledge-intensive firms (KIFs), knowledge-intensive organisations (KIOs), knowledge networks, knowledge-intensive business services firms (KIBS), governments, NGOs and end-users (Figure 9.2). This overview of the sanitation marketplace highlights the businesses, social enterprises and entrepreneurs within the SaniM-KVC and how their interactions within the IFSVC and other external linkages grow and expand the opportunities that exist and could exist. The core players perform different functions along the Sani-KM processes to ensure a continuous stream of new ideas and innovative knowledge to move the sector closer to the SDG sanitation targets, sustainability and profitability.

These functions include activities that produce knowledge content and other products and also provide knowledge services to end-users such as householders, governments, businesses and social enterprises, industry, networks, professionals, students, and so on. The main activities include: knowledge sourcing and acquisition, data and information processing, knowledge storage and retrieval, knowledge (content, product and service) conceptualisation, knowledge design and development, knowledge creation and production, knowledge assets and capital protection, knowledge management systems, knowledge dissemination, knowledge utilisation and evaluation. The sanitation marketplace will be enabled by government at different levels, multilateral organisations and certain non-state-actors (NSAs); and also supported by other sectoral product and service providers and contributors whether as financial donors or investors, clients or contractors, partners or stakeholders, and so on. The circular flow that operates in the marketplace will be driven by the demand, needs and preferences of end-users of the knowledge products and services, and these end-users can also be and/or depend on supply from designers and manufacturers of sanitation management products, provide sanitation management services and facility managers (e.g., treatment plants, disposal sites, etc.). These two groups make up the sanitation knowledge market.

The different aspects of the IFSVC as deliberated upon in the other chapters of this book cut across manufacturing and operate in sectors of the economy at global, national, local and regional levels. They consider value-adding enterprise opportunities that contribute to the sanitation economy via manufacturing and service activities. There are enterprises that manufacture toilets, septic tanks, disposal trucks, treatment and conversion facilities, ancillary items (e.g., pipes, taps, etc.) and other hygiene-related products (e.g. hand-wash basins, diapers, menstrual items, etc.). Then, there are those enterprises involved in construction, installation, collection, emptying, transportation, recovery, recycling and reuse and maintenance. These are mainly service providers and could be individual entrepreneurs and/or workers (e.g., masons, plumbers, etc.) or companies in the formal and informal sectors, and within these companies are managers of operations, administration, finance and human resource management. On the social side, we have enterprises and organisations that attempt to bridge the gaps of access to safe sanitation in urban and rural settings. They could be local facilitators (NGOs, CBOs,

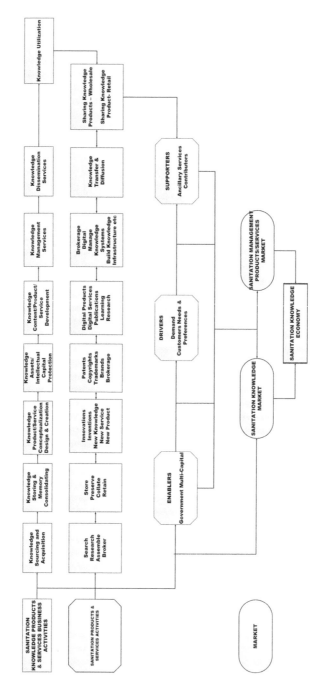

Figure 9.2 Sanitation Knowledge Marketplace. (Source: Authors)

other NSAs or CSR departments of large companies) and/or global interventionists (multilateral organisations, charities, INGOs).

To support this socioeconomic contexts are the investors and funders (financiers, fund managers, etc.), market analysts and advert/promo providers, research and development organisations (e.g., universities, research institutes, public agencies, etc.) that provide novel and innovative solutions; and also professional networks and communities of practices (CoPs) and government institutions that manage and regulate domestic and international interactions within the sanitation market at all levels. These players operate in the sanitation industry that is diverse, complex, dynamic and wicked; adding either economic or social value (or a combination of both) at various degree by supplying services and/or products that are needed in the marketplace or to meet customers' demands and preferences (Figure 9.3).

KM is typically viewed from within organisational/company boundaries, but it can also find calculable value outside these borders. With increasing digital and knowledge interactions among market and sector players, the external marketplace is where KM meets industry economics. Knowledge flows between players (internally and externally) combines the adaptive nature of networks and the tendency of markets to create transactions based on demand and supply; that is assuming that there are buyers (users of such knowledge) and sellers (providers of such knowledge). Buyers of Sani-KRs will be motivated to buy if the knowledge offered is valuable and at a price that is worth their time and effort, but still lower in costs than alternative sources and/or forms (Bryan, 2004). For Sani-K to be valuable though, it cannot be regular, generic and common place, but insightful, relevant, accessible, easy to find and assimilate (Bryan, 2004; Hansen & Haas, 2001; World Bank, 2007), which then determines its value in cost and reputation so that it can be traded in the marketplace (World Bank, 2007). Markets will expectedly

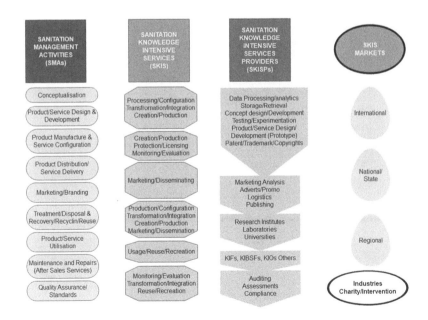

Figure 9.3 Knowledge Intensive Services Activities (KISA) in the IFSVC. (Source: Authors)

form around distinctive knowledge that captures the interest of buyers and sellers (Bryan, 2004), but only Sani-K that creates innovative ideas (i.e., KI) has value enough to be traded. It has been argued that for this to happen, there must be a balance between the economic realm (commercialisation) and the physical realm (laws of nature = reality and practicality), which determines whether people will pay for it and what they are willing to pay (Zubair, 2021).

The value of a Sani-K marketplace depends on the quantity and quality of the available content churned out to buyers and users. And so, the Sani-K marketplace should consist of intra-trading (i.e., Sani-KMart within organisations) and inter-trading (i.e., Sani-KMart in the public space) between seekers and suppliers of Sani-K content, products and services. It is part of the KE whereby knowledge-based resources related to sanitation and its safe management are exchanged whether on a fee-based basis (i.e., purchased at a price) or free-based (i.e., made available for public good) (Simard, 2006; Stewart, 1999). In other words, assuming that there exists Sani-KRs to be transacted upon and there are users and providers to operate such exchanges, then the job of the Sani-KMart is to bring them together. Sani-K is, however, distinctively different from other products and services in the sanitation marketplace because of its unique features. For one, it can be at several places at the same time; never runs out of supply; buyers need only purchase it once (but can be renewed); and it is regenerative (Stewart, 1999). This means that the K-Mart of the sanitation sector will operate differently from other aspects of the value chain, especially as those who provide KRs, sometimes, also use them and so there is a continuous cycle whereby KRs are provided and used at multiple points and by many different agents (i.e., multisectoral and industrial) (OECD, 2006; Simard, 2006) (see Figure 9.4).

9.3.1 Sanitation knowledge-intensive services activities (Sani-KISA)

Enterprises and organisations in the Sani-KMart will require high quality knowledge innovation (KI) to pursue novel creations that target customers' demands and

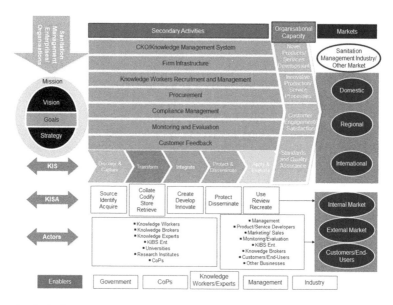

Figure 9.4 Sanitation KIFs/KIOs Value Chain. (Source: Authors)

expectations, foster viable progress and return-on-investments (ROI) and also achieve the SDG 6 Targets on sanitation. The KI will be derived from Knowledge-intensive Services Activities (KISA) (OECD, 2006; Windrum & Tomlinson, 1999). Knowledge-intensive Activities (KIAs) are those tasks, functions and operations that source, acquire, collate, manage and use knowledge resources and assets in the IFSVC. The performance of these activities provides services internally and externally for enterprises and organisations by knowledge workers (KWers), knowledge experts (KExps), knowledge entrepreneurs (KEnts) and knowledge businesses/organisations even governments.

These Knowledge-intensive Services (KIS) are those functions that rely on professionalism and expertise that relate to particular technical or functional domains to provide forms of data, information and knowledge (through reports, manuals, trainings, consultancies, etc.) or major inputs to manufacturing and service delivery processes (Windrum & Tomlinson, 1999). The OECD (2006) also recognised the corresponding activities as KISA that is the regular activities of business and public sector actors to support manufacturing and services and could be initiated, produced and delivered internally (within firms, organisations and government agencies and knowledge workers) or by external KR providers to add capabilities that may not be available internally in a bid to activate creative, fresh and independent perspectives and also possibly provide support for compliance, certifications and in some cases reduced HR costs (OECD, 2006). There are KISA enterprises that support the sanitation industry with knowledge inputs and they also provide Sani-K to other related businesses and organisations, even in different industry sectors, that need Sani-K outputs. Essentially, for innovation to thrive in the sanitation sector, a high demand for KISA is critical, particularly as external providers like safe sanitation management enterprises and organisations need a wide set of skills and knowledge that are often beyond their capabilities (Andreeva & Kianto, 2011; OECD, 2006), for example, delivering new types of faecal sludge management services in a city or producing new types of prefab septic tanks for the contextualities of a particular location or recovering and producing new resources from sanitation matter. These KISA are performed by KWers and KExps for their employers in knowledge-intensive business or social enterprises, multilateral organisations and governments (knowledge-intensive firms and organisations-KIFs/KIOs), but some KIS are also outsourced to KEnts and/or to external parties (knowledge-intensive business services firms-KIBSFs) who provide the needed artefact or deliver the required services on behalf or in conjunction with their clients (EMCC, 2005). They make up the core players of the Sanitation Management Knowledge Value Chain (SaniM-KVC) presented later in this chapter. For this chapter, Sani-KIFs are those sanitation management enterprises that require a high level of knowledge intensity to produce sanitation devices and/or equipment for safe sanitation management and deliver safe sanitation management services through the tacit and explicit knowledge of workers, experts and organisational practices; who may also source for the services of Sani-KIBSFs. On the other hand, Sano-KIOs are those social enterprises and public agencies that deal primarily with sanitation management and its related activities. Sani-KIBSFs provide those knowledge products and services that support the internal operations of Sani-KIFS and Sani-KIOs and their relationships with their customers and users of their products and services.

9.3.2 Sanitation knowledge-intensive firms (Sani-KIFs)/knowledge-intensive organisations (Sani-KIOs)

KIFs and KIOs in the IFSVC are businesses and organisations where knowledge is more important than other inputs (Starbucks, 1992) and that employ mostly highly skilled personnel with the capacity to provide innovation and strategic renewal (Bontis, 1998) as these employee's skills and expertise are very key contributors to the creation of added value, competitive advantage and survival (Starbucks, 1992; Swart & Kinnie, 2003). Expertise in KIFs/KIOs could be considered from the perspective of individual tacit expert knowledge (internally and externally) as well as expertise embedded in machines and routine processes – that is people, technology and practice (Starbucks, 1992). These enterprises and organisations create commercial value by offering knowledge-based products and services through the use of knowledge innovation and highly skilled and knowledgeable employees to develop novel and dynamic solutions to complex problems of their clients and societies (Alvesson, 2004; Cavaliere *et al.*, 2015; Khadir-Poggi & Keating, 2013; Starbucks, 1992; Swart, 2007; Swart & Kinnie, 2003). Alvesson (2004) classifies KIFs into two broad categories: *professional services firms (PSFs)*, which deal predominantly on intangible products (and services) where KWers often have a high and direct contact with the market; and *research and development firms (RDFs)*, which are science-based companies that deal with tangible output and contact between KWers and customers are less direct and minimal. Lowendahl (1997), on the other hand, classifies KIFs as other firms that deal primarily with clients and are individual-controlled (client-based); or provide creative problem-solving and innovative skills for bespoke solutions and usually work in teams (problem-solving); or adapt available solutions to problematic situations and are often controlled by organisations (output-based).

Sani-KIFs and Sani-KIOs will lean heavily on the expertise and know-how (i.e., skills) of their employee-base and their ability to solve problems through creative and innovative solutions (Sveiby, 1997; Swart, 2007), and also generate, distribute and apply knowledge for safe sanitation management product manufacture and service delivery (Defillippi *et al.*, 2006). These KIAs could also serve to initiate, facilitate and develop innovations (for internal activities and client organisations) and also express innovations through the transfer of existing knowledge for new applications among or within organisations, industries, networks, and so on. (OECD, 2006). Some examples of KIFs/KIOs (Khadir-Poggi & Keating, 2013) include ICT and software development (Alvesson, 2004; Chasserio & Legault, 2010; Cleary, 2009; Marks & Baldry, 2009; Rajala *et al.*, 2008; Scarso & Bolisani, 2010; Swart & Kinnie, 2003; Timo & Arto, 2009), research and development (Whelan *et al.*, 2010), engineering (Erhardt, 2011), university and scientific consultancy (Garcia, 2007), Law and legal services (Forstenlechner & Lettice, 2007; Windrum & Tomlinson, 1999) and others such as sanitation management treatment and disposal and recovery and recycling facilities, sanitation-derived products reuse and processing, safe sanitation service delivery, and so on. Table 9.2 highlights the Sani-KIFs/Sani KIOs and related businesses under these classifications.

9.3.3 Sanitation knowledge-intensive business services firms (Sani-KIBSFs)

Sani-KIBSFs are those businesses that provide knowledge-intensive inputs to the business operations of sanitation management enterprises and organisations (EMCC, 2005; Muller & Doloreux, 2009) and other related public and social sector clients (EMCC, 2005) by helping them deal with problems for which external services are required. They are now a prominent part of the KE in global and domestic economies (EMCC, 2005; Muller & Doloreux, 2009). Such KIFs have been referred to as knowledge-intensive business services firms (KIBSFs) and are highly reliant on professional and technological knowledge and expertise that are related to specific disciplines or domains to provide intermediate knowledge-based products and services not available within clients' internal systems (Den Hertog, 2000; Miles, 2005; Toivonen, 2004; Windrum & Tomlinson 1999). They typically employ highly skilled workers and to a larger extent more than other sectors in the economy (EMCC, 2005) and their core tasks involve economic activities that include the accumulation, creation, dissemination, and utilisation of knowledge to develop and produce bespoke (i.e., custom) innovative and novel solutions

Table 9.2 Classifications of sanitation management and related KIFs/KIOs.

	Client-based	Problem-solving-based	Output-based
Professional Services Firms (PSF)	• Providers of sanitation services (collection, emptying, etc.) • Installation and Construction • Marketing and branding • Advert/Promo/PR	• Faecal sludge management laboratories • Facility Management • Advocacy • Intervention • Awareness creation • Suppliers of sanitation management products and services	• Sanitary wares manufacturers • Resource recovery, recycling and reuse Plants • Treatment plants • Disposal sites • Sanitation management equipment manufacturers • Sanitation-related products manufacturers • Providers of sanitation facilities
Research and Development Firms (RDF)	• Research design and development • Data processing and analytics • Information processing and development • Content generation, design and development	• Universities • Research institutions • Testing and experimentation centres	• Public agencies • NGOs/INGOs • Multilateral organisations

to satisfy their clients' needs (Bettencourt *et al.*, 2002; Dobrai & Farkes, 2009; Miles, 2005); mostly working directly with clients to co-produce Sani-K content, processes, products and services (Zieba, 2013). They could be producers and users of knowledge as well as suppliers of knowledge contents, products and services and are of competitive importance to their client base and also typically depend on outsourcing from client organisations/firms (EMCC, 2005). Workers in SaniKIBSFs use their knowledge assets to diagnose the needs of their clients and then determine a solution, propose a course, and sometimes, implement it on behalf of the client (Bettencourt *et al.*, 2002); which could also involve non-human assets such as inventories, technology, installations and equipment (Nordenflycht, 2010). Even when the service is delivered as an artefact (i.e., book, manual, proceeding, app or technology), the knowledge content is often more valuable than the product itself (EMCC, 2005).

Generally, KIBSFs are made up of traditional professional services (e.g., legal, accounting, architectural, management consultancy, marketing, etc.) and technology-based services (e.g., research and development, engineering, ICT, software/app development, construction, etc.). Sani-KIBSF either operate as suppliers of products primarily used to source and manipulate information, data and existing knowledge or suppliers of specialist knowledge that facilitate support for their clients' business processes (clients, which could be other businesses or KIFs/KIOs in the economy (e.g., sanitation management product and services companies), the public sector (governments), social sector (voluntary organisations like NGOs/INGOs, charities, CSOs, multilaterals, etc.) and sometimes, households and individuals that wish to install, maintain or repair sanitation devices. They are only able to serve based on availability and efficacy of knowledge at their disposal; but could also serve as intermediaries between entities that produce knowledge and users of knowledge (Hipp, 1999) such as other KIBSF or KIFs that need salient knowledge for innovation, business activities or compliance requirements – whether as service integrators (EMCC, 2005), coordinators (Toivonen, 2004) or brokers (Bao & Toivonen, 2014; EMCC, 2005) to provide specialised and/or wider set of inputs and operate as suppliers/subcontractors for other firms who wish to sell their in-house Sani-KISA to other organisations (EMCC, 2005).

KIBSFs are a strength of the KE and are growing so rapidly that they outpace all other sectors, particularly in the European Union (EU) as they supply a wider range of services across all industrial and public sectors (Dobrai & Farkes, 2009; EMCC, 2005; Den Hertog, 2000; Makó et al., 2009; Miles, 2005; Miles et al., 2018; Toivonen, 2004). Santos (2020) also points out that KIBSFs are known to foster and generate innovation nationally and regionally (Fischer, 2015; Miozzo et al., 2016) and they are of great significance to emerging economies (Miles et al., 2018; Zieba, 2013). This makes Sani-KIBSFs a critical part of the IFSVC and the sanitation economy as a whole and it is important that studies are conducted to explore their impacts on the management of sanitation globally, nationally, regionally and locally.

9.3.4 Sanitation knowledge workers (Sani-KWers) and experts (Sani-KExps)

A Sani-KWers is someone works primarily with their ability to think (Davenport, 2005) and their work is described as ever-changing, dynamic and autonomous (Drucker, 1959); they are critical to the IFSVC as they support businesses and organisations with problem-solving and innovation creation (Davenport, 2005; Davenport & Prusak, 1998). These workers depend on knowledge capital and employ their brain more than their might by using their intellect and innate skills to translate data, information and knowledge to knowledge resource (KR), and then develop wisdom in expertise to deliver solutions, processes and products and services for the sectoral market (Davenport, 2008; Davenport & Prusak, 1998) to create value-added assets. They primarily engage in thinking, solving complex problems, collaborations and networking (Davenport, 2005; Reinhardt et al., 2011) and although there is no agreement on a precise definition for KWers (De Sordi et al., 2021; Reinhardt et al., 2011). Davenport (2005) describes them as employees with high levels of expertise, education, training and experience focused on tasks that have to do with the creation, distribution, or application of knowledge. They make up a very important ingredient for the success of sanitation knowledge-intensive businesses and organisations as they hold the knowledge abilities that enhance competitive advantage and innovation (Davenport, 2008; Miles, 2005).

In recent times, particularly in the Covid-19 era, KWers have been known to work remotely from locations outside a formal office, and with collaborators and teams across the globe (Moravec, 2013) without being restricted by space and distance. These KWers are referred to as 'Knowmads' or digital nomads (Iliescu, 2021; Makimoto & Manners, 1997; Moravec, 2013), that is, nomadic workers who use their creativity and imagination to do innovative work with almost anybody, anytime and anywhere; and this is creating new opportunities (Iliescu, 2021; Moravec, 2013; Moravec & van den Hoff, 2015). Sanitation Knowledge entrepreneurs (Sani-KEnts) are those dynamic knowledge experts that have specialised knowledge in their field and may work as consultants or maintain a KIBSF where they continue to innovate and serve their clients (Cooke & Porter, 2007). Most Sani-KWers, Sani-KEnts and Sani-KExps are now Knowmads who work from remote locations for clients from across the globe and in all sectors. This indicates that the SaniM-KVC is not static, but dynamic and complex with producers and providers collaborating and working from different points in the world at the same time and on the same project.

The depth of knowledge intensity of any sanitation and related organisation and enterprise is determined by how much they primarily rely on knowledge (intellectual) capital rather than physical and financial capital, and manual labour (Alvesson, 2004; Khadir-Poggi & Keating, 2013; Starbucks, 1992; Swart & Kinnie, 2003). Khadir-Poggi and Keating (2013) suggest that knowledge intensity can be characterised on the use of intellectual and analytical capabilities of KWers acquired through theoretical education and experience (Alvesson, 2004) as their conceptual skills, knowledge expertise and cognitive skills generate substantial added-value that sets such businesses apart (Nordenflycht, 2010). It could also be embedded in the organisation itself, while inclusive of human capital, the organisation serves as the platform in which knowledge can be generated, created and disseminated; and in the relationship between the KWers and their organisations (Khadir-Poggi & Keating, 2013).

9.4 SANITATION MANAGEMENT KNOWLEDGE VALUE CHAIN (SANI-KVC)

In a Sani-KMart, value is embedded into knowledge, that value is then advanced along the stages of a value chain, and then extracted to yield results at different levels and customers (Simard, 2006). The value of any knowledge is based on the degree of usefulness (functionality and importance of the knowledge unit's utility in valuation) and desirability (demand for the knowledge product or service) (Stocker, 2012). This means that value is not just created when Sani-k is produced, but when it is used to solve problems or satisfy the needs of customers and/or society either through artefacts (i.e., products such as VIP latrines, prefab septic tanks, sanitation-derived products like fertiliser, energy, reuse water, disposal trucks, etc.) and/or services (such as emptying, disposal, transportation, training, education, research, repairs, installation, construction, facility management, advocacy, etc.), and such knowledge could also be created, modified, or reconfigured (Stabell & Fjeldstad, 1998 in Stocker, 2012). Therefore, the SaniM-KVC is dependent on the Sani-KMart as it is buyer-driven – that is it rests on the desirability and subsequent demand of customers and society (Simard, 2006).

A value chain (VC) is the range of activities required to bring a product or service from raw material supply to production (conceptualisation) through to final consumption/ end-use/consumption (Lowitt *et al.* 2015; Porter, 1985). It is critical for systematically comprehending the interactions between actors and processes/stages/phases/levels in a market and interpreting the development and innovative possibilities within specific sectoral and locational contexts (Humphrey & Schmitz, 2002; Lowitt *et al.*, 2015). The Sani-KVC should provide products and services that lead to the production and delivery of new and improved solutions (products and services), programmes, processes and interventions for societies, governments and people (Chyi Lee & Yang, 2000; Holsapple & Singh, 2003; Landry *et al.*, 2006). It should also proceed from acquiring knowledge and mapping the interactions through the processes in-between up to the production of new and improved solutions and interventions that add value for people (Landry *et al.*, 2006).

In sanitation management organisations, knowledge is continually sourced, acquired and dispersed within their knowledge management (KM) systems and the set of activities that make up the entire process is referred to as the Knowledge Value Chain (KVC) (Chyi Lee & Yang, 2000; Ermine, 2013; Lee, 2016; Powell, 2001; Wang & Ahmed, 2005; Weggeman, 1997, 2000). The KVC applies Porter's Value chain (Porter, 1985) to knowledge processing and production, and operates as a model for Sani-KM framework of Sani-KIFs/KIOs/KIBSFs (Chyi Lee & Yang, 2000; Ermine, 2013; He & Wong, 2004; Holsapple & Jones, 2004; Lee, 2016; Powell, 2001; Wang & Ahmed, 2005) that organise knowledge activities in a series of intellectual tasks in stages and steps towards the transformation and creation of commercially valuable knowledge products and services (Chyi Lee & Yang, 2000; Ermine, 2013; Lee, 2016; Powell, 2001; Strambach, 2008). It illustrates the processes that Sani-KWers and Sani-KEnts of Sani-KIFs/KIOs/ KIBSFs use to transform data to intelligence and then onwards to contributing to performance outcomes (King & Ko, 2001; Powell, 2001) and enhance their employer's/ clients' competitive advantage, knowledge absorptive capacity, innovation capabilities and socioecological benefits (Lee, 2016). In the SaniM-KVC, the raw material is data, information and existing knowledge; the sequential activities on the chain include value-adding processes at each stage and makes up the building blocks that finally deliver products or services that is valuable to customers/end-users/clients (Probst, 1998) and contribute to the innovative capacity and competitive advantage of enterprises and organisations whether public or private, business or social (Ermine, 2013).

Actors in the SaniM-KVC are delineated into knowledge phases and categories (Table 9.3) and three knowledge bases that serve as the key dimension of knowledge relevant for innovation in specific industries (Malerba & Orsenigo, 2000; Strambach, 2008) as supplier-dominated, production-intensive and science-based; and the knowledge base determines what is produced and provided, and how. There are knowledge categories that work within each knowledge base and sometimes cut across: analytical (i.e., use of science-based deductive knowledge), synthetic (i.e., use of existing knowledge and new knowledge) and symbolic (i.e., use of ideas, symbols, social constructs and culture). Furthermore, Sani-KIFs/KIOs/KIBSFs go through knowledge phases of exploration, examination and exploitation. Knowledge exploration is the search for new products, services, concepts, processes, content, resources, knowledge, competencies, market domains, innovations, technologies, alternatives, possibilities and opportunities (Benner & Tushman, 2002, 2003; Danneels, 2002, 2007; He & Wong, 2004; Katila & Ahuja, 2002; March, 1991; Sinha, 2015; Strambach, 2008). It involves actions and activities that include search, research, risk-taking, experimentation, discovery, variation, flexibility, play and innovations (Li *et al.*, 2008; March, 1991; Popadiuk & Vidal, 2009; Sinha, 2015). Knowledge examination, on the other hand, is where testing, piloting, reviewing, evaluation and validation occur to improve internal knowledge assets and make them appropriate for commercial value adding purposes (Cooke, 2005; Cooke & Leydesdorff, 2006; Cooke & Porter, 2007; Strambach, 2008). The third phase, knowledge exploitation is where existing knowledge and competencies are used to refine existing merchandise and create new products and services, knowledge, resources and competencies in new dimensions and for new markets as well as competitive advantage and market strategies; and involves production, implementation, execution, innovation, efficiency and selection (Benner & Tushman, 2002, 2003; Danneels, 2002; He & Wong, 2004; Katila & Ahuja, 2002; March, 1991; Sinha, 2015; Strambach, 2008). Table 9.3 illustrates the interactions of KIBSFs between knowledge categories and phases.

The knowledge categories (analytical, synthetic and symbolic) and knowledge phases (exploration, examination and exploitation) of KIBSFs enable them to deliver composite knowledge products and services that could complement or even change the knowledge base of their clients through integrated knowledge provision services (Strambach, 2008).

The KVC has been applied in different contexts of research such as e-learning (Wild *et al.*, 2002), competitive tendering (Dewagoda & Perera, 2019), KIBS (Bao & Toivonen, 2014), supply chain management (Lee & Han, 2009), organisational performance (Chyi Lee & Yang, 2000; Lee, 2016; Wang & Ahmed, 2005), new product development (Gurd & Jothidas, 2009; He & Wong, 2004), competitiveness (Holsapple & Jones, 2004), research and development (Un & Asakawa, 2015), work performance (OuYang & Lee, 2019), health (Landry *et al.*, 2006) and government services (Simard, 2006)

Table 9.3 Sani-KIBSFs according to knowledge categories and phases.

	Knowledge Categories		
Knowledge Phases	Analytical	Synthetic	Symbolic
Exploration	• Contract research, design and development • Conceptualisation • Content development	• Information processing and content generation • Website and app design • Architectural design • Content generation and design • Engineering design services	• Market research and analysis • Business and corporate management consultancy • Fund-raising and management services
Examination	• Research design and development • Data processing and analytics • Testing and validation • Auditing	• Proto-type development • Research in natural/applied sciences, technology, social sciences and humanities • Experimentation laboratories • Accounting • Finance Management	• Financial management consultancy • Tax consultancy • Knowledge workers recruitment and management
Exploitation	• Legal services (registrations, patents, copyright, trademarks, agreements, etc.) • Biotechnology • Software development • Website and app development • Computer services • ICT consultancy/supply • Specialist consultancy	• Construction and installation • Facility management • Maintenance and repair	• Biotechnology production and services • Piloting • Publishing • Legal services (litigations) • Advertising/public relations • Promotions/awareness creation • Knowledge management and brokerage services • Insurance

among several others. There are no studies on sanitation management or related KIFs/KIOs/KIBSFs in the sector. With the considerations of this chapter, it seems obvious that an understanding of the KVC within the sanitation economy and corresponding IFSVC will give insights into the knowledge flow, impacts and management and how this affects competitive advantage, performance, knowledge work and innovation in the sector. According to Chyi Lee and Yang (2000), the KVC model indicates the progress of competitive advantage from the KM structure of any KIFs/KIOs/KIBSFs. Thus, this Chapter proposes the Sanitation Management Knowledge Value Chain (SaniM-KVC) as a model that could describe the knowledge flow within sanitation KIFs/KIOs/KIBSFs, which contribute to the core operations that generate value for innovations in products and services in a way that enhances competitive advantage and knowledge diffusion in the industry and communities of practice (CoP). The focus is on transforming sanitation knowledge into new products and services (He & Wong, 2004) with commercial value, practical functionality and consumer satisfaction.

This chapter considers KVC for operations within Sani-KIFs and Sani-KIOs (Figure 9.5) and Sani-KIBSFs (Figure 9.6) and then proposes a creative concept map for the Sanitation Management Knowledge Value Chain (SaniM-KVC) in the sanitation industry. They are adapted from the KVC models of Weggeman (1997, 2000), Chyi Lee and Yang (2000), Wang and Ahmed (2005) and Simard (2006) as well as Porter's value chain (Porter, 1985). These proposed concepts are not tested yet, but should be further explored in the contexts of the different stages of the IFSVC and their impacts on the sanitation economy.

Figure 9.5 illustrates the Sani-KVC for KIFs and KIOs in the SE highlighting the value-adding activities whereby each stage adds value to data and information and existing knowledge that is then translated into products and services. They provide intermediate inputs that add value along the sequential stages to the final products or services and is the major ingredient for innovation (Albors-Garrigos et al., 2009; Santos, 2020). This Sani-KVC considers what happens in sanitation management enterprises and organisations and is driven by their missions, visions, goals and strategies; and the KM system comprises of the Sanitation Knowledge-intensive Services (Sani-KIS) conducted through Sanitation Knowledge-intensive Services Activities (Sani-KISA) and together make up the core activities that provide resources for conceptualisation, design, development, production and delivery. These activities are played out by internal sanitation knowledge workers and experts that coalesce tacit and explicit (analytical, synthetic and symbolic) knowledge through exploration, examination and exploitation phases towards organisational goals and strategies. In addition, Sani-KIS and Sani-KISA could be outsourced to external actors that search, produce and deliver knowledge-intensive business services (KIBS) on behalf of their clients.

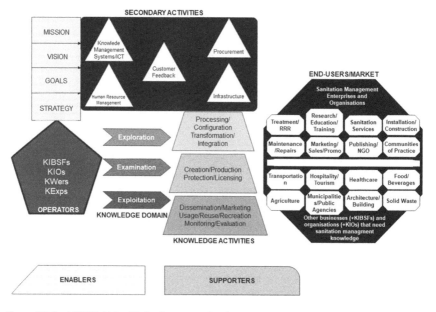

Figure 9.5 Sani-KIBSFs Value Chain. (Source: Authors)

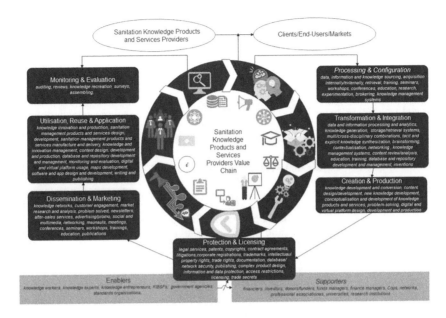

Figure 9.6 Sanitation Management Knowledge Value Chain. (Source: Authors)

However, in the middle of the diagram is the secondary activities, and it is important for the enterprise and/or organisation to build its own Knowledge Management System (KMS) overseen by a Chief Knowledge Officer that manages the process of processing, producing and sharing valuable knowledge across the entire community. The secondary activities support the core activities like hiring and managing knowledge workers (KWers), firm infrastructure (administration, finance, legal, etc.), and even customer feedback and engagement to ensure a seamless transition from the knowledge bank to the area of need like marketing, sales, management for decision making and actions, planning, and so on. At the bottom of that middle phase is the enablers that provide foundation for the sanitation KIFs/KIOs in their search, research, acquisition, transformation, integration, creation, dissemination and protection of their knowledge assets. When all of these are appropriately commissioned then it could positively affect organisational performance and outcomes (hanging on the side of the secondary activities) and then subsequently (as seen on the top right of the diagram) existing products and services are improved and new products and services are introduced to the market place, sometimes at domestic, regional and international levels or all levels. Meanwhile, the Sani-KIS and Sani-KISA produce knowledge products and services that feed the internal knowledge market of the sanitation management KIFs/KIOs and could also be outsourced to Sani-KIBSFs and supplied to other enterprises and organisations that use sanitation and related artefacts and services (e.g., transportation, tourism and hospitality, healthcare, etc.); and sometimes, directly to end-users who need the knowledge for research or home use.

Figure 9.6, on the other hand, highlights the value-adding stages along the Sani-KISA involved with providing Sani-KIS for the clients of Sani-KIBSFs who serve as business enterprises or not-for-profit organisations that provide Sani-K content, products and services to sanitation management entities and other related establishments in the

SE and the structure of their operations is also driven by the mission, vision, goal and strategy designed for the existence from the onset. The difference between the Sani-KIBSFs and Sani-KIFs/KIOs is that Sani-KIBSFs are primarily knowledge providers and their core activities are KISA to produce and provide knowledge products and services to their clients, which could also be other KIBSFs or companies that have sanitation concerns to deal with. The secondary activities here are essentially designed to be able to aptly and accountably serve their clients with their primary activities of producing and providing knowledge in cooperation with their clients (which makes customer feedback and engagement crucial). The Sani-KISA is split according to the phases of knowledge where exploration, examination and exploitation take place. Subsequently, on the right of the diagram, are the end-users or market of the KIBSFs, split into two types: sanitation management enterprises and organisations, (which include academic institutions, research institutions and professional networks as well) and then other businesses (which include other Sani-KIBSFs and Sani-KIFs) and organisations (Sani-KIOs) that need sanitation management knowledge and expertise. Enablers and supporters provide a base for the Sani-KIBSFs by contributing to marketing and financing (banks, investments, grants and sales) and policies, legislation and regulations as well as other government support and incentives.

Figure 9.6 illustrates a creative concept overview map for the value chain activities of the sequential processes that follow sanitation knowledge products and services. These products and services could be provided as composite units by a single KIBSF (or several units of a KIF/KIO), but most often enterprises, knowledge entrepreneurs, knowledge experts, and knowledge workers operate with clients (Santos, 2020) and employers in specialised areas of two or three or in a particular knowledge base and/or phase. The Sani-KISA here operate in a circular flow that indicates the value-adding steps in the process of sourcing, acquiring, and up to using knowledge content, products and services that accrue as the sum of the value-added at each stage.

Table 9.4 shows a description of knowledge products, services and content that are exchanged in the Sani-KMart and Table 9.5 addresses the activities in the Sanitation Management Knowledge Value Chain.

The Sani-KIAs begin at the point of processing and configuration where data, information and existing knowledge are sourced (through research, education, training, search, etc.) by Sani-KIFs/KIOs and even Sani-KIBSFs as well as KEnts acting as brokers or coordinators; and then material relevant to the need is identified (whether available or not within the system); and subsequently acquired through KWers tacit knowledge, codified and stored explicit knowledge and/or purchased from third parties as an artefact in itself or by hiring knowledge experts and or specialised KIBSFs (Lonnqvist and Laihonen, 2017; Rajala et al. 2008). The next stage is when the units acquired and configured are transformed to usable material whereby the different perspectives, levels, phases and categories of knowledge from different sources have been integrated into one composite package for specific purposes (Berends et al., 2011; Gabbay et al., 2020; Holsapple & Singh, 2003; Krome, 2014; Paralič et al., 2013; Schneider, 2012; Welo & Ringin, 2018; Zahra et al., 2020). The composite knowledge drawn from the transformation and integration stage enables the creation and production of knowledge content, products and services in sanitation and its management specific to required and relevant sectoral needs and expectations and novel innovations (Brix, 2017; Ramirez et al., 2012; Yu et al., 2017).

Knowledge protection (KP) may not be so critical for KIBSFs and KIOs, but it is often really important for KIFs (Chyi Lee & Yang, 2000; Probst, 1998; Probst et al., 1999; Simard, 2006); it serves as a means to protect their tangible and intangible assets from

Table 9.4 Sanitation knowledge products, services and contents.

Item	Description	Examples
Sanitation Knowledge Products	Knowledge products are the tangible outputs that could be used to generate, create, store, distribute, diffuse, use, evaluate and transform existing and new knowledge either for or of itself (IUCN, 2004)	Books, Reports, Guidelines, Journals, Maps, Software, Apps, Websites, Podcasts, Media Programmes, Databases, Repositories, Inventories, Virtual Platforms, and so on.
Sanitation Knowledge Services	Knowledge services are the KISA operated internally and externally for clients and users through the tacit and explicit knowledge embedded in knowledge workers and organisational practices and technologies (IUCN, 2004)	Customer Feedback Management, Knowledge Workers/Experts Recruitment, Education and Research, Research and Development, Training and Development, Science and Technology research, Social Science and Humanities research, content generation, design and development
Sanitation Knowledge Content	Sanitation knowledge content generated, designed, developed and produced by KWers, KExps and knowledge entrepreneurs for clients and employers	Manuals, procedures, guidelines, standards, regulations, policies, legislation, training materials, seminar/workshop/conference proceedings, journal articles/publications, books, magazines, newsletters, blogs, podcasts, curricula development, programme design and development. Adverts/promo materials, and so on.

expropriation and imitation (Bolisani *et al.*, 2013; Elliot *et al.*, 2019). Considering that knowledge innovation, particularly for new content, product and service development, is primarily tacit knowledge that exists in the brains of knowledge workers and other experts, and also requires a number of interactions that could unwittingly expose key knowledge assets (Bolisani *et al.*, 2013; de Faria & Sofka, 2010; Elliot *et al.*, 2019; Manhart & Thalmann, 2015; Paallysaho & Kussisto, 2008, 2011), the process of preventing this involves a number of mechanisms and KISA. The outcome of the Covid-19 pandemic has resulted in more people working and interacting remotely and has made KP more dicey as knowledge assets are shared across communication lines that are not completely under the control of organisations, their contractors and collaborators (Bolisani *et al.*, 2013; Elliot *et al.*, 2019; Ha *et al.*, 2021; Paallysaho & Kussisto, 2011). Thus, knowledge-intensive enterprises and organisations in the sanitation management sector need to make use of formal and informal mechanisms to keep their valuable and sensitive knowledge assets vaulted (Bolisani *et al.*, 2013; Elliot *et al.*, 2019; Manhart & Thalmann, 2015; Paallysaho & Kussisto, 2008, 2011), but they are faced with the challenge of balancing knowledge protection and knowledge sharing internally and externally (Bolisani *et al.*, 2013; Elliot *et al.*, 2019; Manhart & Thalmann, 2015; Manhart *et al.*, 2015; Paallysaho & Kussisto, 2011).

The next step is to determine which knowledge is to be freely shared, vaulted, licensed, and how much of it should be shared, and then the manner in which it should be shared. These are activities that fall under the dissemination and marketing stage. Sani-KIOs may mainly share their knowledge units as public goods, but also keep private certain valuable aspects within lock and key as much as possible. But, Sani-KIBSFs and Sani-KIFs (in particular) are stuck with the dilemma of determining what knowledge is available for disseminating, especially as they are much dependent on

Table 9.5 Sanitation knowledge products and services value chain.

Core Activities	Secondary Activities	SKPSPs (Actors: KIFs, KIBSFs, KIOS & OTHERS)	End-users/Market
Processing/Configuration Generating data and information; Sourcing and identifying and acquiring existing data, information and filling knowledge gaps	• Planning and administration • Human Resource Management • Infrastructure • Knowledge Workers/Experts • Recruitment and Management • Knowledge Management Systems • Procurement • Compliance Management • Audit and Certification	• Universities • Research Institutes • Knowledge workers and experts • Testing and Experimentation • Concept Design and Development • Content generators • Data generation and processing • Digital platform development • Software, app and Website development • Marketers, disseminators • Publishers • Operations, maintenance and repair	• Architecture and Building Construction • Transportation • Hospitality and Tourism • Entertainment and Event Management • Sanitary ware manufacturers • Healthcare and Welfare • Sewage and faecal sledge management • Sanitation Services Providers • Advocacy and Intervention organisations • Municipality and City managers • Government agencies
Transformation/Integration Data processing and management; Information verification and management; Collation and Codification; Storage and retrieval			
Creation/Production Content generation and development; Conceptualization; Product and Service design and development			
Protection/Licensing Patent; Trademarks; Copyright; contracts; agreements			
Dissemination/Marketing Branding; Customer Engagement; Multimedia			

- Schools and Public Places (e.g. markets, parks, cinemas, event centres, etc.)

- Engineering consultancy

Usage/Reuse/Recreate
Sanitation Education and training; Sanitation Research and Development; Sanitation Product and Service Design and Development; Sanitation governance instruments and agencies

Monitoring/Evaluation
Auditing; Assessments; Data and information gathering

Supporting Services
Banking; Funds Management; Corporate Management; Administration
- Business and Management Consulting
- Accounting and Tax management
- Financial Management
- Digital and ICT product/service providers
- Publishing

Enablers
Taxes; Policies; Legislations; Regulations; Standards; Public Private Partnerships; Incentives
- Public Agencies and Intervention Agencies
- Communities of Practice
- Management
- Industry
- Knowledge Workers and Experts
- Donors and Funders
- Finance Institutions and Investors

Outcomes
Performance; Competitive Advantage; Improved Capabilities
- Standards and Quality assurance
- Clients' Engagement and Customer Satisfaction
- Innovative Production and Service Delivery Processes
- New Products and Services Design and Development

marketing and collaborations with external parties (Bolisani & Scarso, 2014; Bolisani *et al.*, 2016; Diehr & Wilhelm, 2017; Kohlbacher, 2008). In any case, there are various ways of distributing knowledge content, products and services to the society (public), employees, clients/customers, third parties (e.g., partners, collaborators, etc.), industry, and end-users. When knowledge assets are made available, whether free or fee based, they can be utilised in several ways within the entities that they were created or third-party partners/clients and/or other end-users. They could be used to create sanitation management knowledge content, or specific sanitation products, services, processes, technologies, and so on., in particular as it relates to the circular economy (Ddiba *et al.*, 2020; Mallory *et al.*, 2020; Moya *et al.*, 2019; TBC, 2017) and the COP26 methane mitigation goals (UN, 2021); and they could be used to design sanitation solutions across sectoral industries such as transportation (airlines, ships, buses, trains, etc.), tourism and hospitality, healthcare, and so on. They could also be used to create new knowledge or to create new content, products, services and other commercially valued commodities (Diehr & Gueldenberg, 2017; Diehr & Wilhelm, 2017; Holsapple & Singh, 2003; Simard, 2006; Song *et al.*, 2005).

At this point, the process of monitoring and evaluation becomes vital to determine the efficacy and efficiency of knowledge resources and assets accumulated through all stages of the value chain (as a whole or just for specific stages), how they have added value to the organisations, enterprises, society and industry that they served, and what additional value they might still be able to add (Janus, 2016; UNDP, 2002). The cycle continues from this point as M&E is able to generate new knowledge that could be re-introduced to the value chain system at the various stages and the circular knowledge flow will resume again. At the top and base of the value chain are key stakeholders that contribute to the performance and sustainability of the value chain and market it represents. The providers of the sanitation and vital related knowledge content as well as products, services, processes, and so on., and those that purchase/obtain what they sell and/or final users down the chain. The enablers are those that give authentication and legal backing for the activities and players in the VC; while the supporters are those that provide services that enable and enhance the operations of SaniM-KVC enterprises and organisations.

9.4.1 Key aspects of the SMKPSVC
There are some key aspects of the SKMPSVC that drive the process towards value-adding and creation in the IFSVC and they include CoPs, education, research and training.

9.4.1.1 Communities of practice (CoPs)
Emphasis of KM is to work smarter by acquiring relevant and high-quality knowledge; and this could be achieved through CoP (Venkatraman & Venkatraman, 2018). KM includes motivating individuals to participate in overall goals and create the social processes that will facilitate success. Such social processes include communities of practice and expert networks. This is vital because individual knowledge (and also indigenous knowledge) will suffocate unless it can be shared through groups, teams, networks and associations. CoPs help to manage knowledge assets and resources (Wenger *et al.*, 2002) and their members can work across organisations, sectors and disciplines. They are regarded as an important component of a human-oriented KM (Huysman & Wulf, 2006; Newell *et al.*, 2006) as individual and collective learning take place simultaneously (Lesser & Storck, 2001) and they support learning and knowledge exchanges (Bolisani & Scarso, 2014).

To understand and expand the reach of the social interactions within the SaniM-KVC, especially in the complex and sensitive sanitation sector, may be better explored from the concept of 'communities of practice' (Lave & Wenger, 1991; Lowitt *et al.*, 2015; Wenger, 1998). The transfer of knowledge and learning opportunities between and within CoP in a VC are critical to the productive quality and effectiveness of sanitation management and achieving the SDG 6 (Bammann, 2007). Thus, understanding and developing a knowledge management value chain in sanitation will be effective from the perspectives of sanitation communities of practice (Sani-CoPs) and their related social interactions within and across sectors, spatial scales and landscapes of communities in knowledge transfer and learning (Lowitt *et al.*, 2015). The purpose and aspirations of the Sani-CoPs will drive any SaniM-KVC (Landry *et al.*, 2006), especially as it regards domestic and global expectations such as the SDG 6 and other agenda.

Sani-Cops share a concern or a passion for specific areas in sanitation and its management, and learn how to do things better as they interact regularly (Lave & Wenger, 1991) by focussing on the social relationships that allow people to learn together or from each other. Thus, to understand the operations of social interactions within the SaniM-KVC, the concept of communities of practice is key (Lowit *et al.*, 2015). Multiple Sani-CoPs can come together to form larger 'landscapes' of practice with the potential to support social learning and innovations and also enable cross-learning (Dei & van der Walt, 2020; Wenger-Trayner *et al.*, 2015). In fact, the SaniM-KVC comprises of CoPs within the different stakeholders and organisations it brings together (even across landscapes of practice) (Lowitt *et al.*, 2015) such as sanitation = health, water hygiene, climate, governance, behaviour, economics, and so on.). This highlights the social interactions, knowledge-sharing and learning as vital resources for the IFSVC. They could also be crucial to facilitating coherence and coordination in the knowledge value chain activities of the sanitation management sector (Chisholm & Nielson, 2009; Lowitt *et al.*, 2015; Nahapiet & Ghoshal, 1998). They can provide a pool of knowledge and make it available through education, training, research and archival platforms (e.g., databases, repositories, publications, etc.) and other forms of knowledge-sharing and learning (e.g., conferences, seminars, experts, peers, etc.) for sanitation management problem-solving, decision-making and innovations (Kling & Courtright, 2003; Venkatraman & Venkatraman, 2018); and are involved in creating and sharing knowledge in literature (Hartlung & Oliveira, 2013). They could also add value and contribute to competitive advantage (Kim *et al.*, 2012) by improving old artefacts (i.e. products, services, processes, tools, etc.), creating new knowledge, products, services, and so on., solve problems faster and smarter, disseminate best practices, develop professional skills and support recruitment and talent retention (Wenger & Snyder, 2000). In addition, Sani-CoPs support social capital, which has been described as the structural and cognitive characteristics of social organisations that share values, norms, and trust to facilitate coordination and cooperation for mutual benefits (Nahapiet & Ghoshal, 1998; Putnam, 1995) by providing learning networks of individuals with common interests, drawing people together to generate/ share knowledge and learn in a way that breeds trust and then through created and shared stories about the norms and values, they develop and maintain sector registers and terms that can be transferred to others (Lesser & Storck, 2001; Lowitt *et al.*, 2015).

Members of Sani-CoPs will share experiences within a particular domain of Sani-K/ SaniM-K that allows them to develop perspectives, practices and particular approaches (Wenger *et al.*, 2002) and then engage in collective learning in a subject matter of common interests (with perhaps divergent focus); for example a group of sanitation professionals interested in research towards innovative solutions for contextual safely managed sanitation (Wenger-Trayner & Wenger-Trayner, 2015) that support learning, sharing and stewarding knowledge and could deepen knowledge and expertise (Bolisani & Scarso, 2014; Lesser & Prusak, 1999; Wenger-Trayner & Wenger-Trayner, 2015). However, they do not just share existing knowledge, but also provide innovations, inventions and solutions to problems; and create new knowledge, expand practice, define new territory, introduce new disciplines and develop a collective and strategic voice (Wenger-Trayner & Wenger-Trayner, 2015). These CoPs have the capacity to drive knowledge creation and dissemination since individuals and groups can share and transfer knowledge that improves practice, productivity and fosters innovations within organisations and even personal growth (Aljuwaiber, 2016; Dei & van der Walt, 2020; Hislop, 2003; Wenger, 2004) and also encourage cross-learning amongst landscapes of practice (Probst & Borzillo, 2008; Wenger-Trayner *et al.*, 2015). But, for sanitation practice to be able to properly realise its potential through the knowledge and skills of the Sani-CoPs, a clear picture of the SaniM-KVC is necessary (Wenger & Snyder, 2000). Therefore, research is urgently needed in this area.

9.4.1.2 Sanitation education, research and training (Sani-ERT)

No cause can succeed without first making education its ally (attributed to Victor Hugo) as education provides individuals with knowledge and skills necessary within society and the labour market for their own enlightenment, empowerment and participation; and to increase social advantage in several ways (Exley, 2016). The sanitation knowledge economy (Sani-KE) obviously requires highly qualified and knowledgeable workforce to achieve the SDG targets related to sanitation, so education and professional training is a strong driver while research drives intellectual capital for innovation (Carrillo, 2016; OECD, 2012). To accomplish the goal and targets of SDG 6 and other key related SDGs (e.g., SDG 3, 7, 8, 9 and, 11), and the success of sanitation businesses, investments in human capital and human development are crucial (Jacinto & Garcia de Fanelli, 2014). Strengthening the capacity of domestic citizens and professionals to plan, implement, manage, govern and implement as well as innovate, produce and serve effectively and with enhanced value additions and creations will enable national and local governments in conjunction with global partners to address contextual and interconnected sanitation challenges more successfully. The significance of Sani-ERT in perpetuating safe sanitation management is indisputable – even though this has not yet gained traction in the development and academia arena.

All aspects of the IFSVC are dependent on the learning connections of SERT. There can be no progress without adequately and appropriately trained and equipped human resource in the sanitation sector of any country, in particular, developing countries. Knowledge sharing and creation produces the capacity for innovative solutions and implementation of relevant and effective policies. When ERT is properly designed and positioned in any sector of any given economy, then the sector is strengthened to deliver on expectations and solve contextual and interconnected problems. But, above that, the sector will be equipped to create value-added products and services across the spectrum and beyond. The complex and dynamic nature of the sanitation system and processes make it an interdisciplinary discipline with contextual peculiarities, thus value-added activities are not restricted to technology alone, and this will require capacities at various levels and different aspects.

Knowledge-sharing and transfer rest greatly on education and training while knowledge-creation, translation, exploitation and dissemination through research and development is key for innovation and transformation. The capacity to acquire, source, create, share, transfer and apply knowledge is significant to the ability to tackle sanitation and related problems and to provide sustainable solutions. And so, the instruments of

SERT will be able to round-up all dispersed knowledge and provide tracks of learning that could transform the sanitation management sector and provide effective, knowledgeable and skilled manpower capacity, primarily in developing countries. Sani-CoP are useful to facilitate research, teaching and learning in universities in this regard.

In addition, the capacity that enables individuals, organisations and societies (economies) to identify and understand existing development challenges and trends, solve attendant problems, perform appropriate tasks effectively, efficiently and sustainably through an active learning and knowledge-sharing process is embedded in ERT (Alaerts & Kaspersma, 2009). Weak or inadequate capacity at any level could translate to the inability to achieve goals and targets and even further hamper the progress of any society, economy and/or organisation. For sanitation management capacity to be relevant in any economy or society, however, it must cover the quality and quantity of contextual abilities needed at individual and institutional levels required to match the challenges and expectations; and this capacity should include the combination of attributes that enable the creation and addition of value at different phases and stages of the IFSVC. Providing and strengthening capacity with knowledge transfer (or sharing) and creation happens via ERT.

Capacity development that represents knowing the what (e.g., conceptualisation, creation and integration), knowing the how (e.g., process, procedure, replication), and knowing the where/when (contextualisation, location, climate, time) are primarily learned through education and research. On the other hand, the practice of knowledge that leads to skilling is mainly learned through training (on-the-job, mentorship, apprenticeship, coaching, etc.). now, knowledge is transferred through different levels of education (primary, secondary, technical/vocational and tertiary) while knowledge-sharing and creation comes through research. Skill training can be delivered on the shopfloor or through apprenticeships, mentoring/coaching, internships, seminars, workshops, classes and/or self-taught acquisitions (Alaerts & Kaspersma, 2009).

Knowledge, skills and the ability to understand the nuances in sanitation management and governances are dependent on the agencies of SERT. It is impossible for the sanitation sector to thrive without an above average capacity within its human capital stock to deliver top-notch quality, functional, efficient and sustainable solutions on a continuous basis. SERT are significant mechanisms to provide learning and innovation that improves and delivers home-based solutions and would also contribute to local and national economies with a strong potential for creating a sanitation economy and unique job opportunities. Research and formal higher education drive innovation, design production, finance, technology, management, governance, service and advocacy to meet national and local demands while vocational and technical education and training (formal and informal) provide intermediary level skills for installation, maintenance, operations, service delivery and sales/marketing. In addition, professional and on-the-job trainings provide room for continuous learning to ensure sustainability. Meanwhile, sanitation education at the primary and secondary levels equip citizens and residents with basic critical knowledge about sanitation, which influences a better understanding, appreciative perception and pre-emptive behaviour towards safe sanitation management as adults.

9.5 CONCLUSION

Building the sanitation sector will call for innovations in sanitation products and services, operations/maintenance, installation, design, management/governance and advocacy as well as education, research and training and the interactions of sanitation management communities of practice. The sector will, however, need to develop a strong foundation to carry the knowledge market for the Sani-KMart to survive in the fast-growing general KE. The World Bank (2007) suggests four key pillars to this foundation:

(I)   the availability of highly skilled workforce and a quality education system for sanitation management;
(II)   availability and accessibility to a dense and modern ICT infrastructure and systems;
(III) vibrant and effective inn0vation landscape and interactions between academia, industry, government, public and the environment;
(IV)   institutional support and incentives that target entrepreneurship and the use of knowledge.

The success and acceleration of any knowledge-based growth (whether in North or South countries), however, is dependent on how integrated the two ends of the knowledge spectrum are: that is, a seamless connection flow between the exploration of new knowledge (e.g., research, create, test, and experience) and the diffusion and use of existing knowledge (World Bank, 2007). It is also crucial that government players maintain a strong role in the knowledge market as they also create, use and disseminate data, information and knowledge that support the sanitation industry and governance as well (Simard, 2006). In addition, innovation is central to the future of safe sanitation management and the effectiveness of innovations is dependent on available knowledge capital that is based on a collation of created and shared content within a community (or communities) of practice (e.g., sanitation management) and/or interconnected CoPs (e.g., gender, technology, governance, health, water, hygiene, etc.); and the value of such knowledge capital is determined and weighed by how it is utilized (when and where) (Khadir-Poggi and Keating, 2013; Lee, 2016; Simard, 2006). However, when knowledge content is shared (i.e., knowledge-sharing) amongst individuals or groups in a CoP (and its connections) as relevant ideas, information, suggestions and expertise, it builds and strengthens the practice (Bartol & Srivastva, 2002) and also provides the potential to develop new ideas which is integral to value creation and addition. This leads to complex interactive processes of creating (and recreating), transferring and transforming knowledge from one community of practice to another and then to users in societies, industries, governments, and so on., in a value creating loop, where knowledge is traded as a commodity (Landry et al., 2006).

Thus, promoting the value creation and addition in the sanitation sector required for progress will entail the understanding and managing of relevant knowledge of the complex dynamics in sanitation systems and processes to unlock the value within different stages and interactions. In other words, the process of sourcing, acquiring and storing sanitation information and content need to be coordinated and assembled strategically as knowledge capital available to communities of practice (CoPs) form translation and transformation into practical solutions and exploitation/utilization that will result in wide-reach usage as well as the creation of new knowledge, which could then be transferred, stored and disseminated (Holsapple & Singh, 2003; Lee, 2016) as intellectual capital for policy, implementation, products, services and innovation (Alawneh et al., 2009). This intellectual capital is developed through knowledge innovation, that is, the creation, transformation, transfer, and application of new ideas drawn from old and new knowledge capital pool to develop value in products/services and decision-making with competitive advantage and improved performance and capacities that deliver on expectations and requirements in communities of practice, for clients, societies, economies, governments, and industry (Darroch, 2005; Kostas & John, 2006; Lee, 2016).

All of these activities take place in the sanitation management knowledge value chain (SaniM-KVC) and this is why studies are urgently needed in this area of the IFSVC where there is no previous data or information.

9.6 Take action

(1) Identify the sanitation knowledge intensive business services enterprises and sanitation knowledge intensive firms and organizations in your country and at global levels.
(2) Contact a sanitation knowledge intensive business enterprise in your city and conduct interviews on their operations, challenges and opportunities.

9.7 Journal entry

(1) What is the role of the sanitation management knowledge value chain in the overall IFSVC and its impacts on the sanitation economy?
(2) Why is sanitation no longer just a development concern, but increasingly an integral aspect of enterprises' operations?

9.8 Reflection

(1) Reflect on the sanitation knowledge marketplace and the sanitation management knowledge value chain and how they influence the IFSVC and the sanitation economy.
(2) What is the role of the sanitation knowledge market in the implementation of IFSVC?

REFERENCES

ADBI. (2019). Water insecurity and sanitation in Asia. In: Asian Development Bank Institute, N. Yoshino, E. Araral, and K. E. Seetha Ram (eds), ADBI, Japan, pp. 1–387. ISBN 978-4-89974-114-5. https://www.adb.org/sites/default/files/publication/544131/adbi-water-insecurity-and-sanitation-asia.pdf#page=18 (accessed 03/December/2021)

Albors-Garrigos J., Hervas-Oliver J. L. and Hidalgo A. (2009). Analysing high technology adoption and impact within public supported high tech programs: an empirical case. *Journal of High Technology Management Research*, **20**(2), 153–168, https://doi.org/10.1016/j.hitech.2009.09.006

Alaerts G. J. and Kaspersma J. M. (2009). Progress and challenges in knowledge and capacity development. In: Capacity Development for Improved Water Management, M. W. Blokland, G. J. Alaerts and J. M. Kaspersma (eds), Taylor and Francis, London UK, pp. 3–30.

Alawneh A. A., Abuali A. and Almarebeh T. A. (2009). The role of knowledge management in enhancing the competitiveness of small and medium-size enterprise. *Communication of IBIMA*, **10**, 98–108.

Aljuwaiber A. (2016). Communities of practice as an initiative for knowledge sharing in business organisations: a literature review. *Journal of Knowledge Management*, 20(4), 731–748. https://doi.org/10.1108/JKM-12-2015-0494

Allee V. (2003). Value networks and evolving business models for the knowledge economy. In: Handbook on Knowledge Management 2: Knowledge Directions, C. W. Holsapple (ed.), Springer, Berlin, pp. 605–621.

Alvesson M. (2004). Knowledge Work and Knowledge-Intensive Firms. Oxford University Press, Oxford, UK.

Amidon D. M. (1997). Innovation Strategy for the Knowledge Economy: The Ken Awakening. Routledge, London. ISBN 9780080508795.

Andreeva T. and Kianto A. (2011). Knowledge processes, knowledge-intensity and innovation: a moderated mediation analysis. *Journal of Knowledge Management*, 15(6), 1016–1034. https://doi.org/10.1108/13673271111179343

Argyris C. (1999). On Organisational Learning, 2nd edn. Blackwell Business Publishers, Oxford, UK.

Bammann H. (2007). Participatory value chain analysis for improved farmer incomes, employment and food security. *Pacific Economic Bulletin*, 22(3), 113–125.

Bao S. and Toivonen M. (2014). Finnish knowledge-intensive business services in China: market entry and position in the value chain. *Technology Innovation Management Review*, 4(4), 43–52. http://doi.org/10.22215/timreview/784

Bartol K. M. and Srivastva A. (2002). Encouraging knowledge sharing: the role of organisational reward systems. *Journal of Leadership and Organisation Studies*, 9(1), 64–76, https://doi.org/10.1177/107179190200900105

Becker M. C. (2001). Managing dispersed knowledge: organisational problems, managerial strategies and their effectiveness. *Journal of Management Studies*, 38(7), 1037–1051. https://doi.org/10.1111/1467-6486.00271

Benner M. J. and Tushman M. (2002). Process management and technological innovation: a longitudinal study of the photography and paint industries. *Administrative Quarterly*, 47, 676–706. https://knowledge.wharton.upenn.edu/wp-content/uploads/2013/09/1301.pdf https://doi.org/10.2307/3094913

Benner M. J. and Tushman M. (2003). Exploitation, exploration and process management: the productivity dilemma revisited. *Academy of Management Review*, 28(2), 238–256, https://doi.org/10.5465/amr.2003.9416096

Bereiter C. (2002). Education and Mind in the Knowledge Age. Lawrence Erlbaum Associates, Publishers, New Jersey.

Berends H., Garud R., Debackere K. and Weggeman M. (2011). Thinking along: a process for tapping into knowledge across boundaries. *International Journal of Technology Management*, 53(1), 69–88. https://doi.org/10.1504/IJTM.2011.037238

Bernet T., Devaux A., Ortiz O. and Theile G. (2005). Chapter 23: participatory market chain approach. In: Participatory Research and Development for Sustainable Agriculture and Natural Resource Management: A Sourcebook. Volume 1: Understanding Participatory Research and Development, J. Gonsalves, T. Becker, A. Braun, D. Campilan, H. De Chavez, E. Fajber, M. Kapiriri, J. Rivaca-Caminade and R. Vernooy (eds), International Potato Center-Users' Perspectives with Agricultural Research and Development, Laguna, Philippines and International Development Research Centre, Ottawa, Canada, pp. 1–140.

Bettencourt L., Ostrom A., Brown S. and Roundtree R. (2002). Client co-production in knowledge-intensive business services. *California Management Review*, 44(4), 10–128, https://doi.org/10.2307/41166145

Bhagwath M. V. (2014). Intellectual capital assessment of a management institution – methods and its significance. *International Journal of Advanced Information Science and Technology*, 21(21), 42–46.

Bolisani E. and Scarso E. (2014). Marketing audit for knowledge intensive business services. *The Electronic Journal of Knowledge Management*, 12(1), 3–13. Available online at www.ejkm.com

Bolisani E., Pailoa M. and Scarso E. (2013). Knowledge protection in knowledge intensive business services. *Journal of Intellectual Capital*, 14(2), 192–211, https://doi.org/10.1108/14691931311323841

Bolisani E., Donò A. and Scarso E. (2016). Relational marketing in knowledge-intensive business services: an analysis of the computer services sector. *Knowledge Management Research & Practice*, 14(3), 319–328. https://doi.org/10.1057/kmrp.2014.35

Bontis N. (1998). Intellectual capital: an exploratory study that develops measures and models. *Management Decisions*, **36**(2), 63–76, https://doi.org/10.1108/00251749810204142

Bratianu C. (2015). Organisational knowledge dynamics: managing knowledge creation, acquisition, sharing, and transformation. In: A Volume in the Advances in Knowledge Acquisition, Transfer and Management (AKATM), IGI Global, Hershey, PA, 349 pages.

Bratianu C. and Bejinaru R. (2017). Knowledge strategies for increasing IC of universities. In: Proceddings of the 9th European Conference on Intellectual Capital, Instituto Universitario de Lisboa, Lisbon, Portugal, 6–7 April 2017, I. T. Lopez and R. Serrasqueiro (eds), pp. 34–41.

Braunerhjelm P., Acs Z. J., Audretsch D. B. and Carlsson B. (2010). The missing link: knowledge diffusion and entrepreneurship in endogenous growth. *Small Business Economics*, **34**, 105–125, https://doi.org/10.1007/s11187-009-9235-1

Brix J. (2017). Exploring knowledge creation processes as a source of organizational learning: a longitudinal case study of a public innovation project. *Scandinavian Journal of Management, Procedia CIRP*, **33**(2), 113–127. ISSN 0956-5221, https://doi.org/10.1016/j.scaman.2017.05.001

Bryan L. L. (2004). Making A Market in Knowledge. The McKinsey Quarterly, 3. The McKinsey and Company, New York. https://www.mckinsey.com/~/media/McKinsey/Business%20Functions/Strategy%20and%20Corporate%20Finance/Our%20Insights/Making%20a%20market%20in%20knowledge/Making%20a%20market%20in%20knowledge.pdf

Carrillo F. J. (2016). Knowledge markets: a typology and an overview. *International Journal of Knowledge-Based Development*, **7**(3), 264–289. https://doi.org/10.1504/IJKBD.2016.078540

Cavaliere V., Lombardi S. and Giustiniano L. (2015). Knowledge sharing in knowledge-intensive manufacturing firms. An empirical study of its enablers. *Journal of Knowledge Management*, **19**(6), 1124–1145. https://doi.org/10.1108/JKM-12-2014-0538 https://doi.org/10.1108/JKM-12-2014-0538

Chasserio S. and Legault M. (2010). Discretionary power of project manager in knowledge-intensive firms and gender issues. *Canadian Journal of Administrative Sciences*, **27**(3), 236–248, https://doi.org/10.1002/cjas.147

Chisholm A. M. and Nielson K. (2009). Social capital and the resource-based view of the firm. *International Studies of Management and Organization*, **39**(2), 7–32. https://doi.org/10.2753/IMO0020-8825390201 https://doi.org/10.2753/IMO0020-8825390201

Chyi Lee C. and Yang J. (2000). Knowledge value chain. *Journal of Management Development*, **19**(9), 783–794. https://doi.org/10.1108/02621710010378228 https://doi.org/10.1108/02621710010378228

Cleary P. (2009). Exploring the relationship between management accounting and structural capital in a knowledge-intensive sector. *Journal of Intellectual Capital*, **10**(1), 37–52, https://doi.org/10.1108/14691930910922888

Cooke P. (2005). Regionally asymmetric knowledge capabilities and open innovation exploring 'globalisation 2'- a new model of industry organisation. *Research Policy*, **34**(8), 1128–1148. https://doi.org/10.1016/j.respol.2004.12.005

Cooke P. and Leydesdorff L. (2006). Regional development in the knowledge-based economy: the construction of advantage. *The Journal of Technology Transfer*, **31**(1), 5–15. https://doi.org/10.1007/s10961-005-5009-3

Cooke P. and Porter J. (2007). From seekers to squatters: the rise of knowledge entrepreneurship. *CESifo Forum, ifo Institut für Wirtschaftsforschung an der Universität München, München*, **8**(2), 21–28. ISSN 2190-717X.

Collison C. and Parcell G. (2004). Learning to Fly: Practical Knowledge Management From Leading and Learning Organisations, 2nd edn. Wiley and Sons, UK. ISBN: 978-841-12509-1

Cross R. and Baird L. (2000). Technology is not enough: improving performance by building organisational memory. *Sloan Management Review*, **41**(3), 69–78.

Cowan R. and Nicolas J. (2004). Network structure and the diffusion of knowledge. *Journal of Economic Dynamics ad Control*, **28**(8), 1557–1575, https://doi.org/10.1016/j.jedc.2003.04.002

Danneels E. (2002). The dynamics of product innovation and firm competences. *Strategic Management Journal*, **23**(12), 1095–1121. https://doi.org/10.1002/smj.275

Danneels E. (2007). The process of technological competence leveraging. *Strategic Management Journal*, **23**(5), 511–533. https://doi.org/10.1002/smj.598

Darroch J. (2003). Developing a measure of knowledge management behaviours and practices. *Journal of Knowledge Management*, **7**(5), 41–54, https://doi.org/10.1108/13673270310505377

Darroch J. (2005). Knowledge management, innovation and firm performance. *Journal of Knowledge Management*, **9**(3), 101–115. https://doi.org/10.1108/13673270510602809

Davenport T. H. (2005). Thinking for A Living: How to get Better Performance and Results From Knowledge Workers. Harvard Business School Press, Boston, MA.

Davenport T. H. (2008). Improving knowledge worker performance. In: From Strategy to Execution, D. Pantaleo and N. Pal (eds), Springer, Berlin, Heidelberg, pp. 215–235. https://doi.org/10.1007/978-3-540-71880-2_11

Davenport T. H. and Prusak L. (1998). Working Knowledge: How Organisations Manage What They Know. Harvard Business School Press, Boston, MA. ISBN: 9780585056562.

Ddiba D., Andersson K., Koop S. H. S., Ekener E., Finnveden G. and Dickin S. (2020). Governing the circular economy: assessing the capacity to implement resource-oriented sanitation and waste management systems in low- and middle-income countries. *Earth System Governance*, **4**, 1–11. https://doi.org/10.1016/j.esg.2020.100063

de Faria P. and Sofka W. (2010). Knowledge protection strategies of multinational firms—a cross-country comparison. *Research Policy*, **39**(7), 956–968. ISSN 0048-7333. https://doi.org/10.1016/j.respol.2010.03.005

DeFillippi R. J., Arthur M. B. and Lindsay V. J. (2006). Knowledge at Work: Creative Collaboration in the Global Economy. Blackwell Publishing, USA.

Dei D.-G. J. and van der Walt T. B. (2020). Knowledge management practices in universities: the role of communities of practice. *Social Sciences & Humanities Open*, **2**, 1–8. https://doi.org/10.1016/j.ssaho.2020.100025

Den Hertog P. (2000). Knowledge intensive business services as co-producers of innovation. *International Journal of Innovation Management*, **4**(4), 491–528, https://doi.org/10.1142/S136391960000024X

de Jesus Ginja Antunes H. and Pinheiro P. G. (2020). Linking knowledge management, organizational learning and memory. *Journal of Innovation & Knowledge*, **5**, 140–149. https://doi.org/10.1016/j.jik.2019.04.002

De Sordi J. O., de Azevedo M. C., Bianchi E. M. P. G. and Caradina T. (2021). Defining the term knowledge workers: toward improved ontology and operationalization. *Knowledge and Process Management, in Journal of Corporate Transformation*, **18**(3), 150–174. https://doi.org/10.1002/kpm.378

Dew N., Velamuri S. R. and Venkataraman S. (2004). Dispersed knowledge and an entrepreneurial theory of the firm. *Journal of Business Venturing*, **19**(5), 659–679. https://doi.org/10.1016/j.jbusvent.2003.09.004

Dewagoda K. G. and Perera B. A. K. S. (2019). A conceptual knowledge value chain model for construction organisations engaged in competitive tendering. In: Proceedings of the 8th World Construction Symposium, Colombo, Sri Lanka, 8–10 November 2019, Y. G. Sandanayake, S. Gunatilake and A. Waidyasekara (eds), pp. 2–12. https://doi.org/10.31705/WCS.2019.1. https://2019.ciobwcs.com/papers

Diehr G. and Gueldenberg S. (2017). Knowledge utilisation: an empirical review on processes and factors knowledge utilisation. *Global Business and Economics Review*, **19**(4), 401–419. https://doi.org/10.1504/GBER.2017.085024

Diehr G. and Wilhelm S. (2017). Knowledge marketing: how can strategic customers be utilised for knowledge marketing in knowledge-intensive SMEs?. *Knowledge Management Research & Practice*, **15**, 12–22. https://doi.org/10.1057/s41275-016-0039-1

Dobrai K. and Farkas F. (2009). Knowledge-intensive business services: a brief overview. *Perspectives of Innovations, Economics & Business*, **3**, 15–17, https://doi.org/10.15208/pieb.2009.40

Drucker P. F. (1959). Landmarks of Tomorrow. Harper and Brothers, New York.

Elliot K., Patacconi A., Swierzbinski J. and Williams J. (2019). Knowledge protection in firms: a conceptual framework and evidence from HP labs. *European Management Review*, **16**, 179–193. https://doi.org/10.1111/emre.12336

EMCC. (2005). Sector Futures: The knowledge intensive business services sector. European Foundation for the Improvement of Living and Working Conditions, European Monitoring Centre on Change (EMCC), Ireland. European Foundation for the Improvement of Living and Working Conditions.

Erhardt N. (2011). Is it all about teamwork? Understanding processes in team-based knowledge work. *Managing Learning*, **42**(1), 87–112. https://doi.org/10.1177/1350507610382490

Ermine J. (2013). A knowledge value chain for knowledge management. *Journal of Knowledge & Communication Management*, 3(2), 85–101. https://www.researchgate.net/profile/Jean-Louis-Ermine/publication/324271573_A_Knowledge_Value_Chain/links/5c756979299bf1268d28295e/A-Knowledge-Value-Chain.pdf https://doi.org/10.5958/j.2277-7946.3.2.008

Exley S. (2016). Education and learning. Chapter 6. In: Understanding Social Advantage and Disadvantage, H. Dean and L. Platt (eds), Oxford University Press, Oxford, 391pages. ISBN: 9780198737070.

Fischer B. B. (2015). On the contributions of knowledge-intensive business-services multinationals to laggard innovation systems. *BAR-Brazilian Administrative Review*, 12(2), 150–168. https://doi.or/10.1590/1807-7692bar2015140070

Ford D. P. and Staples S. (2010). Are full and partial knowledge sharing the same? *Journal of Knowledge Management*, 14(3), 394–409, https://doi.org/10.1108/13673271011050120

Forstenlechner I. and Lettice F. (2007). Cultural differences in motivating global knowledge workers. *Equal Opportunities International*, 26(8), 823, https://doi.org/10.1108/02610150710836154

Gabbay J., Le May A., Pope C., Brangan E., Cameron A., Klein J. H. and Wye L. (2020). Uncovering the processes of knowledge transformation: the example of local evidence-informed policy-making in United Kingdom healthcare. *Health Research Policy and Systems*, 18, 110. https://doi.org/10.1186/s12961-020-00587-9

Garcia B. C. (2007) Working and learning in a knowledge city: a multilevel development framework for knowledge workers. *Journal of Knowledge Management*, 11(5), 18, https://doi.org/10.1108/13673270710819771

Giebels D., Carus J., Paul M., Kleyer M., Siebenhüner B., Arns A., Bartholomä A., Carlow V., Jensen J., Tietjen B., Wehrmann A. and Schröder B. (2020). Transdisciplinary knowledge management: a key but underdeveloped skill in EBM decision-making. *Marine Policy*, 109, 1–13. https://doi.org/10.1016/j.marpol.2020.104020

Grant R. (1991). A resource-based perspective of competitive advantage: implications for strategy formulation. *California Management Review*, 33(3), 114–135, https://doi.org/10.2307/41166664

Gunday G., Ulusoy G., Kilic K. and Alpkan L. (2011). Effects of innovation types on firm performance. *International Journal of Production Economics*, 133(2), 552–676, https://doi.org/10.1016/j.ijpe.2011.05.014

Gurd B. and Jothidas A. (2009). Developing the knowledge value chain: a strategy for knowledge sharing in New product development in multinational corporations. *Problems and Perspectives in Management*, 7(2), 26–35. https://www.businessperspectives.org/images/pdf/applications/publishing/templates/article/assets/2621/PPM_EN_2009_02_Gurd.pdf

Ha S. T., Lo M. C., Suaidi M. K., Mohamad A. A. and Razak Z. B. (2021). Knowledge management process, entrepreneurial orientation, and performance in SMEs: evidence from an emerging economy. *Sustainability*, 13(17), 9791. https://doi.org/10.3390/su13179791 https://doi.org/10.3390/su13179791

Hansen M. T. and Haas M. R. (2001). Competing for attention in knowledge markets: electronic document dissemination in a management consulting company. *Administrative Science Quarterly*, 46(1), 1–28. https://doi.org/10.2307/2667123 https://doi.org/10.2307/2667123

Hartlung K. and Oliveira M. (2013). Communities of practice: creating and sharing knowledge. *REGE – Revista de Gestao*, 20(3), 407–422. https://doi.org/10.5700/rege507

He Z.-L. and Wong P.-K. (2004). Exploration vs. Exploitation: an empirical test of the ambidexterity hypothesis. *Organisation Science*, 15(4), 481–494. https://doi.org/10.1287/orsc.1040.0078 https://doi.org/10.1287/orsc.1040.0078

Hipp C. H. (1999). Knowledge intensive business services in the new mode of knowledge production. *AI & Society*, 13, 88–106, https://doi.org/10.1007/BF01205260

Hislop D. (2003). The complex relations between communities of practice and the implementation of technological innovations. *International Journal of Innovation Management*, 7(2), 163–188. https://doi.org/10.1142/S1363919603000775

Holsapple C. W. and Jones K. (2004). Exploring primary activities of the knowledge chain. *Knowledge and Process Management*, 11(3), 155–174. https:doi.org/10.1002/kpm.200

Holsapple C. W. and Singh M. (2003). The knowledge chain model: activities for competitiveness. In: Handbook on Knowledge Management. International Handbooks on Information Systems, Vol 2, C. W. Holsapple (ed.), Springer, Berlin, Heidelberg, pp. 215–251. https://doi.org/10.1007/978-3-540-24748-7_11

Humphrey J. and Schmitz H. (2002). How does insertion in global value chains affect upgrading in industrial clusters? *Regional Studies,* **36**(9), 1017–1027. http://doi.org/10.1080/0034340022000022198

Huysman M. and Wulf V. (2006). IT To support the knowledge-sharing in communities towards a social capital analysis. *Journal of Information Technology,* **21**(1), 40–51, https://doi.org/10.1057/palgrave.jit.2000053

Iliescu A. N. (2021). The emergence of knowmads from the knowledge workers. *Management Dynamics in the Knowledge Economy,* **9**(1), 94–106, https://doi.org/10.2478/mdke-2021-0007

Irani Z., Sharif A. M. and Love P. E. D. (2009). Mapping knowledge management and organizational learning in support of organizational memory. *International Journal of Production Economics,* **122**, 200–215, https://doi.org/10.1016/j.ijpe.2009.05.020

IRC. (2006). Knowledge and information management in the water and sanitation sector: a hard nut to crack. Thematic Overview Paper 14, By: Jan Teun Visscher, Jaap Pels, Viktor Markowski and Sascha de Graaf; Reviewed by: Urs Karl Egger (SKAT) and Ratan Budhathoki (NEWAH). International Water and Sanitation Centre (IRC), Netherlands.

IUCN. (2004). The knowledge products and services study: Addendum to the Review Report. External Review of IUCN Commissions, IUCN-The World Conservation Union. https://www.iucn.org/downloads/knowledge_products_and_services_study_1.pdf

Jacinto C. and Garcia de Fanelli A. (2014). Tertiary technical education and youth integration in Brazil, Colombia and Mexico. *International Development Policy | Revue Internationale de Politique de Développement,* **5**(Part 2), 59–80. https://journals.openedition.org/poldev/1776

Janus S. S. (2016). Monitoring and evaluation. In: Becoming A Knowledge-Sharing Organisation: A Handbook for Scaling-up Solutions Through Knowledge Capturing and Sharing, World Bank.

Katila R. and Ahuja G. (2002). Something old, something new: a longitudinal study of search behaviour and new product introduction. *Academy of Management Journal,* **45**(6), 1183–1194. https://doi.org/10.5465/3069433

Kefela G. T. (2010). Knowledge-based economy and society has become a vital commodity to countries. *International NGO Journal,* **5**(7), 160–166. https://academicjournals.org/article/article1381828238_Kefela.pdf

Khadir-Poggi Y. and Keating M. (2013). Understanding knowledge-intensive organisations within knowledge-based economies: biases and challenges. *International Journal of Knowledge-Based Development,* **4**(1), 64–78. https://doi.org/10.1504/IJKBD.2013.052494

Kim S.-J., Hong J.-H. and Suh E.-H. (2012). A diagnosis framework for identifying the current knowledge sharing activity status in a community of practice. *Expert Systems with Applications: an International Journal,* **39**(18), 13093–13107. https://doi.org/10.1016.j.eswa.2012.092

King W. (2009). Knowledge management and organisational learning. *Annals of Information Systems,* **4**, 3–15, https://doi.org/10.1007/978-1-4419-0011-1_1

King W. and Ko D. (2001). Evaluating knowledge management and the learning organization: an information/knowledge value chain approach. *Communications of the Association for Information Systems,* **5**, 1–28. https://doi.org/10.17705/1CAIS.00514 https://doi.org/10.17705/1CAIS.00514

Kling R. and Courtright C. (2003). Group behaviour and learning in electronic forums: a sociotechnical approach. *Information and Society,* **19**, 221–235. https://doi.org/10.1080/01972240309465

Kohlbacher F. (2008). Knowledge-based marketing: building and sustaining competitive advantage through knowledge co-creation. *International Journal of Management and Decision Making,* **9**(6), 617–645. https://doi.org/10.1504/IJMDM.2008.021218

Kok A. (2007) Intellectual capital management as part of knowledge management initiatives at institutions of higher learning. *The Electronic Journal of Knowledge Management,* **5**(2), 181–192.

Kostas M. and John P. (2006). Analysis of the value of knowledge management learning to innovation. *International Journal of Knowledge Management Studies,* **1**(2), 79–89.

Krome M. A. (2014). Knowledge transformation: a case for workforce diversity. *Journal of Diversity Management (JDM),* **9**(2), 103–110. https://doi.org/10.19030/jdm.v9i2.8975 https://doi.org/10.19030/jdm.v9i2.8975

Landry R., Amara N., Pablos-Mendes A., Shademani R. and Gold I. (2006). The knowledge-value chain: a conceptual framework for knowledge translation in health. *Policy and Practice, Bulletin of the World Health Organization,* **84**, 597–602. https://www.scielosp.org/article/ssm/content/raw/?resource_ssm_path=/media/assets/bwho/v84n8/v84n8a09.pdf https://doi.org/10.2471/BLT.06.031724

Lave J. and Wenger E. (1991). Situated Learning: Legitimate Peripheral Participation. Cambridge University Press, Cambridge UK.

Lee M. C. (2016). Knowledge management and innovation management: best practices in knowledge sharing and knowledge value chain. *International Journal of Innovation and Learning*, 19(2), 206–226, https://doi.org/10.1504/IJIL.2016.074475

Lee M. C. and Han M. W. (2009). Knowledge value chain model implemented for supply chain management performance. Fifth International Joint Conference on INC, IMS and IDC, 2009, pp. 606–611. https://doi.org/10.1109/NCM.2009.302

Lee C. C. and Yang J. (2000). Knowledge value chain. *Journal of Management Development*, 19(9), 783–793. https://www.academia.edu/7829168/Knowledge_value_chain_783_Knowledge_value_chain https://doi.org/10.1108/02621710010378228

Lesser E. and Prusak L. (1999). Communities of practice social capital and organisational knowledge. *Information Systems Review*, 1(1), 1–8. https://www.koreascience.or.kr/article/JAKO199916637999845.pdf

Lesser E. L. and Storck J. (2001). Communities of practice and organisational performance. *IBM Systems Journal*, 40(4), 831–841, https://doi.org/10.1147/sj.404.0831

Lev B. (2001). Intangibles: Management, Measurement and Reporting. The Brookings Institution, Washington D.C.

Li Y., Vanhaverbeke W. and Schoenmakers W. (2008). Exploration and exploitation in innovation: reframing the interpretation. *Creativity and Innovation Management*, 17(2), 107–126, https://doi.org/10.1111/j.1467-8691.2008.00477.x

Liu F. C. and Cheng K. L. (2007). Investigatinh the relationship between knowledge sharing and team innovation climate. Oxford Business and Economic Conference, pp. 1–7.

Lonnqvist A. and Laihonen H. (2017). Management of knowledge-intensive organisations: what do we know after 20 years of research? *International Journal of Knowledge-Based Development*, 8(2), 154–167. https://doi.org/10.1504/IJKBD.2017.085149

Lowendahl B. R. (1997). Strategic Management of Professional Service Firms, 2nd edn. Copenhagen Business Scholl Press, Copenhagen.

Lowitt K., Hickey G. M., Ganpat W. and Phillip L. (2015). Linking communities of practice with value chain development in smallholder farming systems. *World Development*, 74, 363–373. https://doi.org/10.1016/j.worlddev.2015.05.014 https://doi.org/10.1016/j.worlddev.2015.05.014

Lundvall B. A. and Johnson B. (1994). The learning economy. *Journal of Industry*, 1(2), 23–42.

Maclellan E. and Soden R. (2007). The significance of knowledge in learning: a psychologically informed analysis of higher education students' perceptions. *International Journal for the Scholarship of Teaching and Learning*, 1(1), 6. https://doi.org/10.20429/ijsotl.2007.010106

Mallory A., Akrofi D., Dizon J., Mohanty S., Parker A., Vicario D. R., Prasad S., Welivita I., Brewer T., Mekala S., Bundhoo D., Lynch K., Mishra P., Willcock S. and Hutchings P. (2020). Evaluating the circular economy for sanitation: findings from a multi-case approach. *Science of the Total Environment*, 744, 1–10. https://doi.org/10.1016/j.scitotenv.2020.140871

Malik M. A. K., Mukesh K., Xianguang L. and Carl A. (2010). Knowledge management in construction supply chain. *International Journal of Networking and Virtual Organisations*, 7(2/3), 207–221, https://doi.org/10.1504/IJNVO.2010.031218

Makimoto T. and Manners D. (1997). Digital Nomad. Wiley and Sons, New York.

Makó C., Csizmadia P., Illéssy M., Iwasaki I. and Szanyi M. (2009). Organizational Innovation in the Manufacturing Sector and the Knowledge Intensive Business Services. Research Paper Series, Centre for Economic Institutions, Institute of Economic Research, Hitotsubashi University, Japan. https://hermes-ir.lib.hit-u.ac.jp/hermes/ir/re/18025/rp2009-1.pdf

Malerba F. and Orsenigo L. (2000). Knowledge innovation activities and industrial revolution. *Industrial and Corporate Change*, 9(2), 289–313, https://doi.org/10.1093/icc/9.2.289

Manhart M. and Thalmann S. (2015). Protecting organizational knowledge: a structured literature review. *Journal of Knowledge Management*, 19(2), 190–211, https://doi.org/10.1108/JKM-05-2014-0198

Manhart M., Thalmann S. and Maier R. (2015). The ends of knowledge sharing in networks: using information technology to start knowledge protection. Twenty-Third European Conference on Information Systems (ECIS), Münster, Germany.

March J. G. (1991). Exploration and exploitation in organizational learning. *Organization Science* 2(1), 71–87, https://doi.org/10.1287/orsc.2.1.71

Marks J. G. and Baldry C. (2009). Stuck in the middle with who? The class identity of knowledge workers. *Work, Employment & Society*, **23**(1), 49–65. https://doi.org/10.1177/0950017008099777 https://doi.org/10.1177/0950017008099777

Marr B., Gupta O., Pike S. and Roos G. (2003). Intellectual capital and knowledge management effectiveness. *Management Decision*, **41**(8), 771–781. https://doi.org/10.1108/00251740310496288

Miles I., Kastrinos N., Flanagan K., Bilderbeek R., Den Hertog P., Huntik W. and Bouman M. (1995). Knowledge intensive business services: users, carriers and sources of innovation. In: European Innovation Monitoring System (EIMS) 15. European Commission, Brussels, p. 117. https://www.research.manchester.ac.uk/portal/files/32800224/FULL_TEXT.PDF

Miles I. (2005). Knowledge intensive business services: prospects and policies. *Foresight*, **7**(6), 39–63. https://doi.org/10.1108/14636680510630939

Miles I. (2007). Knowledge intensive services and innovation. In: The Handbook of Service Industries, J. Bryson, and P. W. Daniels (eds), Edward Elgar Publishing, Cheltenham, pp. 277–294.

Miles I. D., Belousova V. and Chichkanov N. (2018). Knowledge-intensive business services: ambiguities and continuities. *Foresight*, **20**(1), 1–26. https://doi.org/10.1108/FS-10-2017-0058. https://www.research.manchester.ac.uk/portal/files/65096168/2017_Foresight_submitted_dated_November_24.pdf

Miozzo M., Desyllas P., Lee H.-F. and Miles I. (2016). Innovation collaboration and appropriability by knowledge intensive business services firms. *Research Policy*, **45**(7), 1337–1351. htpps://doi.ord/10.1016/j.respol.2016.03.018

Moya B., Sakrabani R. and Parker A. (2019). Realizing the circular economy for sanitation: assessing enabling conditions and barriers to the commercialization of human excreta derived fertilizer in Haiti and Kenya. *Sustainability*, **11**, 3154. https://doi.org/10.3390/su11113154

Moravec J. W. (2013). Knowmad society: the 'new' work and education. *On the Horizon*, **21**, 79–83, https://doi.org/10.1108/10748121311322978

Moravec J. W. and van den Hoff R. (2015). Higher education 3.0: knowmads create their own value!. In: Transformative Perspectives and Processes in Higher Education, Advances in Business Education and Training Book Series Vol. 6, A. Daily-Herbet and K. S. Dennis (eds), Springer International Publishing, Switzerland, pp. 233–240. https://doi.org/10.1007/978-3-319-09247-8

Muller E. and Doloreux D. (2009). What we should know about knowledge intensive business services. *Technology in Society*, **31**(1), 64–72, https://doi.org/10.1016/j.techsoc.2008.10.001

Nahapiet J. and Ghoshal S. (1998). Social capital, intellectual capital and the organisational advantage. *Academy Management Review*, **23**(2), 242–266. https://doi.org/10.5465/amr.1998.533225 https://doi.org/10.5465/amr.1998.533225

Naughton C. and Mihelcic J. (2017). Introduction to the importance of sanitation. In: Global Water Pathogens Project. Part 1: The Health Hazards of Excreta: Theory and Control, J. B. Rose and B. Jiménez Cisneros (eds), Michigan State University, E. Lansing, MI, UNESCO. http://www.waterpathogens.org/book/introduction http://www.unesco.org/openaccess/terms-use-ccbysa-en.

Newell S., Bresnen M., Edelman L., Scarbrough H. and Swan J. (2006). Sharing knowledge across projects: limits to ICT-led project review practice. *Management Learning*, **37**(2), 167–185, https://doi.org/10.1177/1350507606063441

Nonak I. (1994). A dynamic theory of organisational knowledge creation. *Organisation Science*, **5**(1), 14–37, https://doi.org/10.1287/orsc.5.1.14

Nonaka I. and Konno N. (1998). The concept of 'ba': building a foundation for knowledge creation. *California Management Review*, **40**(3), 40–54, https://doi.org/10.2307/41165942

Nonaka I. and Takeuchi H. (1995). The Knowledge-Creating Company: How Japanese Companies Create the Dynamics of Innovation. Oxford University Press, New York.

Nordenflycht A. V. (2010). What is a professional service firm? Toward a theory of knowledge intensive firms. *Academy of Management Review*, **35**, 155–174.

O'dell C. and Hubert C. (2011). The New Edge in Knowledge: How Knowledge Management is Changing the Way we do Business. Wiley, Hoboken, New Jersey.

OECD. (2006). Innovative and Knowledge-Intensive Service Activities. Organisation for Economic Development (OECD) Publishing, Paris. ISBN: 92-64-02273-2. https://read.oecd-ilibrary.org/science-and-technology/innovation-and-knowledge-intensive-service-activities_9789264022744-en#page1

OECD. (2012). Knowledge Networks and Markets in the Life Sciences. Organisation for Economic Development (OECD), Paris, France. http://doi.org/10.1787/9789264168596-en

OECD. (2013). 'Knowledge Networks and Markets', OECD Science, Technology and Industry Policy Papers, No. 7. OECD Publishing, Paris, https://doi.org/10.1787/5k44wzw9q5zv-en

OuYang Y. and Lee T. (2019). Using knowledge value chain for implementation of work performance. *International Journal of Economics, Business and Management Research*, **3**(1), 14–27. https://ijebmr.com/uploads/pdf/archivepdf/2020/IJEBMR_316.pdf

Paallysaho S. and Kuusisto J. (2008). Intellectual property protection as a key driver of service innovation: an analysis of innovative KIBS businesses in Finland and the UK. *International Journal of Services Technology and Management*, **9**(3–4), 268–284, https://doi.org/10.1504/IJSTM.2008.019707

Paallysaho S. and Kuusisto J. (2011). Informal ways to protect intellectual property (IP) in KIBS businesses. *Innovation: Management, Policy and Practice*, **13**(1), 62–76, https://doi.org/10.5172/impp.2011.13.1.62

Paralič J., Babič F. and Paralic M. (2013). Process-driven approaches to knowledge transformation. *Acta Polytechnica Hungarica*, **10**(5), 125–143. https://doi.org/10.12700/APH.10.05.2013.5.8

Pedler M., Burgoyne J. and Boydell T. (1997). The Learning Company: A Strategy for Sustainable Development, 2nd edn. McGraw Hill Publishers, London.

Pluta-Olearnik M. (2013). Marketing of knowledge based services. *Institute of Aviation, Marketing of Scientific and Research Organisations*, **3**(9), 19. https://doi.org/10.14611/minib.09.03.2013.11

Popadiuk S. and Vidal P. G. (2009). Measuring Knowledge Exploitation and Exploration: An Empirical Application in A Technological Development Center in Brazil. XXXIII Encontro da ANPAD, Sao Paolo. http://www.anpad.org.br/admin/pdf/ADI282.pdf

Porter M. E. (1985). Competitive Advantage: Creating and Sustaining Superior Performance. Free Press, New York.

Probst G. J. B. (1998). Practical Knowledge Management: A Model That Works. Arthur D. Little Prism, Second Quarter, pp. 17–29.

Probst G. and Borzillo S. (2008). Why communities of practice succeed and why they fail. *European Management Journal*, **26**(5), 335–347. https:/doi.org/10.1016/j.emj.2008.05.003

Probst G. J. B., Raub S., Romhardt K. and Doughty H. A. (1999). Managing Knowledge: Building Blocks for Success. Wiley, Chichester.

Powell T. (2001). The knowledge value chain (KVC): How to Fix It when It breaks. Presented at KnowledgeNets 2001, New York city, May 2001. In: Proceedings of the 22nd National Online Meeting, M. E. Williams (ed.), Information Today, Inc., Medford, NJ, p. 14.

Powell W. W. and Snellman K. (2004). The knowledge economy. *Annual Review of Sociology*, **30**, 199–220. https://doi.org/10.1146/annurev.soc.29.010202.100037

Prusak L. (1996). The knowledge advantage. *Planning Review*, **24**(2), 6–8. https://doi.org/10.1108/eb054546 https://doi.org/10.1108/eb054546

Putnam R. (1995). Bowling alone: America's declining social capital. *Journal of Democracy*, **6**(1), 65–78. https://doi.org/10.1007/978-1-349-62397-6_12 https://doi.org/10.1353/jod.1995.0002

Rajala R., Westerlund M., Rajala A. and Leminen S. (2008). Knowledge-intensive service activities in software business. *International Journal of Technology Management*, **41**(3/4), 273–290. https://doi.org/10.1504/IJTM.2008.016784

Ramirez A. M., Vasaukaite J. and Kumpikaite V. (2012). Role of knowledge management within innovation and performance. *Economics and Management*, **17**(1), 381–389. https://doi.org/10.5755/j01.em.17.1.2293 https://doi.org/10.5755/j01.em.17.1.2293

Reinhardt W., Schmidt B., Sloep P. and Drachsler H. (2011). Knowledge worker roles and actions – results of two empirical studies. *Knowledge and Process Management*, **18**(3), 150–174. https://doi.org/10.1002/kpm.378

Revilla M. L. D., Qu F., Seetharam K. and Rao V. V. B. (2021). 'Sanitation' in the Top Development Journals: A Review. ADBI Working Paper Series, No. 1253. Asian Development Bank Institute (ADBI), Tokyo. https://www.econstor.eu/bitstream/10419/238610/1/adbi-wp1253.pdf

Rowley J. (2001). Knowledge management in pursuit of learning: the learning with knowledge cycle. *Journal of Information Science*, **27**(4), 227–237. https://doi.org/10.1177/016555150102700406 https://doi.org/10.1177/016555150102700406

Saliola F. and Zanfei A. (2009). Multinational firms, global value chains and the organization of knowledge transfer. *Research Policy*, **38**(2), 369–381. https://doi.org/10.1016/j.respol.2008.11.003

Santos J. B. (2020). Knowledge intensive business service and innovation performance in Brazil. *Innovation and Management Review*, **17**(1), 58–74. https://doi.org/10.1108INMR-03-2019-0025

Scarso E. and Bolisani E. (2010). Knowledge-based strategies for knowledge intensive business services: a multiple case-study of computer service companies. *Electronic Journal of Knowledge Management*, **8**(1), 151–160.

Sandelin S. K., Hukka J. J. and Katko T. S. (2019). Importance of knowledge management at water utilities. *Public Works Management & Policy*, **00**(0), 1–17. https://doi.org/10.1177/1087724X19870813

Schneider M. (2012). Knowledge integration. In: Encyclopedia of the Sciences of Learning, N. M. Seel (eds), Springer, Boston, MA, pp. 1684–1686. https://doi.org/10.1007/978-1-4419-1428-6_807

Simard A. (2006). Knowledge markets: more than providers and users. *The IPSI BgD Transactions on Advanced Research*, **2**(2), 3–9. http://vipsi.org/ipsi/journals/journals/tar/2006/July/Full%20Journal.pdf#page=3

Sinha S. (2015). The exploration – exploitation dilemma: a review in the context of managing growth of new ventures. *VIKALPA: The Journal for Decision-Makers*, **40**(3), 313–323. https://doi.org/10.1177/0256090915599709

Song M., Van der Bij H. and Weggeman M. (2005). Determinants of the level of knowledge application: a knowledge-based and information processing perspective. *Journal of Product Innovation Management*, **22**, 430–444. https://onlinelibrary.wiley.com/doi/epdf/10.1111/j.1540-5885.2005.00139.x

Springer-Heinz A. (2018a). ValueLinks 2,0: Manual on Sustainable Value Chain Development. Volume 1: Value Chain Analysis, Strategy and Implementation. Deutsche Gesellschaft für Internationale Zusammenarbeit (GIZ) GmbH, Eschborn. https://www.valuelinks.org/material/manual/ValueLinks-Manual-2.0-Vol-1-January-2018.pdf

Springer-Heinz A. (2018b). ValueLinks 2,0: Manual on Sustainable Value Chain Development. Volume 2: Value Chain Solutions. Deutsche Gesellschaft für Internationale Zusammenarbeit (GIZ) GmbH, Eschborn. https://www.valuelinks.org/material/manual/ValueLinks-Manual-2.0-Vol-1-January-2018.pdf

Starbucks W. H. (1992). Learning by knowledge-intensive firms. *Journal of Management Studies*, **3**(4), 262–275.

Stabell C. B. and Fjeldstad Ø. D. (1998). Configuring value for competitive advantage: on chains, shops, and network. *Strategic Management Journal*, **19**, 413–437. https://www.teaching-entrepreneurship.com/uploads/5/2/8/3/5283784/stabellandfjeldstad_config_value.pdf https://doi.org/10.1002/(SICI)1097-0266(199805)19:5<413::AID-SMJ946>3.0.CO;2-C

St. Clair G. and Reich M. J. (2002). Knowledge services: financial strategies and budgeting. *Information Outlook*, **6**(6), 26–30.

Stewart T. A. (1999). Intellectual Capital: the New Wealth of Organisations. Nicholas Brealey Publishers, London, UK.

Stocker M. G. (2012). Value Creation og knowledge intensive companies. PhD Dissertation, Department of Business Studies, Corvinus University of Budapest, Romania. https://web.archive.org/web/20180720050150id_/http://phd.lib.uni-corvinus.hu/690/1/Stocker_Miklos_den.pdf

Strambach S. (2008). Knowledge-Intensive business services (KIBS) as drivers of multilevel knowledge dynamics. *International Journal of Services Technology and Management*, **10**(2/3/4), 152–174, https://doi.org/10.1504/IJSTM.2008.022117

Swart J. (2007). HRM And knowledge workers. In: The Oxford Handbook of Human Resource Management, P. Boxall, J. Purcell and P. Wright (eds), Oxford University Press, New York, pp. 450–468.

Swart J. and Kinnie N. (2003). Sharing in knowledge-intensive firms. *Human Resource Management Journal*, **13**(2), 60–75. https://doi.org/10.1111/j.1748-8583.2003.tb00091.x

Sveiby K. E. (1997). The New Organisational Wealth. Berret-Koehler Publishers Inc., San Francisco.

Sveiby K. E. (2001). A knowledge-based theory of the firm to guide in strategy formulation. *Journal of Intellectual Capital*, **2**(4), 344–358, https://doi.org/10.1108/14691930110409651

TBC. (2017). The Circular Sanitation Economy: New Pathways to Commercial and Societal Benefits Faster at Scale. The Toilet Board Coalition – Toilet Accelerator Series. https://www.toiletboard.org/media/34-The_Circular_Sanitation_Economy.pdf

TBC. (2019a). Scaling Up the Sanitation Economy 2020–2025. The Toilet Board Coalition – Toilet Accelerator Series. https://www.toiletboard.org/wp-content/uploads/2021/03/scaling-up-the-sanitation-economy.png

TBC. (2019b). The Sanitation Economy at Sector scale: transformative solutions for new business vales. The Toilet Board Coalition – Toilet Accelerator Series. http://www.toiletboard.org/wp-content/uploads/2021/03/2019-The_Sanitation_Economy_at_Sector_Scale_FINAL.pdf

Timo P. and Arto O. (2009). International activities of knowledge-intensive small and medium-sized enterprises. *Management Research News*, **32**(7), 645–658, https://doi.org/10.1108/01409170910965242

TNUSSP. (2018). Assessment of Behaviour Change Communication Programme: TNUSSP Phase I. https://doi.org/10.24943/tnusspabc.20181202

Toivonen M. (2004). Expertise as business: long-term development and future prospects of knowledge-intensive business services (KIBS). Helsinki University of Technology Laboratory of Industrial Management Doctoral dissertation series 2004/2. http://lib.tkk.fi/Diss/2004/isbn9512273152/isbn9512273152.pdf

Tsuneo N. (2001). Innovation management using intellectual capital. *International Journal of Entrepreneurship and Innovation Management*, **1**(1), 96–110, https://doi.org/10.1504/IJEIM.2001.000447

UN. (2021). COP26: The Glasgow Climate Pact. UN Climate Change Conference in Glasgow (UK2021). https://ukcop26.org/wp-content/uploads/2021/11/COP26-Presidency-Outcomes-The-Climate-Pact.pdf

Un C. C. and Asakawa K. (2015). Types of R & D collaborations and process innovation: the benefit of collaborating upstream in the knowledge chain. *Journal of Product Innovation Management*, **32**(1), 138–153. https://doi.org/10.1111/jpim.12229 https://doi.org/10.1111/jpim.12229

UNDP. (2002). Knowledge and learning: use of evaluative evidence. In: Handbook of Monitoring and Evaluating Results, Part IV, Chapter 7. UNDP Evaluation Office, New York, pp. 75–89. http://web.undp.org/evaluation/documents/handbook/me-handbook.pdf

UNICEF/WHO. (2020). State of the World's Sanitation: An Urgent Call to Transform Sanitation for Better Health, Environments, Economies and Societies. United Nations Children's Fund (UNICEF) and the World Health Organization, New York.

Venkatraman S. and Venkatraman R. (2018). Communities of practice approach for knowledge management systems. *Systems*, **6**(4), 36. https://doi.org/10.3390/systems6040036

Wang C. L. and Ahmed P. K. (2005). The knowledge value chain: a pragmatic knowledge implementation network. *Handbook of Business Strategy*, **6**(1), 321–326, https://doi.org/10.1108/08944310510558115

Walsh J. P. and Ungson G. R. (1991). Organisational memory. *The Academy of Management Review*, **16**(1), 57–91. https://doi.org/10.2307/258607

Weggeman M. (1997). Knowledge Management Design and Management of Knowledge Intensive Organisation. Scriptum, Schiedam, Netherlands.

Weggeman M. (2000). Kennismanagement: De Praktijk (Knowledge Management: in Practice). Scriptum, Scheidam, Netherlands.

Welo T. and Ringin G. (2018). Investigating organizational knowledge transformation capabilities in integrated manufacturing and product development companies. *Procedia CIRP*, **70**, 150–155. https://doi.org/10.1016/j.procir.2018.03.276

Wenger E. (1998). Communities of Practice: Learning, Meaning and Identity. Cambridge University Press, Cambridge, UK.

Wenger E. (2004). Knowledge Management as A Doughnut: Shaping Your Knowledge Strategy Through Communities of Practice. Ivey Business Journal Online, Ivey Business School, Western University, Canada. https://iveybusinessjournal.com/publication/knowledge-management-as-a-doughnut/

Wenger E. and Snyder W. (2000). Communities of practice: the organisational frontier. Harvard Business Review, 139–145. https://hbr.org/2000/01/communities-of-practice-the-organizational-frontier

Wenger E., McDermott R. A. and Snyder W. (2002). Cultivating Communities of Practice: A Guide to Managing Knowledge. Harvard Business School Press, Boston, MA.

Wenger-Trayner E. and Wenger-Trayner B. (2015). Communities of practice: a brief introduction. https://wenger-trayner.com/wp-content/uploads/2015/04/07-Brief-introduction-to-communities-of-practice.pdf

Wenger-Trayner E., Fenton-O'Çreevy M., Hutchison S., Kubiak C. and Wenger-Trayner B. (2015). Learning in Landscapes of Practice: Boundaries, Identity and Knowledgeability in Practice-Based Learning. Routledge, New York. ISBN: 9781315777122.

Whelan E., Collings D. G. and Donnellan B. (2010). Managing talent in knowledge-intensive settings. *Journal of Knowledge Management*, **14**(3), 486–504, https://doi.org/10.1108/13673271011050175

Wild R. H., Griggs K. A. and Downing T. (2002). A framework for e-learning as a tool for knowledge management. *Industrial Management & Data Systems*, **102**(7), 371–380. https://doi.org/10.1108/02635570210439463

Wilson T. D. (2002). The nonsense of knowledge management. *Information Research*, **8**(1), 144.

Windrum P. and Tomlinson M. (1999). Knowledge-intensive services and international competitiveness: a four country comparison. *Technology Analysis & Strategic Management*, **11**(3), 391–408. https://doi.org/10.1080/095373299107429

World Bank. (2007). Building Knowledge Economies: Advanced Strategies for Development. World Bank Institute Development Studies, World Bank Institute, Washington D.C. https://openknowledge.worldbank.org/bitstream/handle/10986/6853/411720PAPER0Kn101OFFICIAL0USE0ONLY1.pdf?sequence=1&isAllowed=y

Yu C., Zhang Z., Lin C. and Wu Y. J. (2017). Knowledge creation process and sustainable competitive advantage: the role of technological innovation capabilities. *Sustainability*, **9**, 2280. https://doi.org/10.3390/su9122280

Zahra S. A., Neubaum D. O. and Hayton J. C. (2020). What do we know about knowledge integration: fusing micro- and macro-organisational perspectives. *The Academy of Management Annals*, **14**(1), 160–194. https://doi.org/10.5465/annals.2017.0093 https://doi.org/10.5465/annals.2017.0093

Zieba M. (2013). Knowledge-Intensive Business Services (KIBS) and Their Role Based Economy. GUT FME Working Paper Series A, No. 7/2013(7). Gdansk University of Technology, Faculty of Management and Economics, Gdansk (Poland).

Zubair I. G. (2021). Knowledge, ideas and the Marketplace, The News on Sunday (TNS), Political Economy. https://www.thenews.com.pk/tns/detail/807254-knowledge-ideas-and-the-marketplace

doi: 10.2166/9781789061840_0265

Chapter 10

Governance and enabling systems

Peter Emmanuel Cookey, Mayowa Abiodun Peter-Cookey, Thammarat Koottatep and Chongrak Polprasert

Chapter objectives

The aim of this chapter is to help the reader to identify dominant actors, coordination mechanisms and type of governance of IFSVC. It also shows how enterprises and actors in the value chain are making it possible to bring products and services from design and development to the marketplace. Furthermore, it illustrates how formal and informal institutions are used in monitoring and enforcing the norms and rules with which the stakeholders collectively manage their common affairs in the IFSVC.

10.1 INTRODUCTION

Value chains come with a certain degree of coordination or chain governance, particularly when independent firms are linked to each other in a network-like structure to exchange products, services and knowledge for competitiveness (UNIDO, 2011). The IFSVC governance refers to the organization of the actors in the value chain that makes it possible to bring a product from design and development (Chapter 1) to the marketplace (Chapter 7). Governance also is about power and the ability of certain enterprises to exert control along the chain (UNIDO, 2011) as well as the 'official' rules that address output, and the commercial imperatives of competition that influence how IFSVC is structured. This ensures that interactions between actors in the value chain are frequently organized in a system that allows competitive enterprises to meet specific requirements of products, services, processes, and logistics in serving their markets (M4P, 2008). Furthermore, this governance implies the setting, monitoring and enforcing of norms and rules with which the stakeholders collectively manage their common affairs. The collective management can be a value chain (thus value chain governance) or a local, national and/or global community of practice interested in resolving a common problem or promoting a common goal. The basic types of governance include markets, networks and hierarchies (Springer-Heinze, 2018a) and governance instruments range from contracts between value chain participants to government regulatory frameworks and also the unwritten norms that determine who can participate in the market (M4P, 2008).

Requirements for governing the IFSVC could be official and originate within or outside the value chain as requirements could be imposed by any lead member of the value chain in any of the stages (Chapters 2 to 9), which could guarantee that only high-quality products and services are provided to the end-users. Conversely, they may be as complex as government enforcement of rules and regulations regarding quality standards that sanitation infrastructure, products and/or hygiene consumables should meet before they are imported and/or exported into any particular country. Product and service quality is an important aspect in value chain governance, which supports economic growth, environmental and social improvement (see Chapter 3, section 3.4). The quality of a marketable good not only relates to the features of the products and services, but also to the entire process of the IFSVC. The core quality and standards benchmarks for value chain products and services are: (i) legal requirements regulating the minimum level of product safety; (ii) industry-specific technical norms and quality grades facilitating contracts; (iii) quality criteria defined by individual enterprises to position a product and/or service in the market; and (iv) sustainability standards on a wide variety of issues of social and political interest (Springer-Heinze, 2018a). In general, a lead enterprise within the IFSVC should take responsibility for setting, monitoring and facilitating compliance with the rules regarding quality and standards in the value chain.

Leading actors/enterprises that may engage in value chain governance are private (lead) enterprises, government bodies, and (UN or bilateral) development agencies (such as UNIDO, the United Nations Industrial Development Organization). The motivation to enter into IFSVC activities will vary for each type and each will use different programmes and/or projects (Springer-Heinze, 2018a). Relevant legislation may assign the responsibility to some lead actors close to the market who will have to ensure that other actors in the value chain comply with the set rules (Dietz, 2014). For instance, it could also be certain conditionalities and procedures imposed by a multinational sanitaryware firm and/or treatment plant manufacturers or a bulk buyer of sanitation-derived resources as a requirement for participation by a subcontractor in its global value chain (M4P, 2008). Some of these conditions may limit the IFSVC actors from being able (or not) to access services and other forms of support required for meeting the value chain standards, and insufficient support could hinder their participation in higher-value segments of the chain (M4P, 2008). Hence, the IFSVC must operate in a business-enabling environment that includes norms and customs, laws, regulations, policies, and international trade agreements that facilitate or hinder the movement of products or services along the value chain (Market Links, 2021). Also, national and local encompassing policies, administrative procedures, enacted regulations and public infrastructure such as roads, electricity, and so on. (Market Research, 2021) will enhance the ability of the IFSVC to support the sanitation economy and societal requirements for safely managed sanitation.

In the end, appropriate governance of the IFSVC will help to determine the following (Market Links, 2021; Market Research, 2021):

- production and service capabilities for best practices in IFSVC transmitted by lead enterprises through embedded services or provision of hands-on advice on how to improve processes and procedures of small and medium-scale enterprises in the value chain;
- market access by lead enterprises so other actors in the value chain could benefit; small and medium-scale enterprises need to be on the radar of lead enterprises in the IFSVC to ensure that decisions taken are in the best interest of other actors and enterprises in the value chain;
- distribution of gains through engagement in those activities in the IFSVC that bring in the most profitable returns and also help to identify those that engage in these value-adding segments;

- understanding how the governed chain provides small and medium-scale enterprises in the value chain and practitioners with valuable information on how to develop skills and with whom to develop relationships that create flexibility and freedom to undertake additional functions within the IFSVC, thereby attracting extra gains; and
- leverage for policy initiatives that provide opportunities for lead global and/or national enterprises to assist local enterprises for better equitable distributions of gains.

Governance is an important instrument to improve the performance of value chains and sustain increased competitive advantage (Dietz, 2014; M4P, 2008). Key instruments in value chain governance that could support the IFSVC and sanitation economy may include among others: (i) contracts between value chain actors; (ii) standards for products, services and processes; (iii) self-regulatory systems in the value chain; (iv) management enterprises and actors at various stages of the value chain; (v) government regulatory frameworks; and (vi) unwritten norms that determine participation in the market, as well as expectations from the public (Dietz, 2014); see Figure 10.1. Thus, a particular governance system can either help groups of enterprises to grow and develop or it can retard their growth. Understanding how and when lead enterprises set, monitor and enforce rules and standards can help refine integration and coordination (Market Research, 2021).

Drawing on M4P (2008), the key issues that IFSVC governance should attempt to address are:

- The system of coordination that exists for meeting value-chain standards, especially in issues of quality, quantity and consistency; the leading or coordinating enterprises in the value chain; the degree of formalised arrangements (contracts, e.g.) and informal coordination.
- The rules and standards (both official and commercial) that actors must comply with in order to participate, and their origin and enforcement.

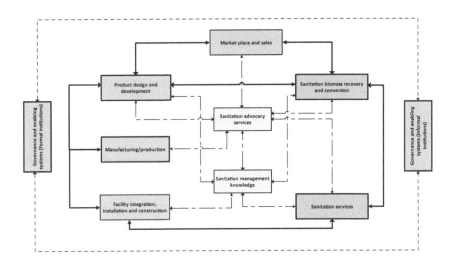

Figure 10.1 Nine stages of the integrated functional sanitation value chain (IFSVC) and linkages between governance and enabling systems (adopted and modified from Koottatep *et al.*, 2019).

- The effects of these rules and regulations on the participation (as economic activities) of poorer actors, especially on actors that enforce these rules, and the system in place for coordination.
- The transmission of information (across the value chain) about applicable rules, standards and services that support compliance, especially through the lead actors or its coordination system.
- The inability of poor actors to comply with these rules, either due to lack of protection information or capacity, and how it limits engagement in higher-value chain activities.

Consequently, the functioning of IFSVC will be governed by the broader enabling business environments, especially how the regulatory environment influences the choices value-chain investors and entrepreneurs make in locating, operating and expanding their businesses. The businesses' ability to access credit and enforce contracts, buy property, process goods through customs, pay taxes and conduct other everyday activities efficiently depends on a governance business environment that protects property rights without unnecessarily burdensome or inappropriate regulations (Market Research, 2021). This means that poor economic choices and onerous regulations can hinder the growth of the value chain and indirectly affect access to safely managed sanitation services even when a country makes significant progress on other development fronts (Market Research, 2021; Springer-Heinze, 2018b). Well functioning and coordinated national policies, regulations and institutional frameworks are crucial to provide a healthy business governance environment. The ability of enterprises to realize their market potential depends on the general conditions of doing business in the economy as a whole and on specific conditions in the IFSVC (Springer-Heinze, 2018b).

Generally, the value chain offers many opportunities for small and/or medium-scale enterprises in the sanitation sector to become more competitive, especially when the rules are transparent and fair. But, if the rules make it difficult for them to get access or benefit from the value-chain activities, they may not be able to comply with them due to lack of knowledge and skills or the cost of compliance may not allow the operation of a profitable business (Dietz, 2014; Gereffi et al., 2005). However, good and effective governance in the value chain could build the capacity of small and/or medium-scale enterprises to strengthen their position in the value chain and lift them out of poverty, which in turn could enhance access to safely managed sanitation services (Dietz, 2014). Also, a well functioning and properly enforced system of sound rules could create an enabling environment for the sanitation business that reduces the risks and lowers the cost of transacting vertically and horizontally within the IFSVC (Agrilinks, 2011). The vertical interactions and/or linkages are between actors that have different market functions, and horizontal linkages exist among the actors who have the same market function in the value chain; linkages within a value chain are mostly business linkages such as contracts between sellers and buyers and can be formal or informal (Dietz, 2014; Gereffi et al., 2005).

10.2 TYPES OF VALUE CHAIN GOVERNANCE

There are different types of value chain governance (VCG) that could be used to manage the IFSVC depending on the environment and existing interactions within the value chain. Connections between value-chain activities and enterprises as well as actors can be described as a long continuum extending from the market, whereby the parties of a transaction are independent and in an equal relationship, to hierarchical value chains that illustrate direct control of the chain by a lead enterprise (or enterprises). Between these two extremes are three network-style forms of governance (Figure 10.1) that represent situations in which the lead firm exercises influence and power by coordinating activities to varying degrees (Dietz, 2014; Gereffi *et al.*, 2005; Market Links, 2021; Market Research, 2021; Springer-Heinze, 2018a, 2018b):

(I) *Market governance* which involves transactions that require little or no formal cooperation between participants and little information is exchanged between enterprises and actors in the chain, where products change hands in a one-off interaction (Springer-Heinze, 2018a). The main interactions between firms in the value chain are price interactions at the point where products and services are sold (UNIDO, 2011). Market-based chains are common when the product is fairly standard and non-differentiated and the price is determined by supply and demand (e.g., sanitaryware products, toilets-related products, plumbing materials, cements and other hygiene products, etc.), and which buyers can easily obtain from the suppliers. In such cases the buyer has no controlling interest because the parameters are defined by each enterprise.

(II) *Modular governance* which ensures that products and services in the value chain are provided to customer specifications. In other words, producers make products and services available according to end-users' specifications (Springer-Heinze, 2018a). These value-chain enterprises tend to take full responsibility for the process technology. The linkages are more substantial than in simple markets, owing to high volumes of information flowing across the links between firms. Classical examples are the delivery of turnkey infrastructure such as faecal sludge treatment plants, sewage treatment plants, sanitation resources recovery and conversion plants, and other technologies used to deliver effective sanitation services by utilities and enterprises (both public and private) to customers.

(III) *Network governance* which could be relational and captive, occurs between enterprises with complementary capabilities and equal say on business decisions. Enterprises that engage in this type of governance are equally skilled in complementary high-value-added activities such as design capabilities, knowledge of the market, a particular technology, and so on. These are captive relations and vertically integrative forms of governance, in which small enterprises depend on a lead enterprise in the value chain (Springer-Heinze, 2018a). Vertical integration means formalizing business linkages between lead enterprises and other smaller enterprises in the chain through written agreements and contracts (Springer-Heinze, 2018a).

(IV) *Hierarchy* which occurs when one enterprise completely dictates the business of another enterprise. Hierarchies can also occur when transnational corporations work with subsidiaries in developing countries. The mother company dictates the subsidiaries' procurement, production and marketing strategies. Market and hierarchy are the extreme opposites in the value-chain governance continuum.

The forms of governance can change as enterprises evolve and mature and also the patterns within enterprises vary from one stage of the chain to another. The dynamic nature of the governance depends on three variables: the complexity of product development and service delivery information (design and process); the ability to codify or systematize the transfer of knowledge to other partnering enterprises in the chain; and the capabilities of partnering enterprises to deliver products and services efficiently and reliably as specified by the lead enterprises (Dietz, 2014). Other influences on the value-chain governance structure are product and service quality, stability, and power of the business-enabling environment and institutions, as well as other sources of power in the value chain such as suppliers and end-users (Dietz, 2014). Studies need to be carried out to determine the most appropriate option(s) for the IFSVC in specific locations and cultures.

10.3 POLICY, REGULATION AND ENABLING BUSINESS ENVIRONMENT

Policy has two sides: one is regulation in the sense of rules made by government or other authorities in order to control the processes and administration of the value chain. The other is the creation of favourable conditions to support enterprise business activities and policies have to be restrictive one way and enabling on the other (Springer-Heinze, 2018b). In most developing countries the main governance challenge that retards sanitation coverage and service improvement is the absence of fit-for-purpose policies and a business-enabling environment. By implication, when supportive policies for planning and implementing sanitation projects and programmes are non-existent, it causes a missing link in the formal recognition and development of an IFSVC (Koottatep *et al.*, 2019). Constraints caused by sanitation policy frameworks and other trade policies could affect the sanitation value chain in most developing countries and could be the major reason for its slow development in comparison with its agricultural counterpart (Springer-Heinze, 2018b). The policy framework needs to promote a business environment that not only makes countries attractive for the location of value chains but facilitates upgrading opportunities over time. This business-enabling environment is an important factor for investment, and subsequent employment and income generation (Springer-Heinze, 2018b). Policies can be regulatory in nature, and at the same time they may promote value chains and guide them into a certain direction. For instance, providing financial incentives for enterprises involved in the circular bioeconomy may trigger actors in the sanitation biomass recovery and conversion value chain (Chapter 6).

Well structured policies, regulation and institutional frameworks are crucial to provide an enabling business environment and for IFSVC enterprises to realize their market potentials, but this depends on the general conditions of doing business in that economy as a whole and on the particular conditions in the value chain (Springer-Heinze, 2018b). The most important factors that set the enabling business environment are the rule of law, infrastructure and an atmosphere of trust and security. Others are social norms, business culture and local expectations, and quite important to value chains enabling environment (Market Links, 2021; Springer-Heinze, 2018b). Apart from the global and lead national value-chain actors and/or enterprises, other actors, especially enterprises at the local levels of the value chains (such as a growth-oriented small business person), have to choose whether to comply with regulations and incur unreasonably high costs that jeopardize business viability; this sometimes makes small enterprises stay in the informal sector for survival (Market Research, 2021; Springer-Heinze, 2018b). Informal rules or regulation found in any business environment do have influence in the value chain of that area. These social institutions exist because they serve a social purpose – which is often to protect the power or privilege of particular groups. An essential part of value chain analysis is the consideration of power, and in particular the power relations that emanate from different social institutions (Market Research, 2021). The business climate (or investment climate) affects value chain governance across all sectors.

Thus, favourable conditions should make investment in the IFSVC easier and safer. Policy design for the IFSVC governance should be guided by the eight principles described by Springer-Heinze (2018b):

(I)   Policies need to address targeted problem because in some cases policies have been created without properly aiming to solve a particular problem. If there are technological solutions available, they should be formulated clearly.

(II) Policies also need to address only one single problem because policies that are design to hit several birds with one stone can be inefficient and contradictory.

It is best to seek a solution of one particular problem for each policy recommendation formulated.

(III) Policies should be implementable; in many countries, policies are formulated, but not implementable. If the chances for implementation are low, then the policy are likely to have no impact and it would be better to look for other interventions outside the policy field;

(IV) Policies should consider different interests to find a political majority because simply offering sound technical solutions without majority support may not be enough to successfully pass. Negotiating the interests of value chain policy actors is a matter of influencing the process between public policy and politics.

(V) Policies should consider the target groups because it is always crucial to identify the possible winners and losers before making recommendations.

(VI) Policy implementation should be affordable because if funds are lacking to support implementation, the suggested policy is most likely to fail.

(VII) Policies should comply with the state's constitution and existing legal framework or other relevant rules and regulations.

(VIII) Institutions to enforce policies should be in place to ensure implementation; and it is important to move within the existing legal frameworks or other relevant rules and regulations.

Thus, improving the business environment – by lifting constraints and filling gaps in regulatory and administrative support mechanisms – is critical for the improvement and upgrading of the IFSVC. Other issues that policies need to address to ensure ease of doing business in a country include but not are limited to (Dietz, 2014; Market Research, 2021; Springer-Heinze, 2018a, 2018b; UNIDO, 2011):

(I) Ease of business registration, as establishing a legal entity makes business ventures less risky and increases their longevity and chance to succeed in several ways.

(II) Ease of business licensing will reduce over-regulation and red-tape associated with lower levels of income and productivity as well as higher levels of informality. In good business licensing regimes licensing is a means to fulfil legitimate regulatory processes such as protection of public health and safety, environmental protection, national security and allocation of scarce resources.

(III) Labour regulations ensure protection of the interests of workers and minimum standards of living for the population as well as basic civil rights protections.

(IV) Ease of property registration ensures that entrepreneurs can obtain mortgages on their homes or land to start or expand businesses. Financial institutions prefer lands and buildings as collateral since they are difficult to move or hide. However, a large proportion of properties in developing countries are not formally registered, which makes it difficult for entrepreneurs to use it as collateral to improve their businesses.

(V) Credit regulations in most developing countries could limit the ability of entrepreneurs with promising business opportunities to obtain loans from banks to expand their businesses. Good credit institutions define property rights for both creditors and debtors. Collateral and insolvency regulations define the rights of creditors to recover their loans. In addition, collateral regulations help debtors by extending the right of property title to the right to use property as security for finance.

(VI) Corporate governance is an essential tool that enhances good business and ethical practices at the national level and is primarily to raise standards and drive reform efforts. Many developed and developing countries have adopted corporate governance codes of best practice to restore and sustain investor confidence in the wake of a financial crisis or corporate scandals.

(VII) Cumbersome tax administrations could be responsible for the reason many businesses operate in the informal economy. Tax compliance costs are often regressive and put a disproportionate burden on small and medium-scale businesses. Since non-compliance is not an option, ease of tax compliance should be a priority of government.

(VIII) Contract enforcement using the court system encourages new business relationships (because partners do not fear being cheated); generates confidence in more complex business transactions by clarifying threat points in the contract and enforcing such threats in the event of default; enables more goods and services to be provided by encouraging investments; and limits injustice and secures social peace. Without a good court system, commercial disputes often end up in feuds, to the detriment of everyone involved.

The appropriate governance and enabling system to guide the IFSVC in any economy or location will determine the effectiveness of the chains, the success levels of the players within each chain, and interactions between chain actors. Governance enablers will resolve all concerns and challenges of the actors and their activities to ensure collaborative and quality management systems along the value chain, and ensure that standard policies, guidelines and sustainability regulating systems are put in place to guarantee effective and efficient service delivery. Public policy regulation of countries' sanitation industries may not be visible in many value-chain maps, but they are important to the system. Thus, typical enabling services for the IFSVC will be provided by relevant government institutions, major providers of public utilities, governance and regulatory services of the value chains and communities of practice CoP) in each chain. While chain supporters will mostly be chain-specific, public agencies do not normally cover specific value chains, but rather oversee entire subsectors (such as agriculture or fisheries) and have regulatory roles over the economy as a whole.

The IFSVC support services would include professional standards setting, provision of information, trade fairs and export marketing, research on generally applicable technical solutions, vocational/professional training, CoP, political advocacy, and knowledge management in the sanitation economy. Support services are often provided by business associations, chambers of commerce and/or by specialized public institutions. This implies that IFSVC governance is a collaborative venture involving multiple chain actors – private enterprises, public and private support service providers, government and public administration – and this constitutes a cooperative system in which the partners perform different functions. It is important for role clarification, and the principle is that those who benefit should get engaged with value chain development and contribute actively. In general terms, the roles of chain actors could be set out as follows (Springer-Heinze, 2018a, 2018b):

(I) *Private sanitation enterprises,* large or small that may be engaged in product design and development (Chapter 2), manufacturing (Chapter 3), facility integration, installation and construction (Chapter 4), sanitation services (Chapter 5), sanitation biomass recovery (Chapter 6), marketing and sales (Chapter 7), sanitation advocacy (Chapter 8) and sanitation management knowledge services (Chapter 9), as well as other commercial business operations; they all assume the risks and pay for the costs. This applies to state-owned enterprises as well. Private operators primarily create benefits for themselves – the viability of their business is the foundation for the competitiveness of the value chain. Large enterprises do have an incentive to contribute to value chains because they benefit from the performance of other value chain actors.

(II) *Sanitation industry associations* and businesses, as well as professional sanitation and related associations and communities of practice, provide support services to groups of value chain operators or the entire value chain and make contributions to regulatory processes and approaches.

(III) *Public and private research, include training institutes*, and specialized units of public administration provide information and other support services to the sanitation and related business community, and also assist enterprises.

(IV) *Government organizations*, from national, state/provincial and local/municipal administrations relevant to the sanitation value chain, regulate and supervise enterprises' activities in the sanitation sector and the value chain for the purpose of employment creation, environmental sustainability, and provide basic infrastructure, such as roads, in the public interest at large.

(V) *Donor agencies and multilateral organizations* are external to the IFSVC system; their main role will or should be to facilitate value chain development and provide support to value chain actors – in the global interest.

Civil societies are stakeholders with great influence upon value chain governance, and they include advocacy groups (Chapter 8). They play an important role in driving the course of the sector and exercise pressure as well as supply energy to the process of change (Springer-Heinze, 2018a). Obviously, governments produce national value-chain strategy documents only if a product and/or service is highly relevant, if the sector has a multiplier effect on other sectors, and if the issues represent a large share in the national economy and could contribute to achieving sustainable development goals and significantly add to foreign currency earnings. Sanitation value chains already exist in some form or the other at local, national, regional and global levels, but mostly with weak governance and enabling business environments in developing countries, and with poor interactions between chain actors. To build and enhance the IFSVC proposed in this book will require some key public promotion strategies (Springer-Heinze, 2018a):

(I) *Support services:* research and technology, trade promotion, professional education, skill development, knowledge management, export promotion;

(II) *Financial incentives:* public co-investment of private productive capital, support to business start-ups and entrepreneurs, especially those that will operate in the circular bioeconomy;

(III) *Public infrastructure investments:* roads, ports, facilities at marketplaces, treatment plant infrastructure, recovery and conversion plants, community gardens, educational institutions;

(IV) *Regulatory interventions:* quality standards, legislations for product safety, legal regulations on labour conditions and the use of technology, taxes and tariffs;

(V) *Coordination and steering*: information, such as value chain data, analyses and studies, facilitation of meetings, organizational strengthening, building strong communities of practice.

The activities mentioned above are just examples. The range of policy instruments is much larger and may include instruments of other policy areas. Ideally, the configurations of activities and instruments are informed by value chain analyses and the strategic considerations covered in this book. Furthermore, this book has shown that value chain growth and indeed improved access to safely managed services can never rely on market processes alone to generate the desired social and environmental outcomes, but acknowledges the explicit role of regulatory improvement, in the economic, environmental, and social development of the value chain (Springer-Heinze, 2018a, 2018b). Adequate policies that will regulate and support the IFSVC are fundamental preconditions for its development, and at the same time sustainably developing the value chain will require strong collective actions of other key players because government alone cannot provide these regulatory services and the requisite leadership (Springer-Heinze, 2018b). Policy failure is as notorious as market failure and value chain development and improvement would be negligent if they simply shifted the responsibility to already overstrained government agencies.

Furthermore, value chains are cooperating systems of private enterprises, and this is visualized by the IFSVC map showing the sequence of value chain operators who are the key stakeholders. The benefits of partnering to develop the value chain will provide better insights to understanding of existing problems; and thus greater relevance of the project, access to the know-how of partners, mobilization of funds, complementary actions and then greater efficiency. For a value chain project to be effective and efficient, the different partners have to deliver their contributions in a coordinated manner and at the right time (Springer-Heinze, 2018a), as vision formulation, strategy generation and specific programme objectives are not enough to create an operational value chain. Vision and programme objectives have to be translated into output and activities to achieve impact. Finally, implementing IFSVC will need a scheduled plan with specific objectives and corresponding activities (Springer-Heinze, 2018a).

10.4 CONCLUSION

There have always been some forms of SVC operating at all levels – locally, regionally, nationally and globally. However, they have not been recognized due to weak linkages and interactions between players, especially in developing countries. In essence, there always has and will be a sanitation market, and sanitaryware manufacturers like those that produce pour flush and toilet systems have existed for centuries and are leading enterprises in the sanitation value chain from the global to the local levels. Going forward, it is key to recognize that the sanitation market has expanded beyond locally made latrines and toilet facilities for rural and semi-urban sanitation programmes to include the higher-level players that can support local-level players in a well developed IFSVC. In addition, the sanitation market is not just for rural areas, but also includes urban players and luxury products which could provide support for interventions for the population at the base of the pyramid (BoP) to increase access to safely managed sanitation services. The IFSVC will still face the same challenges as existing SVCs unless all stakeholders come together to develop a working system with strong linkages and active interactions between all players. This means that governance systems will have to be specific, clear and far-reaching, while enabling business environments will need to be created in such a way that incorporate/aid an intervention for the vulnerable groups of society. Education, promotion and capacity building in the IFSVC are very necessary and studies on comprehending, developing, implementing and analysing the IFSVC in different contexts are required as well. Furthermore, governance and enabling systems have a crucial role to play when it comes to setting the right conditions and are capable of providing some targeted services for IFSVC to flourish.

10.5 Take action

(I)　Review the enabling business environments in your country and how they support the sanitation businesses and enterprises

(II)　Write a letter to your legislator to consider making laws that support the growth of the IFSVC and its effect on achieving the SDG 6 and other related SDGs

10.6 Journal entry

(I)   Identify local and national policies, laws and regulations that affect sanitation businesses and organizations positively and/or negatively.
(II) What could be the best governance options and/or integrated options to manage and enhance the IFSVC for your country and city?

10.7 Reflection

Consider the role of governance and enabling business environment in your city and country, for developing, upgrading and strengthening an existing SVC to IFSVC

10.8 Guiding questions

(I)   What is IFSVC governance?
(II)   List the core quality and standards benchmarks for value chain products and services?
(III)   Identify the leading actors/enterprises that may engage in IFSVC governance?
(IV)   What are the key instruments in value chain governance that could enhance the IFSVC and sanitation economy?
(V)   List the key issues that effective and efficient IFSVC governance should address?
(VI)   Differentiate between the three main types of valuechain governance and how could they be use to manage the IFSVC.
(VII)   What are the two sides of policy in value chain governance?
(VIII) What are the eight principles of IFSVC policy?
(IX)   What are some of the policies required for ease of doing business in a country?

REFERENCES

Agrilinks. (2011). Five Market Systems Enabling Environment Recommendations. https://agrilinks. org/post/5-market-systems-enabling-environment-recommendations (accessed 6 January 2022).

Dietz M. (2014). Value Chain Governance that Benefits the Poor. Working paper. Helvetas Swiss Intercooperation. https://www.shareweb.ch/site/EI/Documents/PSD/Topics/Value%20 Chain%20Development/Helvetas%20-%20Working%20Paper%20-%20Value%20Chain%20 Development%20for%20the%20Poor.pdf (accessed 06 January 2022).

Gereffi G., Humphrey J. and Sturgeon T. (2005). The governance of global value chains. *Review of International Political Economy*, **12**(1), 78–104. https://doi.org/10.1080/09692290500049805, ISSN 0969-2290 print/ISSN 1466-4526, https://www.fao.org/fileadmin/user_upload/ fisheries/docs/GVC_Governance.pdf (accessed 11 January 2022).

Koottatep T., Cookey P. E. and Polprasert C. (eds) (2019). Social-ecological system. In: Regenerative Sanitation: A New Paradigm For Sanitation 4.0. IWA Publishing, London, pp. 106–140. https://doi.org/10.2166/9781780409689_0209

Market Links. (2021). Value Chain Development. https://www.marketlinks.org/good-practice-center/value-chain-wiki/business-enabling-environment-overview (accessed 5 January 2022).

Market Research. (2021). What is Business enabling environment. http://www.marketresearch.mn/attachments/article/269/What%20is%20the%20Business%20Enabling%20Environment.pdf (accessed 5 January 2022).

M4P. (2008). Making Value Chains Work Better for the Poor: A Toolbook for Practitioners of Value Chain Analysis, Version 3. Making Markets Work Better for the Poor (M4P) Project, UK Department for International Development (DFID). Agricultural Development International, Phnom Penh, Cambodia. https://www.fao.org/3/at357e/at357e.pdf (accessed 5 January 2022).

Springer-Heinze A. (2018a). ValueLinks 2.0: Manual on Sustainable Value Chain Development. Vol. 1: Value Chain Analysis, Strategy and Implementation. GIZ, Eschborn, Germany. https://beamexchange.org/uploads/filer_public/f3/31/f331d6ec-74da-4857-bea1-ca1e4e5a43e5/valuelinks-manual-20-vol-1-january-2018_compressed.pdf (accessed 26 April 2021).

Springer-Heinze A. (2018b). ValueLinks 2.0: Manual on Sustainable Value Chain Development. Vol. 2: Value Chain Solutions. GIZ, Eschborn, Germany. https://beamexchange.org/uploads/filer_public/d3/a4/d3a4882e-eb14-4c30-8f7e-6ba4f51f6ec9/valuelinks-manual-20-vol-2-january-2018_compressed.pdf (accessed 26 April 2021).

UNIDO. (2011). Industrial Value Chain Diagnostics: an Integrated Tool. United Nations Industrial Development Organization (UNIDO), Vienna, Austria. https://www.unido.org/sites/default/files/2011-07/IVC_Diagnostic_Tool_0.pdf (accessed 5 January 2022).

Index

ND - #0131 - 031022 - C0 - 234/156/18 - PB - 9781789061833 - Gloss Lamination